精通 Go 语言(影印版)
Mastering Go

Mihalis Tsoukalos 著

南京　东南大学出版社

图书在版编目(CIP)数据

精通 Go 语言:英文/(希)米哈里斯·图卡洛斯(Mihalis Tsoukalos)著. —影印本. —南京:东南大学出版社,2019.5

书名原文:Mastering Go
ISBN 978-7-5641-8322-6

Ⅰ.①精… Ⅱ.①米… Ⅲ.①程序语言-程序设计-英文 Ⅳ.①TP312

中国版本图书馆 CIP 数据核字(2019)第 046189 号
图字:10-2018-497 号

© 2018 by PACKT Publishing Ltd.

Reprint of the English Edition, jointly published by PACKT Publishing Ltd and Southeast University Press, 2019. Authorized reprint of the original English edition, 2018 PACKT Publishing Ltd, the owner of all rights to publish and sell the same.

All rights reserved including the rights of reproduction in whole or in part in any form.

英文原版由 PACKT Publishing Ltd 出版 2018。

英文影印版由东南大学出版社出版 2019。此影印版的出版和销售得到出版权和销售权的所有者 —— PACKT Publishing Ltd 的许可。

版权所有,未得书面许可,本书的任何部分和全部不得以任何形式重制。

精通 Go 语言(影印版)

出版发行:东南大学出版社
地　　址:南京四牌楼 2 号　邮编:210096
出 版 人:江建中
网　　址:http://www.seupress.com
电子邮件:press@seupress.com
印　　刷:常州市武进第三印刷有限公司
开　　本:787 毫米×980 毫米　16 开本
印　　张:37.75
字　　数:739 千字
版　　次:2019 年 5 月第 1 版
印　　次:2019 年 5 月第 1 次印刷
书　　号:ISBN 978-7-5641-8322-6
定　　价:118.00 元

本社图书若有印装质量问题,请直接与营销部联系。电话(传真):025-83791830

mapt.io

Mapt is an online digital library that gives you full access to over 5,000 books and videos, as well as industry leading tools to help you plan your personal development and advance your career. For more information, please visit our website.

Why subscribe?

- Spend less time learning and more time coding with practical eBooks and Videos from over 4,000 industry professionals

- Improve your learning with Skill Plans built especially for you

- Get a free eBook or video every month

- Mapt is fully searchable

- Copy and paste, print, and bookmark content

PacktPub.com

Did you know that Packt offers eBook versions of every book published, with PDF and ePub files available? You can upgrade to the eBook version at www.PacktPub.com and as a print book customer, you are entitled to a discount on the eBook copy. Get in touch with us at service@packtpub.com for more details.

At www.PacktPub.com, you can also read a collection of free technical articles, sign up for a range of free newsletters, and receive exclusive discounts and offers on Packt books and eBooks.

Contributors

About the author

Mihalis Tsoukalos is a technical author, a Unix administrator, a developer, and a mathematician, who enjoys learning new things. He has written more than 250 technical articles for many publications, including *Sys Admin*, *MacTech*, *Linux User and Developer*, *Usenix ;login:*, *Linux Format*, and *Linux Journal*.

Mihalis is also the author of *Go Systems Programming*, by *Packt Publishing*, 2017 and the technical editor for *MongoDB in Action, Second Edition*, by *Manning*. Mihalis' research interests include databases, operating systems, and statistics. You can reach him at `http://www.mtsoukalos.eu/` and `@mactsouk`. He is also a photographer (`http://www.highiso.net/`).

> I would like to thank the people at Packt Publishing for helping me write this book, including Frank Pohlmann and Gary Schwartz, my technical reviewer, Mat Ryer, Radhika Atitkar, for her encouragement and trust, and Kishor Rit, for answering all my questions and encouraging me during the whole process.
>
> For all people everywhere: You will never change your life until you change something you do daily!

About the reviewer

Mat Ryer has been programming computers since he was 6 years old. He would build games and programs, first in BASIC on a ZX Spectrum and then in AmigaBASIC and AMOS on Commodore Amiga with his father. Many hours were spent on manually copying the code from the Amiga Format magazine and tweaking variables or moving GOTO statements around to see what might happen. The same spirit of exploration and obsession with programming led Mat to starting work with a local agency in Mansfield, England, when he was 18, where he started to build websites and other online services.

After several years of working with various technologies and industries in London and around the world, Mat noticed a new systems language called Go that Google was pioneering. Since it addressed very pertinent and relevant modern technical challenges, Mat started using it to solve problems while the language was still in the beta stage. He has used it ever since. Mat contributes to open-source projects and founded Go packages, including Testify, Moq, Silk, and Is, as well as a macOS developer tool called BitBar.

In 2018, Mat co-founded Machine Box and still spends a lot of time speaking at conferences, writing about Go on his blog, and is an active member of the Go community.

Packt is searching for authors like you

If you're interested in becoming an author for Packt, please visit `authors.packtpub.com` and apply today. We have worked with thousands of developers and tech professionals, just like you, to help them share their insight with the global tech community. You can make a general application, apply for a specific hot topic that we are recruiting an author for, or submit your own idea.

Table of Contents

Preface 1

Chapter 1: Go and the Operating System 7
    The structure of the book 8
    The history of Go 8
    Why learn Go? 9
    Go advantages 9
        Is Go perfect? 11
            What is a preprocessor? 11
        The godoc utility 12
    Compiling Go code 13
    Executing Go code 14
    Two Go rules 14
        You either use a Go package or do not include it 15
        There is only one way to format curly braces 16
    Downloading Go packages 17
    Unix stdin, stdout, and stderr 19
    About printing output 19
    Using standard output 21
    Getting user input 23
    About := and = ... 23
        Reading from standard input 24
        Working with command-line arguments 26
    About error output 28
    Writing to log files 30
        Logging levels 31
        Logging facilities 31
        Log servers 31
        A Go program that sends information to log files 32
        About log.Fatal() 35
        About log.Panic() 36
    Error handling in Go 38
        The error data type 38
        Error handling 40
    Additional resources 43
    Exercises 44
    Summary 44

Chapter 2: Understanding Go Internals 45

Table of Contents

The Go compiler — 46
Garbage Collection — 47
    The Tricolor algorithm — 50
    More about the operation of the Go Garbage Collector — 53
    Unsafe code — 55
    About the unsafe package — 57
    Another example of the unsafe package — 57
Calling C code from Go — 59
    Calling C code from Go using the same file — 59
    Calling C code from Go using separate files — 60
    The C code — 61
    The Go code — 62
    Mixing Go and C code — 63
Calling Go functions from C code — 64
    The Go package — 64
    The C code — 66
The defer keyword — 67
Panic and Recover — 69
    Using the panic function on its own — 71
Two handy Unix utilities — 72
    The strace tool — 73
    The dtrace tool — 74
Your Go environment — 76
The Go Assembler — 78
Node Trees — 79
Learning more about go build — 85
General Go coding advices — 86
Additional Resources — 86
Exercises — 87
Summary — 87

Chapter 3: Working with Basic Go Data Types — 89
Go loops — 90
    The for loop — 90
    The while loop — 91
    The range keyword — 91
    Examples of Go for loops — 91
Go arrays — 93
    Multi-dimensional arrays — 94
    The shortcomings of Go arrays — 97
Go slices — 97
    Performing basic operations on slices — 98
    Slices are being expanded automatically — 100
    Byte slices — 102

The copy() function	102
Multidimensional slices	105
Another example of slices	105
Sorting slices using sort.slice()	108
Go maps	110
Storing to a nil map	112
When you should use a map?	113
Go constants	113
The constant generator iota	115
Go pointers	118
Dealing with times and dates	121
Working with times	123
Parsing times	123
Working with dates	125
Parsing dates	125
Changing date and time formats	127
Additional resources	129
Exercises	129
Summary	129
Chapter 4: The Uses of Composite Types	**131**
About composite types	132
Structures	132
Pointers to structures	135
Using the new keyword	137
Tuples	137
Regular expressions and pattern matching	139
Now for some theory	140
A simple example	140
A more advanced example	143
Matching IPv4 addresses	146
Strings	151
What is a rune?	154
The Unicode package	156
The strings package	157
The switch statement	161
Calculating Pi with great accuracy	165
Developing a key/value store in Go	168
Additional resources	173
Exercises	174
Summary	174
Chapter 5: Enhancing Go Code with Data Structures	**175**
About graphs and nodes	176

Algorithm complexity	176
Binary trees in Go	177
Implementing a binary tree in Go	178
Advantages of binary trees	180
Hash tables in Go	181
Implementing a hash table in Go	182
Implementing the lookup functionality	185
Advantages of hash tables	186
Linked lists in Go	186
Implementing a linked list in Go	187
Advantages of linked lists	191
Doubly linked lists in Go	191
Implementing a doubly linked list in Go	193
Advantages of doubly linked lists	196
Queues in Go	196
Implementing a queue in Go	197
Stacks in Go	200
Implementing a stack in Go	200
The container package	203
Using container/heap	204
Using container/list	207
Using container/ring	209
Generating random numbers	211
Generating random strings	214
Additional Resources	217
Exercises	217
Summary	218
Chapter 6: What You Might Not Know About Go Packages	219
About Go packages	220
About Go functions	220
Anonymous functions	221
Functions that return multiple values	221
The return values of a function can be named!	223
Functions with pointer parameters	225
Functions that return pointers	226
Functions that return other functions	228
Functions that accept other functions as parameters	229
Developing your own Go packages	231
Compiling a Go package	233
Private variables and functions	233
The init() function	233
Reading the Go code of a standard Go package	236
Exploring the code of the net/url package	236

Looking at the Go code of the log/syslog package	238
Creating good Go packages	**239**
The syscall package	**241**
Finding out how fmt.Println() really works	244
Text and HTML templates	**246**
Generating text output	247
Constructing HTML output	249
Basic SQLite3 commands	257
Additional resources	**257**
Exercises	**258**
Summary	**258**
Chapter 7: Reflection and Interfaces for All Seasons	**259**
Type methods	**259**
Go interfaces	**262**
About type assertion	**263**
Developing your own interfaces	**265**
Using a Go interface	266
Using switch with interface and data types	268
Reflection	**270**
A simple Reflection example	271
A more advanced reflection example	273
The three disadvantages of reflection	276
Object-oriented programming in Go!	**277**
Additional resources	**281**
Exercises	**281**
Summary	**282**
Chapter 8: Telling a Unix System What to Do	**283**
About Unix processes	**284**
The flag package	**284**
The io.Reader and io.Writer interfaces	**290**
Buffered and unbuffered file input and output	290
The bufio package	**290**
Reading text files	**291**
Reading a text file line by line	291
Reading a text file word by word	293
Reading a text file character by character	295
Reading from /dev/random	297
Reading the amount of data you want from a file	**299**
Why are we using binary format?	**301**
Reading CSV files	**302**
Writing to a file	**305**
Loading and saving data on disk	**308**

The strings package revisited	312
About the bytes package	314
File permissions	316
Handling Unix signals	317
Handling two signals	318
Handling all signals	320
Programming Unix pipes in Go	323
Implementing the cat(1) utility in Go	323
Traversing directory trees	325
Using eBPF from Go	328
About syscall.PtraceRegs	329
Tracing system calls	331
User ID and group ID	336
Additional resources	337
Exercises	338
Summary	339
Chapter 9: Go Concurrency – Goroutines, Channels, and Pipelines	341
About processes, threads, and goroutines	342
The Go scheduler	343
Concurrency and parallelism	343
Goroutines	344
Creating a goroutine	344
Creating multiple goroutines	346
Waiting for your goroutines to finish	348
What if the number of Add() and Done() calls do not agree?	350
Channels	352
Writing to a channel	352
Reading from a channel	354
Channels as function parameters	356
Pipelines	357
Additional resources	361
Exercises	361
Summary	362
Chapter 10: Go Concurrency – Advanced Topics	363
The Go scheduler revisited	364
The GOMAXPROCS environment variable	366
The select keyword	367
Timing out a goroutine	370
Timing out a goroutine – take 1	370
Timing out a goroutine – take 2	372
Go channels revisited	375
Signal channels	376

Table of Contents

- Buffered channels — 376
- Nil channels — 379
- Channel of channels — 380
- Specifying the order of execution for your goroutines — 383
- **Shared memory and shared variables** — 386
  - The sync.Mutex type — 387
    - What happens if you forget to unlock a mutex? — 389
  - The sync.RWMutex type — 391
  - Sharing memory using goroutines — 395
- **Catching race conditions** — 397
- **The context package** — 403
  - An advanced example of the context package — 407
  - Worker pools — 412
- **Additional resources** — 417
- **Exercises** — 418
- **Summary** — 419

Chapter 11: Code Testing, Optimization, and Profiling — 421

- **The Go version used in this chapter** — 422
  - Comparing Go version 1.10 with Go version 1.9 — 422
- **Installing a beta or RC version of Go** — 423
- **About optimization** — 425
- **Optimizing Go code** — 425
- **Profiling Go code** — 426
  - The net/http/pprof standard Go package — 427
  - A simple profiling example — 427
  - A convenient external package for profiling — 435
  - The web interface of the Go profiler — 437
    - A profiling example that uses the web interface — 437
    - A quick introduction to Graphviz — 440
- **The go tool trace utility** — 441
- **Testing Go code** — 447
  - Writing tests for existing Go code — 448
- **Benchmarking Go code** — 452
- **A simple benchmarking example** — 452
  - A wrong benchmark function — 458
- **Benchmarking buffered writing** — 459
- **Finding unreachable Go code** — 464
- **Cross-compilation** — 465
- **Creating example functions** — 467
- **Generating documentation** — 469
- **Additional resources** — 475
- **Exercises** — 476
- **Summary** — 477

Chapter 12: The Foundations of Network Programming in Go — 479
About net/http, net, and http.RoundTripper — 480
- The http.Response type — 480
- The http.Request type — 481
- The http.Transport type — 482

About TCP/IP — 483
About IPv4 and IPv6 — 484
The nc(1) command-line utility — 484
Reading the configuration of network interfaces — 485
Performing DNS lookups — 490
- Getting the NS records of a domain — 492
- Getting the MX records of a domain — 494

Creating a web server in Go — 496
- Profiling an HTTP server — 499
- Creating a website in Go — 504

HTTP tracing — 514
- Testing HTTP handlers — 517

Creating a web client in Go — 520
- Making your Go web client more advanced — 522

Timing out HTTP connections — 526
- More information about SetDeadline — 528
- Setting the timeout period on the server side — 529
- Yet another way to time out! — 531

Wireshark and tshark tools — 533
Additional resources — 533
Exercises — 534
Summary — 535

Chapter 13: Network Programming – Building Servers and Clients — 537
The net standard Go package — 538
A TCP client — 538
- A slightly different version of the TCP client — 540

A TCP server — 542
- A slightly different version of the TCP server — 544

A UDP client — 547
Developing a UDP server — 549
A concurrent TCP server — 551
- A handy concurrent TCP server — 556

Remote Procedure Call (RPC) — 562
- The RPC client — 563
- The RPC server — 564

Doing low-level network programming — 566
- Grabbing raw ICMP network data — 569

Where to go next?	574
Additional resources	574
Exercises	575
Summary	576
Other Books You May Enjoy	577
Index	581

Preface

The book you are reading right now is called *Mastering Go* and is all about helping you become a better Go developer!

I tried to include the right amount of theory and hands on practice, but only you, the reader, can tell if I succeeded or not! Additionally, all presented examples are self-contained, which means that they can be used on their own or as templates for creating more complex applications.

Please try to do the exercises located at the end of each chapter and do not hesitate to contact me with ways to make any future editions of this book even better!

Who this book is for

This book is for amateur and intermediate Go programmers who want to take their Go knowledge to the next level as well as for experienced developers in other programming languages who want to learn Go without learning again how a `for` loop works.

Some of the information found in this book can be also found in my other book, *Go Systems Programming* by *Packt Publishing*. The main difference between these two books is that *Go Systems Programming* is about developing system tools using the capabilities of Go, whereas *Mastering Go* is about explaining the capabilities and the internals of Go in order to become a better Go developer. Both books can be used as a reference after reading them for the first or the second time.

What this book covers

`Chapter 1`, *Go and the Operating System*, begins by talking about the history of Go and the advantages of Go before describing the `godoc` utility and explaining how you can compile and execute Go programs. After that, it talks about printing the output and getting user input, working with the command-line arguments of a program, and using log files. The last topic of the first chapter is error handling, which plays a key role in Go.

Chapter 2, *Understanding Go Internals*, discusses the Go garbage collector and the way it operates. Then it talks about unsafe code and the unsafe package, how to call C code from a Go program, and how to call Go code from a C program. After that, it showcases the use of the `defer` keyword and presents the `strace(1)` and `dtrace(1)` utilities. In the remaining sections of the chapter, you will learn how to find information about your Go environment and the use of the Go assembler.

Chapter 3, *Working with Basic Go Data Types*, talks about the data types offered by Go, which includes arrays, slices, and maps as well as Go pointers, constants, loops, and working with dates and times. You would not want to miss this chapter!

Chapter 4, *The Uses of Composite Types*, begins by teaching you about Go structures and the `struct` keyword before discussing tuples, strings, runes, byte slices, and string literals. The rest of the chapter talks about regular expressions and pattern matching, the switch statement, the strings package, the `math/big` package, and about developing a key-value store in Go.

Chapter 5, *Enhancing Go Code with Data Structures*, is about developing your own data structures when the structures offered by Go do not fit a particular problem. This includes developing binary trees, linked lists, hash tables, stacks, and queues and learning about their advantages. This chapter also showcases the use of the structures found in the container standard Go package. The last topic of this chapter is random number generation.

Chapter 6, *What You Might Not Know About Go Packages*, is all about packages and functions, which also includes the use of the `init()` function, the `syscall` standard Go package, and the `text/template` and `html/template` packages. This chapter will definitely make you a better Go developer!

Chapter 7, *Reflection and Interfaces for All Seasons*, discusses three advanced Go concepts: reflection, interfaces, and type methods. The last part of the chapter is about object oriented programming in Go!

Chapter 8, *Telling a Unix System What to Do*, is about systems programming in Go, which includes subjects such as the `flag` package for working with command-line arguments, handling Unix signals, file input and output, the bytes package, and the `io.Reader` and `io.Writer` interfaces. As I told you before, if you are really into systems programming in Go, then getting *Go Systems Programming* after reading *Mastering Go* is highly recommended!

Chapter 9, *Concurrency in Go – Goroutines, Channels, and Pipelines,* discusses goroutines, channels and pipelines, which is the Go way of achieving concurrency. You will also learn about the differences between processes, threads, and goroutines, and the sync package and the way the Go scheduler operates.

Chapter 10, *Concurrency in Go – Advanced Topics,* will continue from the point where the previous chapter left off and make you a master of goroutines and channels! You will learn more about the Go scheduler, the use of the powerful select keyword and the various types of Go channels as well as shared memory, mutexes, the sync.Mutex type, and the sync.RWMutex type. The last part of the chapter will talk about the context package, worker pools, and how to detect race conditions.

Chapter 11, *Code Testing, Optimization, and Profiling,* discusses code testing, code optimization, and code profiling as well as about cross compilation, creating documentation, benchmarking Go code, creating example function, and finding unreachable Go code.

Chapter 12, *The Foundations of Network Programming in Go,* is all about the net/http package and how you can develop web clients and web servers in Go. This also includes the use of the http.Response, http.Request and http.Transport structures and the http.NewServeMux type. You will even learn how to develop an entire website in Go! Furthermore, in this chapter, you will learn how to read the configuration of your network interfaces and how to perform DNS lookups in Go.

Chapter 13, *Network Programming – Building Your Own Servers and Clients,* talks about creating UDP and TCP servers and clients in Go, using the functionality offered by the net package. Other topics included in this chapter are how to create RPC clients and servers as well as develop a concurrent TCP server in Go and read raw network packages!

To get the most out of this book

This book requires any Unix machine with a relatively recent Go version installed, which includes any machine running Mac OS X, macOS or Linux. Most of the presented code will also work on Microsoft Windows machines.

Preface

To get the most out of this book, you should try to apply the knowledge of each chapter in your own programs as soon as possible, and see what works and what does not! As I told you before, try to solve the exercises found at the end of each chapter, or create your own programming problems.

Download the example code files

You can download the example code files for this book from your account at `www.packtpub.com`. If you purchased this book elsewhere, you can visit `www.packtpub.com/support` and register to have the files emailed directly to you.

You can download the code files by following these steps:

1. Log in or register at `www.packtpub.com`.
2. Select the **SUPPORT** tab.
3. Click on **Code Downloads & Errata**.
4. Enter the name of the book in the **Search** box and follow the onscreen instructions.

Once the file is downloaded, please make sure that you unzip or extract the folder using the latest version of:

- WinRAR/7-Zip for Windows
- Zipeg/iZip/UnRarX for Mac
- 7-Zip/PeaZip for Linux

The code bundle for the book is also hosted on GitHub at `https://github.com/PacktPublishing/Mastering-Go`. In case there's an update to the code, it will be updated on the existing GitHub repository.

We also have other code bundles from our rich catalog of books and videos available at `https://github.com/PacktPublishing/`. Check them out!

Download the color images

We also provide a PDF file that has color images of the screenshots/diagrams used in this book. You can download it here: `https://www.packtpub.com/sites/default/files/downloads/MasteringGo_ColorImages.pdf`.

Conventions used

There are a number of text conventions used throughout this book.

`CodeInText`: Indicates code words in text, database table names, folder names, filenames, file extensions, pathnames, dummy URLs, user input, and Twitter handles. Here is an example: "The first way is similar to using the `man(1)` command, but for Go functions and packages."

A block of code is set as follows:

```
package main

import (
    "fmt"
)

func main() {
    fmt.Println("This is a sample Go program!")
}
```

When we wish to draw your attention to a particular part of a code block, the relevant lines or items are set in bold:

```
package main

import (
    "fmt"
)

func main() {
    fmt.Println("This is a sample Go program!")
}
```

Any command-line input or output is written as follows:

```
$ date
Sat Oct 21 20:09:20 EEST 2017
$ go version
go version go1.9.1 darwin/amd64
```

Bold: Indicates a new term, an important word, or words that you see onscreen. For example, words in menus or dialog boxes appear in the text like this. Here is an example: "Select **System info** from the **Administration** panel."

Warnings or important notes appear like this.

 Tips and tricks appear like this.

Get in touch

Feedback from our readers is always welcome.

General feedback: Email `feedback@packtpub.com` and mention the book title in the subject of your message. If you have questions about any aspect of this book, please email us at `questions@packtpub.com`.

Errata: Although we have taken every care to ensure the accuracy of our content, mistakes do happen. If you have found a mistake in this book, we would be grateful if you would report this to us. Please visit `www.packtpub.com/submit-errata`, selecting your book, clicking on the Errata Submission Form link, and entering the details.

Piracy: If you come across any illegal copies of our works in any form on the Internet, we would be grateful if you would provide us with the location address or website name. Please contact us at `copyright@packtpub.com` with a link to the material.

If you are interested in becoming an author: If there is a topic that you have expertise in and you are interested in either writing or contributing to a book, please visit `authors.packtpub.com`.

Reviews

Please leave a review. Once you have read and used this book, why not leave a review on the site that you purchased it from? Potential readers can then see and use your unbiased opinion to make purchase decisions, we at Packt can understand what you think about our products, and our authors can see your feedback on their book. Thank you!

For more information about Packt, please visit `packtpub.com`.

Go and the Operating System

This chapter will serve as an introduction to various Go topics that may appear slightly ingenuous and naïve at first. The topics contained in this chapter, however, will be used throughout the entire book, so you'll need to make sure that you completely understand them. As happens with most practical subjects, the best way to understand something is to experiment with it. In this case, experimenting means writing Go code on your own, making your own mistakes, and learning from them! Just don't let the error messages discourage you.

In the first chapter, you will learn the following topics:

- The history of the Go programming language
- The reasons that Go is a good choice for developing your applications
- Compiling Go code
- Executing Go code
- Downloading and using external Go packages
- Unix standard input, output, and error
- Printing data on the screen
- Getting user input
- Printing data to standard error
- Working with log files
- Dealing with error handling in Go

The structure of the book

Mastering Go can be divided into three logical parts. The first part consists of four chapters, and it takes a sophisticated look at some important Go concepts, including user input and output, downloading external Go packages, compiling Go code, calling C code from Go, as well as using Go basic types and Go composite types.

The second part consists of three chapters that deal with Go code organization, the design of Go projects, and some advanced features of Go, respectively.

The third part includes the remaining six chapters and deals with the more practical Go topics, including systems programming in Go, concurrency in Go, code testing, optimization, and profiling. The last two chapters of this book will also talk about network programming in Go.

The book will present relatively small, yet complete Go programs that illustrate the concepts presented. This has two main advantages: first, you do not have to look at an endless code listing when trying to learn a single technique, and second, you can use this code as a starting point when creating your own applications and utilities.

Note that the focus of this book is machines that run a variant of the **Unix** operating system; this does not mean that the Go code presented will not run on Microsoft Windows machines—after all Go is portable! It just means that the included Go code has been tested on various Unix variants, mainly on **macOS** High Sierra and Debian **Linux**.

The history of Go

Go is a modern, generic purpose open-source programming language that was officially announced at the end of 2009. It began as an internal Google project, which means that it was started as an experiment, and it is inspired by many other programming languages, including **C**, **Pascal**, **Alef**, and **Oberon**. Its spiritual fathers are *Robert Griesemer*, *Ken Thomson*, and *Rob Pike*, who are professional programmers who designed Go as a language for professional programmers who want to build reliable, robust, and efficient software. Apart from its syntax and its standard functions, Go comes with a pretty rich standard library.

At the time of writing this chapter, the current stable Go version is 1.9.1, but version 1.9.2 is on its way:

```
$ date
Sat Oct 21 20:09:20 EEST 2017
$ go version
go version go1.9.1 darwin/amd64
```

I am pretty confident that by the time this book is published, the output of the `go version` command will be different! The good news is that due to the way that Go progresses, this book will remain relevant for many years!

If you are installing Go for the first time, you can start by visiting https://golang.org/dl/. However, there is a good chance that your Unix variant already has a ready-to-install package for the Go programming language, so you might want to get Go, using your favorite package manager.

Why learn Go?

Go is a modern programming language that allows you to write safe code without silly bugs—do not worry, you can still create complex bugs! Most of all, though, Go wants to have happy developers; therefore, by design, Go code looks attractive and familiar, and it is easy to write.

The next section talks more analytically about the advantages of Go.

Go advantages

Go has many advantages—some of them are unique to Go, while others are shared with other programming languages.

The most significant Go advantages and features are as follows:

- Go is a modern programming language that was created by experienced developers.
- Go release candidates are used first by Google staff for production use!
- Go code is easy to read and easy to understand.

- Go wants happy developers, because a happy developer writes better code!

- The Go compiler prints practical warning and error messages that help you solve the actual problem. Put simply, the Go compiler is here to help you, and not to make your life miserable by printing pointless output!

- Go code is portable, especially between Unix machines.

- Go has support for procedural, concurrent, and distributed programming.

- Go supports **Garbage Collection**, so you do not have to deal with memory allocation and deallocation.

- Go does not have a **preprocessor**. It does high-speed compilation. As a consequence, Go can also be used as a scripting language.

- Go can build web applications, and it provides a simple web server for testing purposes.

- The standard Go library offers many packages that simplify the work of the developer. Additionally, the functions found in the standard Go library are tested and debugged in advance by the people who develop Go, which means that most of the time these functions come without bugs.

- Go uses **static linking** by default, which means that the binary files produced can be easily transferred to other machines with the same OS. As a consequence, once a Go program is compiled successfully and an executable file is generated, the developer does not need to worry about libraries, dependencies, and different library versions anymore.

- You will not need a GUI for developing, debugging, and testing Go applications, as Go can be used from the command-line, which many Unix people prefer.

- Go supports **Unicode**, which means that you do not need any extra code for printing characters from multiple human languages.

- Go keeps concepts orthogonal, because a few orthogonal features work better than many overlapping ones.

Is Go perfect?

There is no such thing as the perfect programming language, and Go is no exception to this rule. However, some programming languages are better at some areas of programming, or we just like them more than other programming languages. Personally, I do not like Java, and while I used to like C++, I do not like it anymore. This is mainly because I find the look of Java and C++ code to be unpleasant.

Some of the disadvantages of Go are as follows:

- Go does not have direct support for **object-oriented programming** (**OOP**), which can be a problem for programmers who are used to writing code in an object-oriented manner. Nevertheless, you can use composition in Go to mimic inheritance.
- For some people who still prefer C, Go will never replace C!
- C is still faster than any other programming language for systems programming, mainly because Unix is written in C.

Nevertheless, Go is a pretty decent and modern programming language that will not disappoint if you find the time to learn it and program in it.

What is a preprocessor?

Earlier, I said that Go does not have a preprocessor, and that this is a good thing. A **preprocessor** is a program that processes your input data and generates output that will be used as the input to another program. In the context of programming languages, the input of a preprocessor is source code that will be processed by the preprocessor before given as input to the compiler of the programming language. The biggest disadvantage of a preprocessor is that it knows nothing about the underlying language or its syntax!

Put simply, this means that when a preprocessor is used, you cannot be certain that the final version of your code will do what you really want it to do because the preprocessor might alter the logic as well as the semantics of your original code!

The list of programming languages with a preprocessor includes, but is not limited to C, C++, Ada, and PL/SQL. The infamous C preprocessor processes lines that begin with # and are called **directives** or **pragmas**. Directives and pragmas are not part of the C programming language!

The godoc utility

The Go distribution comes with a plethora of tools that can simplify your life as a programmer. One of these tools is the `godoc` utility, which allows you to see the documentation of existing Go functions and packages without needing an internet connection.

The `godoc` utility can be executed either as a normal command-line application that displays its output on a Terminal window, or as a command-line application that starts a web server. In the latter case, you will need a web browser to look at the Go documentation.

If you type `godoc` without any command-line parameters, you will get the list of the command-line options supported by `godoc`.

The first way of executing godoc is similar to using the `man(1)` command, but for Go functions and packages. So, in order to find out information about the `Printf()` function of the `fmt` package, you should execute the following command:

```
$ godoc fmt Printf
```

Similarly, you can find out information about the entire `fmt` package by running the next command:

```
$ godoc cmd/fmt
```

The second way requires executing `godoc` with the `-http` parameter:

```
$ godoc -http=:8001
```

The numeric value used in the preceding command, `8001`, is the port number to which the HTTP server will listen. You can choose any port number that is available provided that you have the right privileges. However, note that port numbers `0-1023` are restricted and can only be used by the root user. Thus, it is better to avoid choosing one of them and to pick something else provided that it is not already in use by a different process.

You can omit the equal sign in the command presented and put a space character in its place. So, the next command is the complete equivalent of the previous one:

```
$ godoc -http :8001
```

After that, you should point your web browser to the `http://localhost:8001/pkg/` URL in order to get the list of the available Go packages and to browse their documentation.

Compiling Go code

In this section, you will learn how to compile Go code. The good news is that you can compile your Go code from the command-line without needing a graphical application.

Furthermore, Go does not care about the name of the source file of an autonomous program as long as the package name is `main` and there is a single `main()` function in it, because the `main()` function is where the program execution begins. As a result, you cannot have multiple `main()` functions in the files of a single project.

We will start our first Go program compilation with a program named `aSourceFile.go` that contains the next Go code:

```
package main

import (
    "fmt"
)

func main() {
    fmt.Println("This is a sample Go program!")
}
```

So, in order to compile `aSourceFile.go` and create a **statically linked** executable file, you will need to execute this command:

```
$ go build aSourceFile.go
```

After that, you will have a new executable file named `aSourceFile`:

```
$ file aSourceFile
aSourceFile: Mach-O 64-bit executable x86_64
$ ls -l aSourceFile
-rwxr-xr-x  1 mtsouk  staff  1933104 Oct 14 21:50 aSourceFile
$ ./aSourceFile
This is a sample Go program!
```

The main reason that aSourceFile is that big is because it is statically linked, which means that it does not require any external libraries in order to run.

Executing Go code

There is another way to execute your Go code that does not create any permanent executable files—it just generates some intermediate files that are automatically deleted afterwards.

The method presented in this chapter allows you to use Go as if it were a scripting language like **Python**, **Ruby**, and **Perl**.

In order to run aSourceFile.go without creating an executable file, you will need to execute the next command:

```
$ go run aSourceFile.go
This is a sample Go program!
```

As you can see, the output of the preceding command is exactly the same as before.

With go run, the Go compiler still needs to create an executable file. The fact that you will not see it, that it is automatically executed, and that it is automatically deleted after the program has finished might make you think that there is no need for an executable file!

This book mainly uses go run to execute the example code, primarily because it is simpler than running go build and then running the executable file. Additionally, go run does not leave any files on your hard drive after the program has finished its execution.

Two Go rules

Go has strict coding rules that are there to help you avoid silly errors and bugs in your code, as well as to make your code easier to read among the Go community. This section will present two such Go rules that you need to know.

Please remember that the Go compiler is here to help and not make your life miserable. As a result, the main purpose of the Go compiler is to compile and increase the quality of your Go code.

You either use a Go package or do not include it

Go has strict rules about package usage. Therefore, you cannot just include any package that you might think you will need and not use it afterwards. You will learn more about Go packages in Chapter 6, *What You Might Not Know About Go Packages*.

Look at the following naïve program, which is saved as packageNotUsed.go:

```
package main

import (
    "fmt"
    "os"
)

func main() {
    fmt.Println("Hello there!")
}
```

In this book, you are going to see lots of error messages, error situations, and warnings. I believe that examining code that fails to compile is useful and sometimes even more valuable than just looking at Go code that compiles without any errors. The Go compiler usually displays useful error messages and warnings that will most likely help you resolve an erroneous situation, so do not underestimate these error messages and warnings.

If you execute packageNotUsed.go, you will get the next error message from Go and the program will not be executed:

```
$ go run packageNotUsed.go
# command-line-arguments
./packageNotUsed.go:5:2: imported and not used: "os"
```

If you remove the os package from the import list of the program, packageNotUsed.go will compile just fine—try it on your own.

Although, this is not the perfect time to start talking about breaking Go rules, there is a way to bypass this restriction, which is showcased in the next Go code listing that is saved in the `packageNotUsedUnderscore.go` file:

```
package main

import (
    "fmt"
    _ "os"
)

func main() {
    fmt.Println("Hello there!")
}
```

Using an underscore character in front of a package name in the import list will not create an error message in the compilation process, even if that package is not used in the program:

```
$ go run packageNotUsedUnderscore.go
Hello there!
```

The reason that Go allows you to bypass this rule will become more evident in Chapter 6, *What You Might Not Know About Go Packages*.

There is only one way to format curly braces

Look at the next Go program, which is named `curly.go`:

```
package main

import (
    "fmt"
)

func main()
{
    fmt.Println("Go has strict rules for curly braces!")
}
```

Although it looks just fine, if you try to execute it, you will be fairly disappointed because you will get the next **syntax error** message, and the code will not compile and therefore not run:

```
$ go run curly.go
# command-line-arguments
./curly.go:7:6: missing function body for "main"
./curly.go:8:1: syntax error: unexpected semicolon or newline before {
```

The official explanation for this error message is that Go requires the use of semicolons as statement terminators in many contexts, and the compiler automatically inserts the required semicolons when it thinks that they are necessary. Therefore, putting the opening brace (`{`) in its own line will make the Go compiler insert a semicolon at the end of the previous line (`func main()`), which produces the error message.

Downloading Go packages

Although the standard Go library is very rich, there are times when you will need to download external Go packages in order to use their functionality. This section will teach you how to download an external package and where it will be placed on your Unix machine.

Look at the next simple Go program that is saved as `getPackage.go`:

```
package main

import (
    "fmt"
    "github.com/mactsouk/go/simpleGitHub"
)

func main() {
    fmt.Println(simpleGitHub. AddTwo (5, 6))
}
```

This program uses an external package, because one of the import commands uses an internet address. In this case, the external package is called `simpleGitHub` and is located at https://github.com/mactsouk/go.

Go and the Operating System

If you try to execute `getPackage.go` right away, you will be disappointed:

```
$ go run getPackage.go
getPackage.go:5:2: cannot find package
"github.com/mactsouk/go/simpleGitHub" in any of:
    /usr/local/Cellar/go/1.9.1/libexec/src/github.com/
mactsouk/go/simpleGitHub (from $GOROOT)
    /Users/mtsouk/go/src/github.com/mactsouk/go/
simpleGitHub (from $GOPATH)
```

The thing is that you will need to get the missing package onto your computer. In order to download it, you will need to execute the following command:

```
$ go get -v github.com/mactsouk/go/simpleGitHub
github.com/mactsouk/go (download)
github.com/mactsouk/go/simpleGitHub
```

After that, you can find the downloaded files at the following directory:

```
$ ls -l ~/go/src/github.com/mactsouk/go/simpleGitHub/
total 8
-rw-r--r--  1 mtsouk  staff  66 Oct 17 21:47 simpleGitHub.go
```

However, the `go get` command also compiles the package. The relevant files can be found at the following place:

```
$ ls -l ~/go/pkg/darwin_amd64/github.com/mactsouk/go/simpleGitHub.a
-rw-r--r--  1 mtsouk  staff  1050 Oct 17 21:47 /Users/mtsouk/go/pkg/
darwin_amd64/github.com/mactsouk/go/simpleGitHub.a
```

You are now ready to execute `getPackage.go` without any problems:

```
$ go run getPackage.go
11
```

You can delete the intermediate files of a downloaded Go package as follows:

```
$ go clean -i -v -x github.com/mactsouk/go/simpleGitHub
cd /Users/mtsouk/go/src/github.com/mactsouk/go/simpleGitHub
rm -f simpleGitHub.test simpleGitHub.test.exe
rm -f /Users/mtsouk/go/pkg/darwin_amd64/github.com/mactsouk/
go/simpleGitHub.a
```

Similarly, you can delete an entire Go package that you have downloaded locally using the rm(1) Unix command to delete its Go source after using go clean:

```
$ go clean -i -v -x github.com/mactsouk/go/simpleGitHub
$ rm -rf ~/go/src/github.com/mactsouk/go/simpleGitHub
```

After executing the former commands, you will need to download the Go package again.

You will learn a lot more about Go packages in Chapter 6, *What You Might Not Know About Go Packages*.

Unix stdin, stdout, and stderr

Every Unix operating system has three files open all the time for its processes. As you know, Unix considers everything a file, even a printer or your mouse. Unix uses **file descriptors**, which are positive integer values, as an internal representation for accessing all of its open files, which is much more convenient than using long paths.

By default, all Unix systems support three special and standard filenames: /dev/stdin, /dev/stdout, and /dev/stderr, which can also be accessed using file descriptors 0, 1, and 2, respectively. These three file descriptors are also called **standard input**, **standard output**, and **standard error**, respectively. Additionally, file descriptor 0 can be accessed as /dev/fd/0 on a macOS machine and as both /dev/fd/0 and /dev/pts/0 on a Debian Linux machine.

Go uses os.Stdin for accessing standard input, os.Stdout for accessing standard output, and os.Stderr for accessing standard error. Although you can still use /dev/stdin, /dev/stdout, and /dev/stderr, or the related file descriptor values for accessing the same devices, it is better, safer, and more portable to stick with the os.Stdin, os.Stdout, and os.Stderr standard filenames that Go offers.

About printing output

As is the case with Unix and C, Go also offers a variety of ways for printing your output on the screen. All of the printing functions in this section require the use of the fmt Go standard package and are illustrated in the printing.go program, which will be presented in two parts.

Go and the Operating System

The simplest way to print something in Go is by using the `fmt.Println()` and `fmt.Printf()` functions. The `fmt.Printf()` function has many similarities to the C `printf(3)` function. You can also use the `fmt.Print()` function instead of `fmt.Println()`. The main difference between `fmt.Print()` and `fmt.Println()` is that the latter automatically adds a newline character each time you call it, whereas the biggest difference between `fmt.Println()` and `fmt.Printf()` is that the latter requires a **format specifier** for each *thing* that you want to print, just like the C `printf(3)` function, which means that you have better control over what you are doing, though you have to write more code. Go calls these format specifiers **verbs**. You can find more information about verbs at https://golang.org/pkg/fmt/. If you have to perform any formatting before printing something, or you have to arrange multiple variables, then using `fmt.Printf()` might be a better choice. However, if you only have to print a single variable, then you might need to choose either `fmt.Print()` or `fmt.Println()`, depending on whether you need the newline character or not.

The first part of `printing.go` contains the next Go code:

```go
package main

import (
    "fmt"
)

func main() {
    v1 := "123"
    v2 := 123
    v3 := "Have a nice day\n"
    v4 := "abc"
```

In this part, you can see the import of the `fmt` package and the definition of four Go variables. The `\n` used in `v3` is the line break character-if you just want to insert a line break in your output, however, you can call `fmt.Println()` without any arguments instead of using something like `fmt.Print("\n")`.

The second part follows:

```go
    fmt.Print(v1, v2, v3, v4)
    fmt.Println()
    fmt.Println(v1, v2, v3, v4)
    fmt.Print(v1, " ", v2, " ", v3, " ", v4, "\n")
    fmt.Printf("%s%d %s %s\n", v1, v2, v3, v4)
}
```

In this part, you print the four variables using `fmt.Println()`, `fmt.Print()`, and `fmt.Printf()` in order to understand their differences better.

If you execute `printing.go`, you will get the following output:

```
$ go run printing.go
123123Have a nice day
abc
123 123 Have a nice day
abc
123 123 Have a nice day
abc
123123 Have a nice day
abc
```

As you can see in the preceding output, the `fmt.Println()` function also adds a space character between its parameters, which is not the case with `fmt.Print()`. As a result, a statement like `fmt.Println(v1, v2)` is equivalent to `fmt.Print(v1, " ", v2, "\n")`.

Apart from `fmt.Println()`, `fmt.Print()`, and `fmt.Printf()`, which are the simplest functions that can be used for generating output on the screen, there is also the `S` family of functions that includes `fmt.Sprintln()`, `fmt.Sprint()`, and `fmt.Sprintf()`, which are used for creating strings based on the given format and the `F` family of functions. This includes `fmt.Fprintln()`, `fmt.Fprint()` and `fmt.Fprintf()`, which are used for writing to files using an `io.Writer`.

> You will learn more about the `io.Writer` and `io.Reader` interfaces in Chapter 8, *Telling a Unix System What to Do*.

The next section will teach you how to print your data using standard output, which is pretty common in the Unix world.

Using standard output

Standard output is more or less equivalent to printing on the screen. However, using **standard output** might require the use of functions that do not belong to the `fmt` package, which is why it is presented in its own section.

Go and the Operating System

The relevant technique will be illustrated in `stdOUT.go`, which will be offered in three parts. The first part of the program follows:

```
package main

import (
    "io"
    "os"
)
```

Here `stdOUT.go` uses the `io` package instead of the `fmt` package. The `os` package is used for reading the command-line arguments of the program and for accessing `os.Stdout`.

The second portion of `stdOUT.go` contains the next Go code:

```
func main() {
    myString := ""
    arguments := os.Args
    if len(arguments) == 1 {
        myString = "Please give me one argument!"
    } else {
        myString = arguments[1]
    }
```

The `myString` variable holds the text that will be printed on the screen, which is either the first command-line argument of the program or, if the program was executed without any command-line arguments, a hard-coded text message.

The third part of the program is as follows:

```
    io.WriteString(os.Stdout, myString)
    io.WriteString(os.Stdout, "\n")
}
```

In this case, the `io.WriteString()` function works in the same way as the `fmt.Print()` function; however, it takes only two parameters. The first parameter is the file to which you want to write, which in this case is `os.Stdout`, and the second parameter is a `string` variable.

> **NOTE**: Strictly speaking, the type of the first parameter of the `io.WriteString()` function should be an `io.Writer` interface, which requires a **slice** of bytes as the second parameter. However, in this case, a `string` does the job just fine. You will learn more about slices in Chapter 3, *Working with Basic Go Data Types*.

Executing `stdOUT.go` will produce the following output:

```
$ go run stdOUT.go
Please give me one argument!
$ go run stdOUT.go 123 12
123
```

The preceding output verifies that the `io.WriteString()` function sends the contents of its second parameter on the screen when its first parameter is `os.Stdout`.

Getting user input

There are three main ways to acquire user input:

1. By reading the command-line arguments of a program
2. By asking the user for input
3. By reading external files

This section will present the first two ways. Should you wish to learn how to read an external file, visit `Chapter 8`, *Telling a Unix System What to Do*.

About := and =

Before continuing, it would be very useful to talk about the use of `:=` and how it differs from `=`. The official name for `:=` is the **short assignment statement**. The short assignment statement can be used in place of a `var` declaration with an implicit type.

The `var` keyword is mostly used for declaring global variables in Go programs as well as for declaring variables without an initial value. The reason for the former is that every statement that exists outside of the code of a function must begin with a keyword, such as `func` or `var`. This means that the short assignment statement cannot be used outside of a function because it is not available there.

The `:=` operator works as follows:

```
m := 123
```

The result of the preceding statement is a new integer variable named m with a value of 123.

However, if you try to use := on an already declared variable, the compilation will fail with the next error message, which will make perfect sense:

```
$ go run test.go
# command-line-arguments
./test.go:5:4: no new variables on left side of :=
```

Now you might ask what will happen if you are expecting two or more values from a function, and you want to use an existing variable for one of them. Should you use := or =? The answer is simple: you should use := as shown in the next code example:

```
i, k := 3, 4
j, k := 1, 2
```

As the j variable is used for the first time in the second statement, you should use := even though k has already been defined in the first statement.

Although it seems boring to talk about such insignificant things, knowing them will save you from various types of errors in the long run!

Reading from standard input

The reading of data from the standard input will be illustrated in stdIN.go, which you will see in two parts. The first part follows:

```
package main

import (
    "bufio"
    "fmt"
    "os"
)
```

In the preceding code, you see the use of the bufio package for the first time in this book.

You will learn more about the bufio package, which is related to file input and output, in Chapter 8, *Telling a Unix System What to Do*.

Although the `bufio` package is mostly used for file input and output, you will keep seeing the `os` package all of the time in this book because it contains many handy functions. Its most common functionality is that it provides you with a way to access the command-line arguments of a Go program (`os.Args`). The official description of the `os` package tells us that it offers functions that perform operating system operations. This includes functions for creating, deleting, and renaming files and directories, as well as functions for learning the Unix permissions and other characteristics of files and directories. The main advantage of the `os` package is that it is platform independent. Put simply, its functions will work on both Unix and Microsoft Windows machines!

The second part of `stdIN.go` contains the following Go code:

```go
func main() {
    var f *os.File
    f = os.Stdin
    defer f.Close()

    scanner := bufio.NewScanner(f)
    for scanner.Scan() {
        fmt.Println(">", scanner.Text())
    }
}
```

Here there is a call to `bufio.NewScanner()` using standard input (`os.Stdin`) as its parameter. This call returns a `bufio.Scanner` variable, which is then used with the `Scan()` function for reading from `os.Stdin` line by line. Each line that is read is printed on the screen before getting the next one. Note that each line printed by the program begins with the > character.

The execution of `stdIN.go` will produce the following type of output:

```
$ go run stdIN.go
21
> 21
This is Mihalis!
> This is Mihalis!
```

In Unix, you can tell a program to stop reading data from standard input by pressing *Ctrl + D*.

The Go code of `stdIN.go` and `stdOUT.go` will be very useful when we talk about Unix **pipes** in Chapter 8, *Telling a Unix System What to Do*, so do not underestimate their simplicity.

Working with command-line arguments

The technique covered in this section will be illustrated by using the Go code of `cla.go`, which will be presented in three parts. The program will find the minimum and the maximum of its command-line arguments.

The first part of the program is as follows:

```
package main

import (
    "fmt"
    "os"
    "strconv"
)
```

What is important here is to realize that obtaining the command-line arguments requires the use of the `os` package. Additionally, you need another package, named `strconv`, in order to be able to convert a command-line argument, which is given as a string, into an arithmetical data type.

The second part of the program is as follows:

```
func main() {
    if len(os.Args) == 1 {
        fmt.Println("Please give one or more floats.")
        os.Exit(1)
    }

    arguments := os.Args
    min, _ := strconv.ParseFloat(arguments[1], 64)
    max, _ := strconv.ParseFloat(arguments[1], 64)
```

Here, `cla.go` checks whether you have any command-line arguments or not by examining the length of `os.Args`, because the program needs at least one command-line argument to operate. Note that `os.Args` is a Go slice with `string` values. The first element in the slice is the name of the executable program. Therefore, in order to initialize the `min` and `max` variables, you will need to use the second element of the `os.Args` string slice that has an index value of 1.

The important point here is that the fact that you are expecting one or more floats does not necessarily mean that the user will give you valid floats, either by accident or on purpose. However, as we have not talked about error handling in Go so far, cla.go assumes that all command-line arguments are in the right format and therefore will be acceptable. As a result, cla.go ignores the error value returned by the strconv.ParseFloat() function using the following statement:

```
n, _ := strconv.ParseFloat(arguments[i], 64)
```

The preceding statement tells Go that you only want to get the first value returned by strconv.ParseFloat() and that you are not interested in the second value, which in this case is an error variable by assigning it to the underscore character. The underscore character, which is called a **blank identifier**, is the Go way of discarding a value. If a Go function returns multiple values, you can use the blank identifier multiple times.

WARNING: Ignoring all or some of the return values of a Go function, especially the error values, is a very dangerous practice that should not be used in production code!

The third part comes with the following Go code:

```
    for i := 2; i < len(arguments); i++ {
        n, _ := strconv.ParseFloat(arguments[i], 64)
        if n < min {
            min = n
        }
        if n > max {
            max = n
        }
    }

    fmt.Println("Min:", min)
    fmt.Println("Max:", max)
}
```

Here you use a `for` loop that will help you visit all of the elements of the os.Args slice, which was previously assigned to the arguments variable.

Executing `cla.go` will create the next type of output:

```
$ go run cla.go -10 0 1
Min: -10
Max: 1
$ go run cla.go -10
Min: -10
Max: -10
```

As you might expect, the program does not behave well when it receives erroneous input. Worst of all, the program does not generate any warnings to inform the user that there were one or more errors while processing the command-line arguments:

```
$ go run cla.go a b c 10
Min: 0
Max: 10
```

About error output

This section presents a technique for sending data to **Unix standard error**, which is the Unix way of differentiating between actual values and error output.

The Go code for illustrating the use of standard error in Go is included in `stdERR.go` and will be presented in two parts. As writing to standard error requires the use of the file descriptor related to standard error, the Go code of `stdERR.go` will be based on the Go code of `stdOUT.go`.

The first part of the program follows:

```
package main

import (
    "io"
    "os"
)
func main() {
    myString := ""
    arguments := os.Args
    if len(arguments) == 1 {
        myString = "Please give me one argument!"
    } else {
        myString = arguments[1]
    }
```

So far, stdERR.go is almost identical to stdOUT.go.

The second part of the program, stdERR.go, is as follows:

```
io.WriteString(os.Stdout, "This is Standard output\n")
io.WriteString(os.Stderr, myString)
io.WriteString(os.Stderr, "\n")
```

Here you call `io.WriteString()` two times to write to standard error (`os.Stderr`) and one more time to write to standard output (`os.Stdout`).

Executing stdERR.go will create the following output:

```
$ go run stdERR.go
This is Standard output
Please give me one argument!
```

The preceding output cannot help you differentiate between data written to standard output and data written to standard error, which could be very useful at times. However, if you are using the bash(1) shell, there is a trick that you can use in order to distinguish between standard output data and standard error data. Almost all Unix shells offer this functionality in their own way.

Thus, when using bash(1), you can redirect the standard error output to a file as follows:

```
$ go run stdERR.go 2>/tmp/stdError
This is Standard output
$ cat /tmp/stdError
Please give me one argument!
```

The number after the name of a Unix program or system call refers to the section of the manual to which its page belongs. Although most names can be found only once in the manual pages, which means that indicating the section number is not required, there are names that can be located in multiple sections because they have multiple meanings, such as crontab(1) and crontab(5). Therefore, if you try to retrieve the manual page of a name with multiple meanings without stating its section number, you will get the entry that has the smallest section number.

Similarly, you can discard error output by redirecting it to the /dev/null device, which is like telling Unix to ignore it completely:

```
$ go run stdERR.go 2>/dev/null
This is Standard output
```

Go and the Operating System

What we did in the two preceding examples is to redirect the file descriptor of standard error into a file and /dev/null, respectively. If you want to save both standard output and standard error to the same file, you can redirect the file descriptor of standard error (2) to the file descriptor of standard output (1)! The following command shows this technique, which is pretty common in Unix systems:

```
$ go run stdERR.go >/tmp/output 2>&1
$ cat /tmp/output
This is Standard output
Please give me one argument!
```

Last, you can send both standard output and standard error to /dev/null as follows:

```
$ go run stdERR.go >/dev/null 2>&1
```

Writing to log files

The log package allows you to send log messages to the system logging service of your Unix machine, whereas the syslog Go package, which is part of the log package, allows you to define the **logging level** and the **logging facility** that your Go program will use.

Usually, most system log files on a Unix operating system can be found under the /var/log directory. However, the log files of many popular services, such as Apache and Nginx, can be found elsewhere, depending on their configuration.

Generally speaking, using a log file to write some information is considered a better practice than writing the same output on the screen for two reasons:

1. The output does not get lost as it is stored in a file
2. You can search and process log files using Unix tools such as grep(1), awk(1), and sed(1), which cannot be done when messages are printed on a Terminal window

The log package offers many functions for sending output to the syslog server of a Unix machine. The list of function includes log.Printf(), log.Print(), log.Println(), log.Fatalf(), log.Fatalln(), log.Panic(), log.Panicln(), and log.Panicf().

Logging functions can be extremely handy for debugging your programs, especially server processes written in Go, so you should not undervalue their power.

[30]

Logging levels

The **logging level** is a value that specifies the severity of the log entry. Various logging levels exist including `debug`, `info`, `notice`, `warning`, `err`, `crit`, `alert`, and `emerg` (in reverse order of severity).

Logging facilities

A **logging facility** is a category used for logging information. The value of the logging facility part can be one of `auth`, `authpriv`, `cron`, `daemon`, `kern`, `lpr`, `mail`, `mark`, `news`, `syslog`, `user`, `UUCP`, `local0`, `local1`, `local2`, `local3`, `local4`, `local5`, `local6`, and `local7`. It is defined inside `/etc/syslog.conf`, `/etc/rsyslog.conf`, or another appropriate file depending on the server process used for system logging on your Unix machine. This means that if a logging facility is not defined and thus handled, the log messages you send to it might get ignored and therefore lost.

Log servers

All Unix machines have a separate server process that is responsible for receiving logging data and writing it to log files. Various log servers exist that work on Unix machines; however, only two of them are used on most Unix variants: `syslogd(8)` and `rsyslogd(8)`.

On macOS machines, the name of the process is `syslogd(8)`. On the other hand, most Linux machines use `rsyslogd(8)`, which is an improved and more reliable version of `syslogd(8)`, which was the original Unix system utility for message logging.

However, despite the Unix variant you are using or the name of the server process used for logging, logging works the same way on every Unix machine and therefore does not affect the Go code that you will write.

The configuration file of `rsyslogd(8)` is usually named `rsyslog.conf` and is located in `/etc`. The contents of a `rsyslog.conf` configuration file, without the lines with comments and lines starting with $, might look like the following:

```
$ grep -v '^#' /etc/rsyslog.conf | grep -v '^$' | grep -v '^\$'
auth,authpriv.*                 /var/log/auth.log
*.*;auth,authpriv.none          -/var/log/syslog
daemon.*                        -/var/log/daemon.log
kern.*                          -/var/log/kern.log
```

```
lpr.*                              -/var/log/lpr.log
mail.*                             -/var/log/mail.log
user.*                             -/var/log/user.log
mail.info                          -/var/log/mail.info
mail.warn                          -/var/log/mail.warn
mail.err                            /var/log/mail.err
news.crit                           /var/log/news/news.crit
news.err                            /var/log/news/news.err
news.notice                        -/var/log/news/news.notice
*.=debug;\
    auth,authpriv.none;\
    news.none;mail.none            -/var/log/debug
*.=info;*.=notice;*.=warn;\
    auth,authpriv.none;\
    cron,daemon.none;\
    mail,news.none                 -/var/log/messages
*.emerg                            :omusrmsg:*
daemon.*;mail.*;\
    news.err;\
    *.=debug;*.=info;\
    *.=notice;*.=warn              |/dev/xconsole
local7.* /var/log/cisco.log
```

In order to send your logging information to /var/log/cisco.log, you will need to use the local7 logging facility. The star character after the name of the facility tells the logging server to catch every logging level that goes to the local7 logging facility and write it to /var/log/cisco.log.

The syslogd(8) server has a pretty similar configuration file that is usually /etc/syslog.conf. On macOS High Sierra, the /etc/syslog.conf file is almost empty and has been replaced by /etc/asl.conf. Nevertheless, the logic behind the configuration of /etc/syslog.conf, /etc/rsyslog.conf, and /etc/asl.conf is the same.

A Go program that sends information to log files

The Go code of logFiles.go will explain the use of the log and log/syslog packages.

The log/syslog package is not implemented on the Microsoft Windows version of Go.

The first part of `logFiles.go` follows:

```
package main

import (
    "fmt"
    "log"
    "log/syslog"
    "os"
    "path/filepath"
)

func main() {
    programName := filepath.Base(os.Args[0])
    sysLog, err := syslog.New(syslog.LOG_INFO|syslog.LOG_LOCAL7,
 programName)
```

The first parameter of the `syslog.New()` function is the priority, which is a combination of the logging facility and the logging level. Therefore, a priority of `LOG_NOTICE | LOG_MAIL`, which is mentioned as an example, will send notice logging-level messages to the `MAIL` logging facility.

As a result, the preceding code sets the default logging to the `local7` logging facility using the `info` logging level. The second parameter of the `syslog.New()` function is the name of the process that will appear on the logs as the sender of the message. Generally speaking, it is considered a good practice to use the real name of the executable in order to be able to find the information you want easily in the log files at another time.

The second part of the program contains the following Go code:

```
    if err != nil {
        log.Fatal(err)
    } else {
        log.SetOutput(sysLog)
    }
    log.Println("LOG_INFO + LOG_LOCAL7: Logging in Go!")
```

After the call to `syslog.New()`, you will have to check the error variable that it returns so that you can make sure that everything is fine. If everything is OK, which means that the value of the error variable is equal to `nil`, you call the `log.SetOutput()` function. This sets the output destination of the default logger, which in this case is the logger you created earlier on (`sysLog`). Then you can use `log.Println()` to send information to the log server.

Go and the Operating System

The third part of `logFiles.go` comes with the following code:

```
sysLog, err = syslog.New(syslog.LOG_MAIL, "Some program!")
if err != nil {
    log.Fatal(err)
} else {
    log.SetOutput(sysLog)
}

log.Println("LOG_MAIL: Logging in Go!")
fmt.Println("Will you see this?")
}
```

The last part shows that you can change the logging configuration in your programs as many times as you want, and that you can still use `fmt.Println()` for printing output on the screen.

The execution of `logFiles.go` will create the following output on the screen on a Debian Linux machine:

```
$ go run logFiles.go
Broadcast message from systemd-journald@mail (Tue 2017-10-17 20:06:08 EEST):
logFiles[23688]: Some program![23688]: 2017/10/17 20:06:08 LOG_MAIL: Logging in Go!
Message from syslogd@mail at Oct 17 20:06:08 ...
Some program![23688]: 2017/10/17 20:06:08 LOG_MAIL: Logging in Go!
Will you see this?
```

Executing the same Go code on a macOS High Sierra machine generated the following output:

```
$ go run logFiles.go
Will you see this?
```

Keep in mind that most Unix machines store logging information in more than one log file, which is also the case with the Debian Linux machine used in this section. As a result, `logFiles.go` sends its output to multiple log files, which can be verified by the output of the following shell commands:

```
$ grep LOG_MAIL /var/log/mail.log
Oct 17 20:06:08 mail Some program![23688]: 2017/10/17 20:06:08 LOG_MAIL: Logging in Go!
$ grep LOG_LOCAL7 /var/log/cisco.log
Oct 17 20:06:08 mail logFiles[23688]: 2017/10/17 20:06:08 LOG_INFO + LOG_LOCAL7: Logging in Go!
$ grep LOG_ /var/log/syslog
```

```
Oct 17 20:06:08 mail logFiles[23688]: 2017/10/17 20:06:08 LOG_INFO +
LOG_LOCAL7: Logging in Go!
Oct 17 20:06:08 mail Some program![23688]: 2017/10/17 20:06:08
LOG_MAIL: Logging in Go!
```

The preceding output shows that the message of the `log.Println("LOG_INFO + LOG_LOCAL7: Logging in Go!")` statement was written on both `/var/log/cisco.log` and `/var/log/syslog`, whereas the message of the `log.Println("LOG_MAIL: Logging in Go!")` statement was written on both `/var/log/syslog` and `/var/log/mail.log`.

The important thing to remember from this section is that if the logging server of a Unix machine is not configured to catch all logging facilities, some of the log entries that you send to it might get discarded without any warnings.

About log.Fatal()

In this section, you will see the `log.Fatal()` function in action. The `log.Fatal()` function is used when something really bad has happened, and you just want to exit your program as fast as possible after reporting the bad situation. The use of `log.Fatal()` is illustrated in the `logFatal.go` program, which contains the following Go code:

```
package main

import (
    "fmt"
    "log"
    "log/syslog"
)

func main() {
    sysLog, err := syslog.New(syslog.LOG_ALERT|syslog.LOG_MAIL,
 "Some program!")
    if err != nil {
        log.Fatal(err)
    } else {
        log.SetOutput(sysLog)
    }

    log.Fatal(sysLog)
    fmt.Println("Will you see this?")
}
```

Executing `log.Fatal()` will create the following output:

```
$ go run logFatal.go
exit status 1
```

As you can easily understand, the use of `log.Fatal()` terminates a Go program at the point where `log.Fatal()` was called, which is the reason that you did not see the output from the `fmt.Println("Will you see this?")` statement.

However, because of the parameters of the `syslog.New()` call, a log entry has been added to the log file that is related to mail, which is `/var/log/mail.log`:

```
$ grep "Some program" /var/log/mail.log
Oct 17 20:20:29 iMac Some program![4663]: 2017/10/17 20:20:29 &{17
Some program! iMac.local    {0 0} 0xc42000c220}
```

About log.Panic()

There are situations where a program will fail for good, and you want to have as much information about the failure as possible. In such difficult circumstances, you might consider using `log.Panic()`, which is the logging function that is illustrated in this section using the Go code of `logPanic.go`.

The Go code of `logPanic.go` follows:

```
package main

import (
    "fmt"
    "log"
    "log/syslog"
)

func main() {
    sysLog, err := syslog.New(syslog.LOG_ALERT|syslog.LOG_MAIL,
 "Some program!")
    if err != nil {
        log.Fatal(err)
    } else {
        log.SetOutput(sysLog)
    }

    log.Panic(sysLog)
    fmt.Println("Will you see this?")
}
```

Executing `logPanic.go` on macOS High Sierra will produce the following output:

```
$ go run logPanic.go
panic: &{17 Some program! iMac.local    {0 0} 0xc42000c220}
goroutine 1 [running]:
log.Panic(0xc42004ff50, 0x1, 0x1)
        /usr/local/Cellar/go/1.9.1/libexec/src/log/log.go:330 +0xc0
main.main()
        /Users/mtsouk/Desktop/masterGo/ch/ch1/code/logPanic.go:17 +0xea
exit status 2
```

Running the same program on a Debian Linux machine with Go version 1.3.3 will generate the following output:

```
$ go run logPanic.go
panic: &{17 Some program! mail    {0 0} 0xc2080400e0}
goroutine 16 [running]:
runtime.panic(0x4ec360, 0xc208000320)
        /usr/lib/go/src/pkg/runtime/panic.c:279 +0xf5
log.Panic(0xc208055f20, 0x1, 0x1)
        /usr/lib/go/src/pkg/log/log.go:307 +0xb6
main.main()
        /home/mtsouk/Desktop/masterGo/ch/ch1/code/logPanic.go:17 +0x169
goroutine 17 [runnable]:
runtime.MHeap_Scavenger()
        /usr/lib/go/src/pkg/runtime/mheap.c:507
runtime.goexit()
        /usr/lib/go/src/pkg/runtime/proc.c:1445
goroutine 18 [runnable]:
bgsweep()
        /usr/lib/go/src/pkg/runtime/mgc0.c:1976
runtime.goexit()
        /usr/lib/go/src/pkg/runtime/proc.c:1445
goroutine 19 [runnable]:
runfinq()
        /usr/lib/go/src/pkg/runtime/mgc0.c:2606
runtime.goexit()
        /usr/lib/go/src/pkg/runtime/proc.c:1445
exit status 2
```

The output of `log.Panic()` includes additional low-level information that will hopefully help you resolve difficult and rare situations that happen in your Go code.

Analogous to the `log.Fatal()` function, the use of the `log.Panic()` function will add an entry to the proper log file and will immediately terminate the Go program.

Error handling in Go

Errors and error handling are two very important Go topics. Go likes error messages so much that it has a separate data type for errors, named `error`! This also means that you can easily create your own **error messages** if you find that what Go gives you is inadequate. You will most likely need to create and handle your own errors when you are developing your own Go packages.

Note that having an error condition is one thing, while deciding how to react to an error condition is a totally different thing. Put simply, not all error conditions are created equal, which means that some error conditions might require that you immediately stop the execution of a program, whereas other error situations might require printing a warning message for the user and continuing with the execution of the program. It is up to the developer to use common sense and decide what to do with each `error` value that the program might get.

The error data type

Many occasions exist where you might end up having to deal with a new error case while developing your own Go application. The `error` data type is here to help you define your own errors.

This subsection will teach you how to create your own `error` variables. As you will see in a while, in order to create a new `error` variable, you will need to call the `New()` function of the `errors` standard Go package.

The example Go code illustrating this process can be found in `newError.go`, and it will be presented in two parts. The first part of the program follows next:

```
package main

import (
    "errors"
    "fmt"
)

func returnError(a, b int) error {
```

```
        if a == b {
            err := errors.New("Error in returnError() function!")
            return err
        } else {
            return nil
        }
    }
```

There are many interesting things happening here. First of all, you can see the definition of a Go function other than `main()` for the first time in this book. The name of this new unsophisticated function is `returnError()`. Additionally, you can see the `errors.New()` function in action, which takes a `string` value as its parameter. Last, if a function should return an `error` variable but there is no error to report, it returns `nil` instead.

 You will learn more about the various types of Go functions in *Chapter 6, What You Might Not Know About Go Packages*.

The second part of `newError.go` is as follows:

```
    func main() {
        err := returnError(1, 2)
        if err == nil {
            fmt.Println("returnError() ended normally!")
        } else {
            fmt.Println(err)
        }

        err = returnError(10, 10)
        if err == nil {
            fmt.Println("returnError() ended normally!")
        } else {
            fmt.Println(err)
        }

        if err.Error() == "Error in returnError() function!" {
            fmt.Println("!!")
        }
    }
```

As the code illustrates, most of the time, you need to check whether an `error` variable is equal to `nil` or not and then act accordingly. Also presented here is the use of the `err.Error()` method, which allows you to convert an `error` variable into a `string` variable. This function lets you compare an `error` variable with a `string`.

Sending your error messages to the logging service of your Unix machine, especially when a Go program is a server or some other critical application. However, the code presented in this book will not follow this principle everywhere in order to avoid filling your log files with unnecessary data.

Executing `newError.go` will produce the following output:

```
$ go run newError.go
returnError() ended normally!
Error in returnError() function!
!!
```

If you try to compare an `error` variable with a `string` variable without converting the `error` variable to a `string` first, the Go compiler will create the following error message:

```
# command-line-arguments
./newError.go:33:9: invalid operation: err == "Error in returnError() function!" (mismatched types error and string)
```

Error handling

Error handling is a very important feature of Go because almost all Go functions return an error message or `nil`, which is the Go way of saying whether there was an error condition while executing a function or not. You will most likely get tired of seeing the following Go code, not only in this book but also in every other Go program you can find on the Internet:

```
if err != nil {
    fmt.Println(err)
    os.Exit(10)
}
```

Do not confuse error handling with printing to error output, because they are two totally different things. The former has to do with Go code that handles error conditions, whereas the latter has to do with writing something to the standard error file descriptor.

The preceding code prints the error message on the screen and exits using `os.Exit()`. Should you wish to send the error message to the logging service instead of the screen, use the following variation of the preceding Go code:

```
if err != nil {
    log.Println(err)
    os.Exit(10)
}
```

Last, there is another variation of the preceding code that is used when something really bad has happened and you want to terminate the program:

```
if err != nil {
    panic(err)
    os.Exit(10)
}
```

The `panic` function is a built-in Go function that stops the execution of a program and starts panicking! If you find yourself using `panic` too often, you might want to reconsider your Go implementation. As you will see in Chapter 2, *Understanding Go Internals*, Go also offers the `recover` function, which might be able to save you when you're in some bad situations. For now, you will need to wait for the next chapter to learn more about the power of the `panic` and `recover` function pair.

It's now time to see a Go program that not only handles error messages generated by standard Go functions, but one that also defines its own error message. The name of the program is `errors.go`, and it will be presented to you in five parts. As you will see, the `errors.go` utility tries to improve the functionality of the `cla.go` program that you saw earlier in this chapter by examining whether its command-line arguments are acceptable floats or not.

The first part of the program follows:

```
package main

import (
    "errors"
    "fmt"
    "os"
    "strconv"
)
```

This part of `errors.go` contains the expected `import` statements.

The second portion of errors.go comes with the following Go code:

```go
func main() {
    if len(os.Args) == 1 {
        fmt.Println("Please give one or more floats.")
        os.Exit(1)
    }

    arguments := os.Args
    var err error = errors.New("An error")
    k := 1
    var n float64
```

Here you create a new error variable named err in order to initialize it with your own value.

The third part of the program comes next:

```go
    for err != nil {
        if k >= len(arguments) {
            fmt.Println("None of the arguments is a float!")
            return
        }
        n, err = strconv.ParseFloat(arguments[k], 64)
        k++
    }

    min, max := n, n
```

This is the trickiest part of the program because, if the first command-line argument is not a proper float, you will need to check the next one and keep checking until you find a suitable command-line argument. If none of the command-line arguments are in the correct format, errors.go will terminate and print a message on the screen. All this checking happens by examining the error value that is returned by strconv.ParseFloat(). All of this code is there just for the accurate initialization of the min and max variables.

The fourth part of the program comes with the following Go code:

```go
    for i := 2; i < len(arguments); i++ {
        n, err := strconv.ParseFloat(arguments[i], 64)
        if err == nil {
            if n < min {
                min = n
            }
            if n > max {
                max = n
            }
```

            }
        }

Here you just process all of the right command-line arguments in order to find the minimum and maximum floats among them.

Finally, the last code portion of the program just deals with printing out the current values of the `min` and `max` variables:

        fmt.Println("Min:", min)
        fmt.Println("Max:", max)
}

As you can see from the Go code in `errors.go`, the biggest part of the code is about error handling than about the actual functionality of the program. Unfortunately, this is the case for software developed in most modern programming languages, and Go is no exception.

If you execute `errors.go`, you will get the following output:

```
$ go run errors.go a b c
None of the arguments is a float!
$ go run errors.go b c 1 2 3 c -1 100 -200 a
Min: -200
Max: 100
```

Additional resources

Have a look at the following resources:

- Visit the Go website at `https://golang.org/`
- Browse the Go documentation site at `https://golang.org/doc/`
- Visit the documentation page of the `log` package at `https://golang.org/pkg/log/`
- Visit the documentation of the `log/syslog` package at `https://golang.org/pkg/log/syslog/`
- Visit the documentation page of the `os` package at `https://golang.org/pkg/os/`
- Have a look at `https://golang.org/cmd/gofmt/`, which is the documentation page of the `gofmt` tool that is used for formatting Go code

- If you are working on a Mac, check the **TextMate** editor at `http://macromates.com/` as well as **BBEdit** at `https://www.barebones.com/products/bbedit/`

- Visit the documentation page of the `fmt` package at `https://golang.org/pkg/fmt/` to learn more about Go verbs and the available functions

Exercises

- Write a Go program that finds the sum of all of its numeric command-line arguments
- Write a Go program that finds the average value of all of its float command-line arguments
- Write a Go program that keeps reading integers until it gets the word STOP as input

Summary

This chapter addressed many interesting Go topics including compiling Go code, working with standard input, accessing standard output and standard error in Go, processing command-line arguments, printing on the screen, and using the logging service of a Unix system as well as error handling and some general information about Go. You should consider all of these topics as foundational information about Go.

The next chapter is all about the internals of Go, which includes learning about **Garbage Collection**, working with the Go compiler, calling C code from Go, using the `defer` keyword, and working with the Go assembler as well as the `panic` and `recover` function pair.

2
Understanding Go Internals

All of the Go features that you learned in the previous chapter are extremely handy, and you will be using them all the time. However, there is nothing more rewarding than being able to see and understand what is going on in the background and how Go operates behind the scenes.

In this chapter, you will learn about the Go garbage collector and how it works. Additionally, you will find out how to call C code from your Go programs, which you might find indispensable in certain situations. However, you will not need to use this capability too often, because Go is a very capable programming language. Likewise, you will also understand how to call Go code from your C programs and how to use the `panic()` and `recover()` functions and the `defer` keyword.

In this chapter of Mastering Go, you will learn the following topics:

- The Go compiler
- How garbage collection works in Go
- How to check the operation of the garbage collector
- Calling C code from your Go programs
- Calling Go code from a C program
- The `panic()` and `recover()` functions
- The `unsafe` package
- The handy, yet tricky `defer` keyword
- The `strace(1)` Linux utility
- The `dtrace(1)` utility that can be found in FreeBSD systems including macOS High Sierra
- Finding out information about your Go environment
- Node trees
- The Go assembler

The Go compiler

The **Go compiler** is executed with the help of the `go` tool. This tool does many more things than just generating executable files.

 The `unsafe.go` file used in this section does not contain any special code-the commands presented will work on every valid Go source file. You will see the contents of `unsafe.go` in a short while.

You can compile a Go source file using the `go tool compile` command. What you will get is an **object file**, which is a file with the `.o` file extension. This is illustrated in the output of the following commands, which were executed on a macOS High Sierra machine:

```
$ go tool compile unsafe.go
$ ls -l unsafe.o
-rw-r--r--  1 mtsouk  staff  5495 Oct 30 19:51 unsafe.o
$ file unsafe.o
unsafe.o: data
```

An object file is a binary file that contains **object code**, which is machine code in a relocatable format that, most of the time is not directly executable. The biggest advantage of the relocatable format is that it requires as little memory as possible during the linking phase.

If you use the `-pack` command-line flag when executing `go tool compile`, you will get an **archive file** instead of an object file:

```
$ go tool compile -pack unsafe.go
$ ls -l unsafe.a
-rw-r--r--  1 mtsouk  staff  5680 Oct 30 19:52 unsafe.a
$ file unsafe.a
unsafe.a: current ar archive
```

An archive file is a binary file that contains one or more files that is primarily used for grouping multiple files into a single file. The archive format used by Go is called **ar**.

You can list the contents of an `.a` archive file as follows:

```
$ ar t unsafe.a
__.PKGDEF
_go_.o
```

Another truly valuable command-line flag of the `go tool compile` command is `-race`, which allows you to detect **race conditions**. You will learn more about race conditions and why you want to avoid them in `Chapter 10`, *Go Concurrency – Advanced Topics*.

You will learn the additional uses of the `go tool compile` command toward the end of this chapter, when we talk about assembly language and node trees. However, in order to tease you a little, try executing the following command:

```
$ go tool compile -S unsafe.go
```

The preceding command generates lots of output that you might find difficult to understand, which means that Go does a pretty good job on hiding any unnecessary complexities, unless you ask for them!

Garbage Collection

Garbage Collection is the process of freeing memory space that is not being used. In other words, the garbage collector sees which objects are out of scope and can no longer be referenced, and it frees the memory space they consume. This process happens in a concurrent manner while a Go program is running, not before or after the execution of a Go program. The documentation of the Go garbage collector implementation states the following:

The GC runs concurrently with mutator threads, is type accurate (aka precise), allows multiple GC threads to run in parallel. It is a concurrent mark and sweep that uses a write barrier. It is non-generational and non-compacting. Allocation is done using size segregated per P allocation areas to minimize fragmentation while eliminating locks in the common case.

There is a lot of terminology here that I will explain in a while. First, however, I will show you a way to look at some parameters of the garbage collection process. Fortunately, the Go standard library offers functions that allow you to study the operation of the garbage collector and learn more about what the garbage collector does secretly. The relevant code is saved as `gColl.go`, and it will be presented in three parts.

The first code segment of `gColl.go` is as follows:

```
package main

import (
    "fmt"
    "runtime"
    "time"
```

```go
)

func printStats(mem runtime.MemStats) {
    runtime.ReadMemStats(&mem)
    fmt.Println("mem.Alloc:", mem.Alloc)
    fmt.Println("mem.TotalAlloc:", mem.TotalAlloc)
    fmt.Println("mem.HeapAlloc:", mem.HeapAlloc)
    fmt.Println("mem.NumGC:", mem.NumGC)
    fmt.Println("-----")
}
```

Note that each time that you need to retrieve the more recent garbage collections statistics, you will need to call the `runtime.ReadMemStats()` function. The purpose of the `printStats()` function is to avoid writing the same Go code all of the time.

The second part of the program is as follows:

```go
func main() {
    var mem runtime.MemStats
    printStats(mem)

    for i := 0; i < 10; i++ {
        s := make([]byte, 50000000)
        if s == nil {
            fmt.Println("Operation failed!")
        }
    }
    printStats(mem)
```

The `for` loop creates multiple, big Go **slices** in order to allocate large amounts of memory and trigger the garbage collector.

The last part of `gColl.go` has the following Go code, which does more memory allocations using Go slices:

```go
    for i := 0; i < 10; i++ {
        s := make([]byte, 100000000)
        if s == nil {
            fmt.Println("Operation failed!")
        }
        time.Sleep(5 * time.Second)
    }
    printStats(mem)
}
```

The output of `gColl.go` on a macOS High Sierra machine is as follows:

```
$ go run gColl.go
mem.Alloc: 66024
mem.TotalAlloc: 66024
mem.HeapAlloc: 66024
mem.NumGC: 0
-----
mem.Alloc: 50078496
mem.TotalAlloc: 500117056
mem.HeapAlloc: 50078496
mem.NumGC: 10
-----
mem.Alloc: 76712
mem.TotalAlloc: 1500199904
mem.HeapAlloc: 76712
mem.NumGC: 20
-----
```

Although you will not examine the operation of the Go garbage collector all of the time, being able to watch the way the Go garbage collector operates on a slow application can save you a lot of time in the long run—I can assure you that you will not regret the time you spend learning about garbage collection in general and, more specifically, about the way the Go garbage collector works.

There is a trick that allows you to get even more detailed output about the way the Go garbage collector operates, which is illustrated by the next command:

```
$ GODEBUG=gctrace=1 go run gColl.go
```

So, if you put `GODEBUG=gctrace=1` in front of any `go run` command, Go will print analytical data about the operation of the garbage collector. The generated data will have the following format:

```
gc 4 @0.025s 0%: 0.002+0.065+0.018 ms clock, 0.021+0.040/0.057/0.003+0.14 ms cpu, 47->47->0 MB, 48 MB goal, 8 P
gc 17 @30.103s 0%: 0.004+0.080+0.019 ms clock, 0.033+0/0.076/0.071+0.15 ms cpu, 95->95->0 MB, 96 MB goal, 8 P
```

The preceding output gives you more information about the heap sizes during the garbage collection process. Let's take the `47->47->0 MB` trinity of values as an example. The first number is the heap size when the garbage collector is about to run. The second value is the heap size when the garbage collector ends its operation. The last value is the size of the live heap.

The Tricolor algorithm

The operation of the Go garbage collector is based on the **tricolor algorithm**, which is the subject of this subsection.

The tricolor algorithm is not unique to Go, and it can be used in other programming languages as well.

Strictly speaking, the official name for the algorithm used in Go is the **tricolor mark-and-sweep algorithm**. It can work concurrently with the program and uses a **write barrier**. This means that when a Go program runs, the Go scheduler is responsible for the scheduling of the application and the garbage collector as if the Go scheduler had to deal with a regular application with multiple **goroutines**! You will learn more about goroutines and the Go scheduler in Chapter 9, *Go Concurrency – Goroutines, Channels, and Pipelines*.

The core idea behind this algorithm is that of **Edsger W. Dijkstra**, **Leslie Lamport**, **A. J. Martin**, **C. S. Scholten**, and **E. F. M. Steffens**. It was first illustrated on a paper, *On-the-fly Garbage Collection: An Exercise in Cooperation*. The primary principle behind the tricolor mark-and-sweep algorithm is that it divides the objects of the heap into three different sets according to their color, which is assigned by the algorithm. I will address the mark-and-sweep algorithm further in the *More about the operation of the Go Garbage Collector* section of this chapter.

Now let's talk about the meaning of each color set. The objects of the **black set** are guaranteed to have no pointers to any object of the white set. However, an object in the **white set** can have a pointer to an object of the black set, because this has no effect on the operation of the garbage collector! The objects of the **grey set** might have pointers to some objects of the white set. Also, the objects of the white set are candidates for garbage collection.

Note that no object can go directly from the black set to the white set, which allows the algorithm to operate and be able to clear the objects in the white set. Additionally, no object of the black set can directly point to an object of the white set.

When the garbage collection begins, all objects are white and the garbage collector visits all of the root objects and colors them grey. The **roots** are the objects that can be directly accessed by the application, which includes global variables and other things on the stack. These objects mostly depend on the Go code of a particular program. After this, the garbage collector picks a grey object, makes it black, and starts searching to determine if that object has pointers to other objects of the white set. This means that when a grey object is being scanned for pointers to other objects, it is colored black. If that scan discovers that this particular object has one or more pointers to a white object, it puts that white object in the grey set. This process keeps going for as long as objects exist in the grey set. After that, the objects in the white set are unreachable and their memory space can be reused. Therefore, at this point, the elements of the white set are said to be garbage collected.

If an object of the grey set becomes unreachable at some point in a garbage collection cycle, it will not be collected in that garbage collection cycle but rather in the next one! Although this is not an optimal situation, it is not that bad.

During this process, the running application is called the **mutator**. The mutator runs a small function named **write barrier** that is executed each time a pointer in the heap is modified. If the pointer of an object in the heap is modified, which means that this object is now reachable, the write barrier colors it grey and puts it in the grey set.

The mutator is responsible for the invariant that no element of the black set has a pointer to an element of the white set. This is accomplished with the help of the write barrier function. Failing to accomplish this invariant will ruin the garbage collection process, and it will most likely crash your program in an ugly and undesired way.

As a result, the heap can be viewed as a graph of connected objects, as shown in the following diagram, which demonstrates a single phase of a garbage collection cycle:

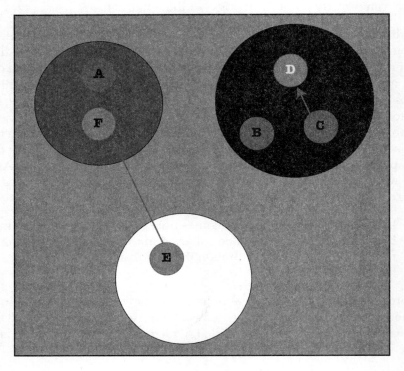

The Go garbage collector represents the heap of a program as a graph

Thus, there are three different colors: black, white, and grey. When the algorithm begins, all objects are colored white. As the algorithm continues, white objects are moved into one of the other two sets. The objects that are left in the white set are the ones that will be cleared at some point.

In the preceding graph, you can see that while object **E**, which is in the white set, can access object **F**, it cannot be accessed by any other object because no other object points to object **E**, which makes it a perfect candidate for garbage collection! Additionally, objects **A**, **B**, and **C** are root objects and are always reachable and therefore cannot be garbage collected.

Can you guess what will happen next in that graph? Well, it is not that difficult to fathom that the algorithm will have to process the remaining elements of the grey set, which means that objects **A** and **F** will go into the black set. Object **A** will go into the black set because it is a root element, and **F** will go into the black set because it does not point to any other object while it is in the grey set. After object **A** is garbage collected, object **F** will become unreachable and will be garbage collected in the next cycle of the garbage collector, as an unreachable object cannot magically become reachable in the next iteration of the garbage collection cycle.

The Go garbage collection can also be applied to variables such as **channels**! When the garbage collector finds out that a channel is unreachable and that the channel variable can no longer be accessed, it will free its resources even if the channel has not been closed! You will learn more about channels in `Chapter 9`, *Go Concurrency – Goroutines, Channels and Pipelines*.

Go allows you to initiate a garbage collection manually by putting a `runtime.GC()` statement in your Go code. However, keep in mind that `runtime.GC()` will block the caller, and it might block the entire program, especially if you are running a very busy Go program with many objects. This happens mainly because you cannot perform garbage collections while everything else is rapidly changing, as this will not give the garbage collector the opportunity to identify clearly the members of the white, black, and grey sets! This garbage collection status is also called the **garbage collection safe-point**.

You can find the long and relatively advanced Go code of the garbage collector at `https://github.com/golang/go/blob/master/src/runtime/mgc.go`. You can study this if you want to learn even more about the garbage collection operation. You can even make changes to that code if you are brave enough!

The Go garbage collector is always being improved by the Go team, mainly by trying to make it faster by lowering the number of scans it needs to perform over the data of the three sets . However, despite the various optimizations, the general idea behind the algorithm remains the same.

More about the operation of the Go Garbage Collector

This section will explore the Go garbage collector further and present additional information about its activities.

The main concern of the Go garbage collector is low latency, which basically means short pauses in its operation in order to have a real-time operation. On the other hand, what a program does all the time is to create new objects and manipulate existing objects with pointers. This process can end up creating objects that cannot be accessed any longer because no pointers exist that point to these objects. Such objects are now garbage waiting for the garbage collector to clean them up and free their memory space. After that, the memory space that has been freed is ready to be used again.

The classic algorithm for garbage collection is called mark-and-sweep, and it is the simplest algorithm in use. The way that the **mark-and-sweep algorithm** works is pretty simple and easy to understand: the algorithm stops the program execution (**stop-the-world garbage collector**) in order to visit all of the accessible objects of the heap of a program and *marks* them. After that, it *sweeps* the inaccessible objects. During the mark phase of the algorithm, each object is marked as white, grey, or black. The children of a grey object are colored grey, whereas the original grey object is now colored black. The sweep phase begins when there are no more grey objects to examine. This technique works because there are no pointers from the black set to the white set, which is a fundamental invariant of the algorithm.

Although the mark-and-sweep algorithm is simple, it suspends the execution of the program while it is running, which means that it adds **latency** to the actual process. Go tries to lower that particular latency by running the garbage collector as a concurrent process and using the tricolor algorithm described in the previous section. However, other processes can move pointers or create new objects while the garbage collector runs concurrently. This fact can make things pretty difficult for the garbage collector. As a result, the principal point that allows the tricolor algorithm to run concurrently is to be able to maintain the fundamental invariant of the mark-and-sweep algorithm-no object of the black set can point to an object of the white set.

The solution to this problem is to fix all of the cases that can cause a problem for the algorithm! Therefore, new objects must go to the grey set, because this way the fundamental invariant of the mark-and-sweep algorithm cannot be altered. Additionally, when a pointer of the program is moved, you color the object to which the pointer points as grey. You can say that the grey set acts like a barrier between the white set and the black set. Last, each time a pointer is moved, some Go code gets automatically executed, which is the **write barrier** mentioned earlier that does some recoloring.

The **latency** introduced by the execution of the write barrier code is the price you have to pay for being able to run the garbage collector concurrently.

Note that the **Java** programming language has many garbage collectors that are highly configurable with the help of multiple parameters. One of these Java garbage collectors is called G1, and it is recommended for low-latency applications.

 It is really important to remember that the Go garbage collector is a real-time garbage collector, which runs concurrently with the other **goroutines** of a Go program and only optimizes for low latency.

In Chapter 11, *Code Testing, Optimization and Profiling*, you will learn how to represent graphically the performance of a program. This chapter also includes information about the operations of the Go garbage collector.

That's enough about garbage collection. The next topic covered will be **unsafe code** and the unsafe standard Go package.

Unsafe code

Unsafe code is Go code that bypasses the type safety and the memory security of Go. Most of the time, unsafe code is related to pointers. However, keep in mind that using unsafe code can be dangerous for your programs, so if you are not completely sure that you need to use unsafe code in one of your programs, do not use it!

The use of unsafe code will be illustrated in the unsafe.go program, which is presented in three parts.

The first part of unsafe.go is as follows:

```
package main

import (
    "fmt"
    "unsafe"
)
```

Note that in order to use unsafe code, you will need to import the unsafe standard Go package.

The second part of the program occurs with the following Go code:

```
func main() {
    var value int64 = 5
    var p1 = &value
    var p2 = (*int32)(unsafe.Pointer(p1))
```

Note the use of the unsafe.Pointer() function here which allows you, at your own risk, to create a int32 pointer named p2 that points to a int64 variable named value, which is accessed using the p1 pointer. Any Go pointer can be converted to unsafe.Pointer.

A pointer of the unsafe.Pointer type can override the type system of Go. This is unquestionably fast, but it can also be dangerous if used incorrectly or carelessly. Additionally, it gives developers more control over data.

The last part of unsafe.go contains the following Go code:

```
    fmt.Println("*p1: ", *p1)
    fmt.Println("*p2: ", *p2)
    *p1 = 5434123412312431212
    fmt.Println(value)
    fmt.Println("*p2: ", *p2)
    *p1 = 54341234
    fmt.Println(value)
    fmt.Println("*p2: ", *p2)
}
```

You can **dereference a pointer** and get, use, or set its value using the star character (*).

If you execute unsafe.go, you will get the following output:

```
$ go run unsafe.go
*p1:  5
*p2:  5
5434123412312431212
*p2:  -930866580
54341234
*p2:  54341234
```

What does this output tell us? It tells us that a 32-bit pointer cannot store a 64-bit integer!

As you will see in the next section, the functions of the unsafe package can do many more interesting things with memory.

About the unsafe package

Now that you have seen the unsafe package in action, it's a good time to talk about what makes it a special kind of a package!

First of all, if you look at the source code of the unsafe package, you might be a little surprised. On a macOS High Sierra system with Go version 1.9.1 that is installed using Homebrew (https://brew.sh/), the source code of the unsafe package is located at /usr/local/Cellar/go/1.9.1/libexec/src/unsafe/unsafe.go and its contents without the comments are as follows:

```
$ cd /usr/local/Cellar/go/1.9.1/libexec/src/unsafe/
$ grep -v '^//' unsafe.go | grep -v '^$'
package unsafe
type ArbitraryType int
type Pointer *ArbitraryType
func Sizeof(x ArbitraryType) uintptr
func Offsetof(x ArbitraryType) uintptr
func Alignof(x ArbitraryType) uintptr
```

OK. Where is the rest of the Go code of the unsafe package? The answer to that question is relatively simple: the Go compiler implements the unsafe package when you import it into your programs.

Many low-level packages such as runtime, syscall, and os constantly use the unsafe package.

Another example of the unsafe package

In this subsection, you will learn more about the unsafe package and its capabilities with the help of another small Go program named moreUnsafe.go. This program will be presented in three parts. What moreUnsafe.go does is to access all the elements of an array using pointers.

Understanding Go Internals

The first part of the program is as follows:

```
package main

import (
    "fmt"
    "unsafe"
)
```

The second part of `moreUnsafe.go` comes with the following Go code:

```
func main() {
    array := [...]int{0, 1, -2, 3, 4}
    pointer := &array[0]
    fmt.Print(*pointer, " ")
    memoryAddress := uintptr(unsafe.Pointer(pointer)) + unsafe.Sizeof(array[0])

    for i := 0; i < len(array)-1; i++ {
            pointer = (*int)(unsafe.Pointer(memoryAddress))
            fmt.Print(*pointer, " ")
            memoryAddress = uintptr(unsafe.Pointer(pointer)) + unsafe.Sizeof(array[0])
    }
```

At first, the `pointer` variable points to the memory address of `array[0]`, which is the first element of the array of integers. Next, the `pointer` variable that points to an integer value is converted to an `unsafe.Pointer()` function and then to `uintptr`. The result is stored in `memoryAddress`.

The value of `unsafe.Sizeof(array[0])` is what gets you to the next element of the array, because this is the memory occupied by each array element. This value is added to the `memoryAddress` variable in each iteration of the `for` loop, which allows you to get the memory address of the next array element. The `*pointer` notation dereferences the pointer and returns the stored integer value.

The third part is as follows:

```
    fmt.Println()
    pointer = (*int)(unsafe.Pointer(memoryAddress))
    fmt.Print("One more: ", *pointer, " ")
    memoryAddress = uintptr(unsafe.Pointer(pointer)) + unsafe.Sizeof(array[0])
    fmt.Println()
}
```

In the last part, you are trying to access an element of the array that does not exist using pointers and memory addresses. The Go compiler cannot catch such a logical error due to the use of the `unsafe` package and therefore will return something inaccurate.

Executing `moreUnsafe.go` will create the following output:

```
$ go run moreUnsafe.go
0 1 -2 3 4
One more:  842350722816
```

You have now accessed all of the elements of a Go array using pointers! However, the real problem here is that when you tried to access an invalid array element, the program did not complain and returned a random number instead.

Calling C code from Go

Although the intention of Go is to make your programming experience better and save you from having to deal with the quirks of C, C remains a very capable programming language that is still useful. This means that there are situations, such as using a database or a device driver written in C, which still require the use of C. This means that you will need to work with C code in your Go projects.

 If you find yourselves using this capability several times in the same project, you might need to reconsider your approach or your choice of programming language!

Calling C code from Go using the same file

The simplest way to call C code from a Go program is to include the C code in your Go source file. This requires a special treatment, but it is pretty fast and not that difficult to do.

The name of the Go source file that contains both C and Go code will be `cGo.go`, and it will be presented in three parts.

The first part of the Go source file is as follows:

```
package main

//#include <stdio.h>
//void callC() {
//    printf("Calling C code!\n");
//}
import "C"
```

 As you can see, the C code is included in the comments of the Go program. However, the go tool knows what to do with these kinds of comments because of the use of the C Go package.

The second part of the program contains the following Go code:

```
import "fmt"

func main() {
```

All of the other packages should be imported separately.

The last part of cGo.go contains the following code:

```
    fmt.Println("A Go statement!")
    C.callC()
    fmt.Println("Another Go statement!")
}
```

Thus, in order to execute the callC() C function, you will need to call it as C.callC().

Executing cGo.go will create the following output:

```
$ go run cGo.go
A Go statement!
Calling C code!
Another Go statement!
```

Calling C code from Go using separate files

Now let's continue learning how to call C code from a Go program when the C code is located in a separate file.

Chapter 2

First, let me explain the imaginary problem that we will solve with our program. We will need to use two C functions that we have implemented in the past and that we do not want or cannot rewrite in Go.

The C code

This subsection will present the C code for the example which comes in two files: `callC.h` and `callC.c`. The include file (`callC.h`) contains the following code:

```c
#ifndef CALLC_H
#define CALLC_H

void cHello();
void printMessage(char* message);

#endif
```

The C source file (`callC.c`) contains the following C code:

```c
#include <stdio.h>
#include "callC.h"

void cHello() {
    printf("Hello from C!\n");
}

void printMessage(char* message) {
    printf("Go send me %s\n", message);
}
```

Both the `callC.c` and `callC.h` files are stored in a separate directory, which in this case is `callClib`. You can use any directory name you want, however.

The actual C code is not important as long as you call the right C functions with the correct type and number of parameters. There is nothing in the C code that tells you that it will be used from a Go program. You should look at the Go code for the juicy part.

The Go code

This subsection will address the Go source code of the example, which will be called `callC.go` and will be presented in three parts.

The first part of `callC.go` includes the following Go code:

```
package main

// #cgo CFLAGS: -I${SRCDIR}/callClib
// #cgo LDFLAGS: ${SRCDIR}/callC.a
// #include <stdlib.h>
// #include <callC.h>
import "C"
```

The single most important Go statement of the entire Go source file is the inclusion of the C package using a separate `import` statement. However, C is a virtual Go package that just tells `go build` to preprocess its input file using the `cgo` tool before the Go compiler processes the file! You can still see that you need to use comments to inform the Go program about the C code. In this case, you tell `callC.go` where to find the `callC.h` file as well as where to find the `callC.a` library file that we will create in a little while—such lines begin with `#cgo`.

The second part of the program is as follows:

```
import (
    "fmt"
    "unsafe"
)

func main() {
    fmt.Println("Going to call a C function!")
    C.cHello()
```

The last part of `callC.go` follows:

```
    fmt.Println("Going to call another C function!")
    myMessage := C.CString("This is Mihalis!")
    defer C.free(unsafe.Pointer(myMessage))
    C.printMessage(myMessage)

    fmt.Println("All perfectly done!")
}
```

Chapter 2

In order to pass a string to a C function from Go, you will need to create a C string using `C.CString()`. Additionally, you will need a `defer` statement in order to free the memory space of the C string when it is no longer needed. The `defer` statement includes a call to `C.free()` and another one to `unsafe.Pointer()`.

In the next section, you will see how to compile and execute `callC.go`.

Mixing Go and C code

Now that you have the C and the Go code, it is time to learn what to do next in order to execute the Go file that calls the C code.

The good news is that you do not need to do anything particularly difficult because all of the critical information is contained in the Go file. The only critical thing that you will need to do is to compile the C code in order to create a library, which requires the execution of the following commands:

```
$ ls -l callClib/
total 16
-rw-r--r--@ 1 mtsouk    staff    162 Oct 31 18:14 callC.c
-rw-r--r--@ 1 mtsouk    staff     89 Oct 31 18:14 callC.h
$ gcc -c callClib/*.c
$ ls -l callC.o
-rw-r--r--  1 mtsouk    staff    944 Oct 31 18:14 callC.o
$ file callC.o
callC.o: Mach-O 64-bit object x86_64
$ ar rs callC.a *.o
ar: creating archive callC.a
$ ls -l callC.a
-rw-r--r--  1 mtsouk    staff   1152 Oct 31 18:15 callC.a
$ file callC.a
callC.a: current ar archive random library
$ rm callC.o
```

After that, you will have a file named `callC.a` located in the same directory as the `callC.go` file. The `gcc` executable is the name of the **C compiler**.

Now you are ready to compile the file with the Go code and create a new executable file:

```
$ go build callC.go
$ ls -l callC
-rwxr-xr-x  1 mtsouk    staff  2322000 Oct 31 18:16 callC
$ file callC
callC: Mach-O 64-bit executable x86_64
```

[63]

Understanding Go Internals

Executing the `callC` executable file will create the following output:

```
$ ./callC
Going to call a C function!
Hello from C!
Going to call another C function!
Go send me This is Mihalis!
All perfectly done!
```

If you will call a small amount of C code, then using a single Go file for both the C and Go code is highly recommended because of its simplicity. However, if you will do something more complex and advanced, creating a static C library should be your preferred method.

Calling Go functions from C code

It is also possible to call a Go function from your C code. Therefore, this section will present a small example where two Go functions will be called from a C program. The Go package will be converted into a **C shared library** that will be used in the C program.

The Go package

This subsection will present you with the code of the Go package that will be used in a C program. The name of the Go package needs to be `main`, but its filename can be anything you want. In this case, the filename will be `usedByC.go`, and it will be presented in three parts.

You will learn more about Go packages in `Chapter 6`, *What You Might Not Know About Go Packages*.

The first part of the code of the Go package is as follows:

```
package main

import "C"

import (
    "fmt"
)
```

[64]

As I mentioned before, it is mandatory that you name the Go package as `main`. You will also need to import the `C` package in your Go code.

The second part of the program contains the following Go code:

```
//export PrintMessage
func PrintMessage() {
    fmt.Println("A Go function!")
}
```

Each Go function that will be called by the C code needs to be exported first. This means that you should put a comment line starting with `//export` before its implementation. After `//export`, you will need to put the name of the function because this is what the C code will use.

The last part of `usedByC.go` is as follows:

```
//export Multiply
func Multiply(a, b int) int {
    return a * b
}

func main() {
}
```

The `main()` function of `usedByC.go` needs no code because it will not be exported and therefore used by the C program. Additionally, as you also want to export the `Multiply()` function, you will need to put `//export Multiply` before its implementation.

After this, you will need to generate a C shared library from the Go code by executing the following command:

```
$ go build -o usedByC.o -buildmode=c-shared usedByC.go
```

The preceding command will generate two files named `usedByC.h` and `usedByC.o`:

```
$ ls -l usedByC.*
-rw-r--r--@ 1 mtsouk  staff        204 Oct 31 20:37 usedByC.go
-rw-r--r--  1 mtsouk  staff       1365 Oct 31 20:40 usedByC.h
-rw-r--r--  1 mtsouk  staff    2329472 Oct 31 20:40 usedByC.o
$ file usedByC.o
usedByC.o: Mach-O 64-bit dynamically linked shared library x86_64
```

You should not make any changes to usedByC.h.

The C code

The relevant C code can be found in the willUseGo.c source file, which will be presented in two parts. The first part of willUseGo.c is as follows:

```
#include <stdio.h>
#include "usedByC.h"

int main(int argc, char **argv) {
   GoInt x = 12;
   GoInt y = 23;

   printf("About to call a Go function!\n");
   PrintMessage();
```

If you already know C, you should understand why you need to include usedByC.h. This is the way that the C code knows about the available functions of a library.

The second part of the C program follows next:

```
   GoInt p = Multiply(x,y);
   printf("Product: %d\n",(int)p);
   printf("It worked!\n");
   return 0;
}
```

The GoInt p variable is needed for getting an integer value from a Go function, which is converted to a C integer using the (int) p notation.

Compiling and executing willUseGo.c on a macOS High Sierra machine will create the following output:

```
$ gcc -o willUseGo willUseGo.c ./usedByC.o
$ ./willUseGo
About to call a Go function!
A Go function!
Product: 276
It worked!
```

The defer keyword

The `defer` keyword postpones the execution of a function until the surrounding function returns. It is widely used in file input and output operations because it saves you from having to remember when to close an opened file: the `defer` keyword allows you to put the function call that closes an opened file near to the function call that opened it. As you will learn about the use of `defer` in file-related operations in Chapter 8, *Telling a Unix System What to Do*, this section will present a different usage of `defer`. You will also see `defer` in action in the section that talks about the `panic()` and `recover()` built-in Go functions.

It is very important to remember that **deferred functions** are executed in **Last In First Out (LIFO)** order after the return of the surrounding function. Put simply, this means that if you `defer` function `f1()` first, function `f2()` second, and function `f3()` third in the same surrounding function, when the surrounding function is about to return, function `f3()` will be executed first, function `f2()` will be executed second, and function `f1()` will be the last one to get executed.

As this definition of `defer` is a little unclear, I think that you will understand the use of `defer` a little better by looking at the Go code and the output of the `defer.go` program, which will be presented in three parts.

The first part of the program is as follows:

```
package main

import (
    "fmt"
)

func d1() {
    for i := 3; i > 0; i-- {
        defer fmt.Print(i, " ")
    }
}
```

Apart from the `import` block, the preceding Go code implements a function named `d1()` with a `for` loop and a `defer` statement that will be executed three times.

The second part of `defer.go` contains the following Go code:

```
func d2() {
    for i := 3; i > 0; i-- {
        defer func() {
            fmt.Print(i, " ")
```

```
            }()
        }
        fmt.Println()
}
```

In this part of the code, you can see the implementation of another function that is named `d2()`. The `d2()` function also contains a `for` loop and a `defer` statement that will be also executed three times. However, this time the `defer` keyword is applied to an **anonymous function** instead of a single `fmt.Print()` statement. Additionally, the anonymous function takes no parameters.

The last part of the Go code is as follows:

```
func d3() {
    for i := 3; i > 0; i-- {
        defer func(n int) {
            fmt.Print(n, " ")
        }(i)
    }
}

func main() {
    d1()
    d2()
    fmt.Println()
    d3()
    fmt.Println()
}
```

Apart from the `main()` function that calls the `d1()`, `d2()`, and `d3()` functions, you can also see the implementation of the `d3()` function, which has a `for` loop that uses the `defer` keyword on an anonymous function. However, this time the anonymous function requires one integer parameter named `n`. The Go code tells us that the `n` parameter takes its value from the `i` variable used in the `for` loop.

Executing `defer.go` will create the following output:

```
$ go run defer.go
1 2 3
0 0 0
1 2 3
```

You will most likely find the generated output complicated and challenging to understand. This underscores the fact that the operation and the results of the use of `defer` can be tricky if your code is not clear and unambiguous.

Let's examine the results in order to get a better idea of how tricky `defer` can be if you do not pay close attention to your code. We will start with the first line of the output (1 2 3), which is generated by the `d1()` function. The values of `i` in `d1()` are 3, 2, and 1 in that order. The function that is deferred in `d1()` is the `fmt.Print()` statement. As a result, when the `d1()` function is about to return, you get the three values of the `i` variable of the `for` loop in reverse order because deferred functions are executed in LIFO order.

Now, let's inspect the second line of the output that is produced by the `d2()` function. It is really strange that we got three zeros instead of 1 2 3 in the output. The reason for this, however, is relatively simple. After the `for` loop has ended, the value of `i` is 0, because it is that value of `i` that made the `for` loop terminate. However, the tricky part here is that the deferred anonymous function is evaluated after the `for` loop ends, because it has no parameters. This means that is evaluated three times for an `i` value of 0, hence the generated output! This kind of confusing code is what might lead to the creation of nasty bugs in your projects, so try to avoid it!

Also, we will talk about the third line of the output, which is generated by the `d3()` function. Due to the parameter of the anonymous function, each time the anonymous function is deferred, it gets and uses the current value of `i`. As a result, each execution of the anonymous function has a different value to process, thus the generated output.

After this, it should be clear that the best approach to the use of `defer` is the third one, which is exhibited in the `d3()` function. This is so because you intentionally pass the desired variable in the anonymous function in an easy to understand way.

Panic and Recover

This section will present you with a tricky technique that was first mentioned in Chapter 1, *Go and the Operating System*. This technique involves the use of the `panic()` and `recover()` functions, and it will be presented in `panicRecover.go`, which you will review in three parts.

Strictly speaking, `panic()` is a built-in Go function that terminates the current flow of a Go program and starts panicking! On the other hand, the `recover()` function, which is also a built-in Go function, allows you to take back the control of a **goroutine** that just panicked using `panic()`.

The first part of the program follows:

```
package main

import (
   "fmt"
)

func a() {
   fmt.Println("Inside a()")
   defer func() {
        if c := recover(); c != nil {
              fmt.Println("Recover inside a()!")
        }
   }()
   fmt.Println("About to call b()")
   b()
   fmt.Println("b() exited!")
   fmt.Println("Exiting a()")
}
```

Apart from the `import` block, this part includes the implementation of the `a()` function. The most important part of the `a()` function is the `defer` block of code, which implements an anonymous function that will be called when there is a call to `panic()`.

The second code segment of `panicRecover.go` follows next:

```
func b() {
   fmt.Println("Inside b()")
   panic("Panic in b()!")
   fmt.Println("Exiting b()")
}
```

The last part of the program, which illustrates the `panic()` and `recover()` functions, is as follows:

```
func main() {
    a()
    fmt.Println("main() ended!")
}
```

Executing `panicRecover.go` will create the following output:

```
$ go run panicRecover.go
Inside a()
About to call b()
Inside b()
Recover inside a()!
main() ended!
```

What just happened here is really impressive! However, as you can see from the output, the `a()` function did not end normally, because its last two statements did not get executed:

```
fmt.Println("b() exited!")
fmt.Println("Exiting a()")
```

Nevertheless, the good thing is that `panicRecover.go` ended according to our will without panicking because the anonymous function used in `defer` took control of the situation! Also note that function `b()` knows nothing about function `a()`. However, function `a()` contains Go code that handles the panic condition of the `b()` function!

Using the panic function on its own

You can also use the `panic()` function on its own without any attempt to **recover**, and this subsection will show you its results using the Go code of `justPanic.go`, which will be presented in two parts.

The first part of `justPanic.go` follows next:

```
package main

import (
    "fmt"
    "os"
)
```

As you can see, the use of `panic()` does not require any extra Go packages.

The second part of `justPanic.go` is shown in the following Go code:

```
func main() {
    if len(os.Args) == 1 {
        panic("Not enough arguments!")
    }

    fmt.Println("Thanks for the argument(s)!")
}
```

If your Go program does not have at least one command-line argument, it will call the `panic()` function. The `panic()` function takes one parameter, which is the error message that you want to print on the screen.

Executing `justPanic.go` on a macOS High Sierra machine will create the following output:

```
$ go run justPanic.go
panic: Not enough arguments!
goroutine 1 [running]:
main.main()
        /Users/mtsouk/ch2/code/justPanic.go:10 +0x9e
exit status 2
```

Thus, using the `panic()` function on its own will terminate the Go program without giving you the opportunity to recover! Therefore use of the `panic()` and `recover()` pair is much more practical and professional than just using `panic()` alone.

> The output of the `panic()` function looks like the output of the `Panic()` function from the `log` package. However, the `panic()` function sends nothing to the logging service of your Unix machine.

Two handy Unix utilities

There are times that a Unix program fails for some unknown reason or does not perform well, and you want to find out why without having to rewrite your code and add a plethora of debugging statements.

Chapter 2

Consequently, this section presents two command-line utilities that allow you to see the C system calls executed by an executable file. The names of the two tools are strace(1) and dtrace(1), and they allow you to inspect the operation of a program.

Remember that, at the end of the day, all programs that work on Unix machines end up using C system calls to communicate with the Unix kernel and perform most of their tasks.

Although both tools can work with the go run command, you will get less unrelated output if you first create an executable file using go build and use that file. This occurs mainly because, as you already know, go run creates various temporary files before actually running your Go code, and both tools will see that and try to display information about the temporary files, which is not what you want.

The strace tool

The strace(1) command-line utility allows you to trace system calls and signals. As strace(1) is not available on macOS, this section will use a Debian Linux machine to showcase strace(1).

The strace(1) tool only works on Linux machines.

The output that strace(1) generates looks like the following:

```
$ strace ls
execve("/bin/ls", ["ls"], [/* 15 vars */]) = 0
brk(0)                                  = 0x186c000
fstat(3, {st_mode=S_IFREG|0644, st_size=35288, ...}) = 0
```

The strace(1) output displays each system call with its parameters as well as its return value. Note that in the Unix world, a return value of 0 is a good thing!

[73]

In order to process a binary file, you will need to put the `strace(1)` command in front of the executable that you want to process. However, you will need to interpret the output on your own in order to use it to make useful conclusions. The good thing is that tools like `grep(1)` can get you the output that you are actually seeking:

```
$ strace find /usr 2>&1 | grep ioctl
ioctl(0, SNDCTL_TMR_TIMEBASE or SNDRV_TIMER_IOCTL_NEXT_DEVICE or TCGETS,
0x7ffe3bc59c50) = -1 ENOTTY (Inappropriate ioctl for device)
ioctl(1, SNDCTL_TMR_TIMEBASE or SNDRV_TIMER_IOCTL_NEXT_DEVICE or TCGETS,
0x7ffe3bc59be0) = -1 ENOTTY (Inappropriate ioctl for device)
```

The `strace(1)` tool can count time, calls, and errors for each system call when used with the `-c` command-line option:

```
$ strace -c find /usr 1>/dev/null
% time     seconds  usecs/call     calls    errors syscall
------ ----------- ----------- --------- --------- ----------------
 82.88    0.063223           2     39228           getdents
 16.60    0.012664           1     19587           newfstatat
  0.16    0.000119           0     19618        13 open
```

As the normal program output is printed in standard output form, and the output of `strace(1)` is printed in standard error form, the previous command discards the output of the command that is examined and shows the output of `strace(1)`. As you can see from the last line of the output, the `open(2)` system call was called 19,618 times, generated 13 errors, and took about 0.16% percent of the execution time of the entire command, or about 0.000119 seconds.

The dtrace tool

Although debugging utilities such as `strace(1)` and `truss(1)` can trace system calls produced by a process, they can be slow and therefore not appropriate for solving performance problems on busy Unix systems. Another tool named **DTrace** allows you to see what happens behind the scenes on a system-wide basis without the need to modify or recompile anything. It also allows you to work on production systems and watch running programs or server processes dynamically without introducing a big overhead.

Although there is a version of `dtrace(1)` that works on Linux, the `dtrace(1)` tool works best on macOS and the other FreeBSD variants.

This subsection will use the dtruss(1) command-line utility that comes with macOS, which is just a dtrace(1) script that shows the system calls of a process and saves you from having to write dtrace(1) code. Note that both dtrace(1) and dtruss(1) need root privileges to run.

The output that dtruss(1) generates looks like the following:

```
$ sudo dtruss godoc
ioctl(0x3, 0x80086804, 0x7FFEEFBFEC20)      = 0 0
close(0x3)           = 0 0
access("/AppleInternal/XBS/.isChrooted\0", 0x0, 0x0)   = -1 Err#2
thread_selfid(0x0, 0x0, 0x0)         = 1895378 0
geteuid(0x0, 0x0, 0x0)           = 0 0
getegid(0x0, 0x0, 0x0)           = 0 0
```

So, dtruss(1) works in the same way as the strace(1) utility. Analogously to strace(1), dtruss(1) will print system call counts when used with the -c parameter:

```
$ sudo dtruss -c go run unsafe.go 2>&1
CALL                                    COUNT
access                                  1
bsdthread_register                      1
getuid                                  1
ioctl                                   1
issetugid                               1
kqueue                                  1
write                                   1
read                                    244
kevent                                  474
fcntl                                   479
lstat64                                 553
```

The preceding output will quickly inform you about potential bottlenecks in your Go code or allow you to compare the performance of two different command-line programs.

You need to get used to using utilities such as strace(1), dtrace(1), and dtruss(1), but such tools can make your lives so much easier and better. I strongly suggest that you start learning at least one such tool right now!

Understanding Go Internals

You can learn more about the `dtrace(1)` utility by reading *DTrace: Dynamic Tracing in Oracle Solaris, Mac OS X and FreeBSD* by **Brendan Gregg** and **Jim Mauro** (Prentice Hall, 2011) and by visiting `http://dtrace.org/`. Please bear in mind that `dtrace(1)` is much more powerful than `strace(1)` because it has its own programming language. However, `strace(1)` is more versatile when all you want to do is to watch the system calls of an executable file.

Your Go environment

This section will help you find out information about your current Go environment using the functions and properties of the `runtime` package. The name of the program that will be developed in this section is `goEnv.go`, and it will be presented in two parts.

The first part of `goEnv.go` follows:

```
package main

import (
    "fmt"
    "runtime"
)
```

As you will see shortly, the `runtime` package contains functions and properties that will reveal the desired information. The second portion of the code of `goEnv.go` contains the implementation of the `main()` function:

```
func main() {
    fmt.Print("You are using ", runtime.Compiler, " ")
    fmt.Println("on a", runtime.GOARCH, "machine")
    fmt.Println("Using Go version", runtime.Version())
    fmt.Println("Number of CPUs:", runtime.NumCPU())
    fmt.Println("Number of Goroutines:", runtime.NumGoroutine())
}
```

Executing `goEnv.go` on a macOS High Sierra machine with Go version 1.9.2 will create the following output:

```
$ go run goEnv.go
You are using gc on a amd64 machine
Using Go version go1.9.2
Number of CPUs: 8
Number of Goroutines: 1
```

The same program generates the following output on a Debian Linux machine with Go version 1.3.3:

```
$ go run goEnv.go
You are using gc on a amd64 machine
Using Go version go1.3.3
Number of CPUs: 1
Number of Goroutines: 4
```

The real benefit you can get from being able to find information about your Go environment, however, is illustrated in the next program, named `requiredVersion.go`, which tells you if you are using Go version 1.8 or higher:

```
package main

import (
    "fmt"
    "runtime"
    "strconv"
    "strings"
)

func main() {
    myVersion := runtime.Version()
    major := strings.Split(myVersion, ".")[0][2]
    minor := strings.Split(myVersion, ".")[1]
    m1, _ := strconv.Atoi(string(major))
    m2, _ := strconv.Atoi(minor)

    if m1 == 1 && m2 < 8 {
        fmt.Println("Need Go version 1.8 or higher!")
        return
    }

    fmt.Println("You are using Go version 1.8 or higher!")
}
```

The `strings` Go standard package is used for splitting the Go version string you get from `runtime.Version()` in order to get its first two parts, whereas the `strconv.Atoi()` function is used for converting a string to an integer.

Executing `requiredVersion.go` on the macOS High Sierra machine will create the following output:

```
$ go run requiredVersion.go
You are using Go version 1.8 or higher!
```

If you run `requiredVersion.go` on the Debian Linux machine, however, it will generate the following output:

```
$ go run requiredVersion.go
Need Go version 1.8 or higher!
```

Thus using the Go code of `requiredVersion.go`, you will be able to identify whether your Unix machine has the required Go version or not.

The Go Assembler

This section will briefly talk about the assembly language and the **Go assembler**, which is a Go tool that allows you to see the assembly language used by the Go compiler.

As an example, you can see the assembly language of the `goEnv.go` program from the previous section by executing the following command:

```
$ GOOS=darwin GOARCH=amd64 go tool compile -S goEnv.go
```

The value of the `GOOS` variable defines the name of the target operating system, whereas the value of the `GOARCH` variable defines the compilation architecture. The preceding command was executed on a macOS High Sierra machine, hence the use of the `darwin` value for the `GOOS` variable.

The output of the preceding command is pretty large, even for a small program such as `goEnv.go`. Some of the output is shown next:

```
"".main STEXT size=859 args=0x0 locals=0x118
    0x0000 00000 (goEnv.go:8)       TEXT      "".main(SB), $280-0
    0x00be 00190 (goEnv.go:9)       PCDATA    $0, $1
    0x0308 00776 (goEnv.go:13)      PCDATA    $0, $5
    0x0308 00776 (goEnv.go:13)      CALL      runtime.convT2E64(SB)
"".init STEXT size=96 args=0x0 locals=0x8
    0x0000 00000 (<autogenerated>:1) TEXT     "".init(SB), $8-0
    0x0000 00000 (<autogenerated>:1) MOVQ     (TLS), CX
    0x001d 00029 (<autogenerated>:1) FUNCDATA          $0, gclocals
d4dc2f11db048877dbc0f60a22b4adb3(SB)
```

```
    0x001d 00029 (<autogenerated>:1) FUNCDATA        $1, gclocals
33cdeccccebe80329f1fdbee7f5874cb(SB)
```

The lines that contain the FUNCDATA and PCDATA directives are read and used by the Go **garbage collector** and are automatically generated by the Go compiler.

An equivalent variant of the preceding command follows:

```
$ GOOS=darwin GOARCH=amd64 go build -gcflags -S goEnv.go
```

The list of valid GOOS values includes android, darwin, dragonfly, freebsd, linux, nacl, netbsd, openbsd, plan9, solaris, windows, and zos. On the other hand, the list of valid GOARCH values includes 386, amd64, amd64p32, arm, armbe, arm64, arm64be, ppc64, ppc64le, mips, mipsle, mips64, mips64le, mips64p32, mips64p32le, ppc, s390, s390x, sparc, and sparc64.

> If you are really interested in the Go assembler and you want to learn more, visit https://golang.org/doc/asm.

Node Trees

Now it's time for something completely different and low level. You are free to skip this section if you like, but I bet that you are not reading this book for its easy topics but for the difficult ones, such as **node trees**!

> The go tool 6g -W test.go command does not work on newer Go versions. You should use go tool compile -W test.go instead.

A Go node is a struct with a large number of properties. You will learn more about defining and using Go **structures** in Chapter 4, *The Uses of Composite Types*. Everything in a Go program is being parsed and analyzed by the modules of the Go compiler according to the grammar of the Go programming language. The final product of this analysis is a **tree** that is specific to the provided Go code, and it represents the program in a different way that is suited for the compiler rather than for the developer.

Understanding Go Internals

This section will start with the following Go code, which is saved as `nodeTree.go`. It is presented here as an example of the kind of low-level information that the `go` tool can provide:

```
package main

import (
    "fmt"
)

func main() {
    fmt.Println("Hello there!")
}
```

The Go code of `nodeTree.go` is pretty easy to understand, so you will not be surprised by its output, which comes next:

```
$ go run nodeTree.go
Hello there!
```

Now it's time to see some of the internal workings of Go by executing the following command:

```
$ go tool compile -W nodeTree.go
before main
.   CALLFUNC l(8) tc(1) STRUCT-(int, error)
.   .   NAME-fmt.Println a(true) l(4) x(0) class(PFUNC) tc(1) used FUNC-func(...interface {}) (int, error)
.   .   DDDARG l(8) esc(no) PTR64-*[1]interface {}
.   CALLFUNC-list
.   .   CONVIFACE l(8) esc(h) tc(1) implicit(true) INTER-interface {}
.   .   .   NAME-main.statictmp_0 a(true) l(8) x(0) class(PEXTERN) f(1) tc(1) used string
.   VARKILL l(8) tc(1)
.   .   NAME-main..autotmp_0 a(true) l(8) x(0) class(PAUTO) esc(N) used ARRAY-[1]interface {}
after walk main
.   CALLFUNC-init
.   .   AS l(8) tc(1)
.   .   .   NAME-main..autotmp_0 a(true) l(8) x(0) class(PAUTO) esc(N) tc(1) addrtaken assigned used ARRAY-[1]interface {}
.   .   AS l(8) tc(1)
.   .   .   NAME-main..autotmp_2 a(true) l(8) x(0) class(PAUTO) esc(N) tc(1) assigned used PTR64-*[1]interface {}
.   .   .   ADDR l(8) tc(1) PTR64-*[1]interface {}
.   .   .   .   NAME-main..autotmp_0 a(true) l(8) x(0) class(PAUTO) esc(N) tc(1) addrtaken assigned used ARRAY-[1]interface {}
```

```
      .   .       BLOCK l(8)
      .   .       BLOCK-list
      .   .   .     AS l(8) tc(1) hascall
      .   .   .         INDEX l(8) tc(1) assigned bounded hascall INTER-interface
{}
      .   .   .   .       IND l(8) tc(1) implicit(true) assigned hascall ARRAY-
[1]interface {}
      .   .   .   .   .     NAME-main..autotmp_2 a(true) l(8) x(0) class(PAUTO)
esc(N) tc(1) assigned used PTR64-*[1]interface {}
      .   .   .   .   .     LITERAL-0 a(true) l(8) tc(1) int
      .   .   .   .       EFACE l(8) tc(1) INTER-interface {}
      .   .   .   .   .     ADDR a(true) l(8) tc(1) PTR64-*uint8
      .   .   .   .   .   .   NAME-type.string a(true) x(0) class(PEXTERN) tc(1)
uint8
      .   .   .   .   .     ADDR l(8) tc(1) PTR64-*string
      .   .   .   .   .   .   NAME-main.statictmp_0 a(true) l(8) x(0)
class(PEXTERN) f(1) tc(1) addrtaken used string
      .   .       BLOCK l(8)
      .   .       BLOCK-list
      .   .   .     AS l(8) tc(1) hascall
      .   .   .   .   NAME-main..autotmp_1 a(true) l(8) x(0) class(PAUTO) esc(N)
tc(1) assigned used SLICE-[]interface {}
      .   .   .   .   SLICEARR l(8) tc(1) hascall SLICE-[]interface {}
      .   .   .   .   .   NAME-main..autotmp_2 a(true) l(8) x(0) class(PAUTO)
esc(N) tc(1) assigned used PTR64-*[1]interface {}
      .   CALLFUNC l(8) tc(1) hascall STRUCT-(int, error)
      .   .   NAME-fmt.Println a(true) l(4) x(0) class(PFUNC) tc(1) used FUNC-
func(...interface {}) (int, error)
      .   .   DDDARG l(8) esc(no) PTR64-*[1]interface {}
      .   CALLFUNC-list
      .   .   AS l(8) tc(1)
      .   .   .   INDREGSP-SP a(true) l(8) x(0) tc(1) addrtaken main.__ SLICE-
[]interface {}
      .   .   .   NAME-main..autotmp_1 a(true) l(8) x(0) class(PAUTO) esc(N)
tc(1) assigned used SLICE-[]interface {}
      .   VARKILL l(8) tc(1)
      .   .   NAME-main..autotmp_0 a(true) l(8) x(0) class(PAUTO) esc(N) tc(1)
addrtaken assigned used ARRAY-[1]interface {}
before init
      .   IF l(1) tc(1)
      .   .   GT l(1) tc(1) bool
      .   .   .   NAME-main.initdone· a(true) l(1) x(0) class(PEXTERN) tc(1)
assigned used uint8
      .   .   .   LITERAL-1 a(true) l(1) tc(1) uint8
      .   IF-body
      .   .   RETURN l(1) tc(1)
      .   IF l(1) tc(1)
      .   .   EQ l(1) tc(1) bool
```

```
.   .   .       NAME-main.initdone   a(true) l(1) x(0) class(PEXTERN) tc(1)
assigned used uint8
.   .   .       LITERAL-1 a(true) l(1) tc(1) uint8
.   IF-body
.   .   CALLFUNC l(1) tc(1)
.   .   .       NAME-runtime.throwinit a(true) x(0) class(PFUNC) tc(1) used
FUNC-func()
.   AS l(1) tc(1)
.   .       NAME-main.initdone   a(true) l(1) x(0) class(PEXTERN) tc(1) assigned
used uint8
.   .       LITERAL-1 a(true) l(1) tc(1) uint8
.   CALLFUNC l(1) tc(1)
.   .       NAME-fmt.init a(true) l(4) x(0) class(PFUNC) tc(1) used FUNC-func()
.   AS l(1) tc(1)
.   .       NAME-main.initdone   a(true) l(1) x(0) class(PEXTERN) tc(1) assigned
used uint8
.   .       LITERAL-2 a(true) l(1) tc(1) uint8
.   RETURN l(1) tc(1)
after walk init
.   IF l(1) tc(1)
.   .   GT l(1) tc(1) bool
.   .   .       NAME-main.initdone   a(true) l(1) x(0) class(PEXTERN) tc(1)
assigned used uint8
.   .   .       LITERAL-1 a(true) l(1) tc(1) uint8
.   IF-body
.   .   RETURN l(1) tc(1)
.   IF l(1) tc(1)
.   .   EQ l(1) tc(1) bool
.   .   .       NAME-main.initdone   a(true) l(1) x(0) class(PEXTERN) tc(1)
assigned used uint8
.   .   .       LITERAL-1 a(true) l(1) tc(1) uint8
.   IF-body
.   .   CALLFUNC l(1) tc(1) hascall
.   .   .       NAME-runtime.throwinit a(true) x(0) class(PFUNC) tc(1) used
FUNC-func()
.   AS l(1) tc(1)
.   .       NAME-main.initdone   a(true) l(1) x(0) class(PEXTERN) tc(1) assigned
used uint8
.   .       LITERAL-1 a(true) l(1) tc(1) uint8
.   CALLFUNC l(1) tc(1) hascall
.   .       NAME-fmt.init a(true) l(4) x(0) class(PFUNC) tc(1) used FUNC-func()
.   AS l(1) tc(1)
.   .       NAME-main.initdone   a(true) l(1) x(0) class(PEXTERN) tc(1) assigned
used uint8
.   .       LITERAL-2 a(true) l(1) tc(1) uint8
.   RETURN l(1) tc(1)
```

As you can appreciate, the Go compiler and its tools do many things behind the scenes, even for a small program such as nodeTree.go.

The -W parameter tells the go tool compile command to print the **debug parse tree** after the type checking.

Look at the output of the next two commands:

```
$ go tool compile -W nodeTree.go | grep before
before main
before init
$ go tool compile -W nodeTree.go | grep after
after walk main
after walk init
```

As you can see, the before keyword is about the beginning of the execution of a function. If your program had more functions, you would have gotten more output. This is illustrated in the following example:

```
$ go tool compile -W defer.go | grep before
before d1
before d2
before d3
before main
before d2.func1
before d3.func1
before init
before type..hash.[2]interface {}
before type..eq.[2]interface {}
```

The preceding example uses the Go code of defer.go, which is much more complicated than nodeTree.go. However, it should be obvious than the init() function is automatically created by Go as it exists in both examples (nodeTree.go and defer.go).

I will now present you with a juicier version of nodeTree.go, named nodeTreeMore.go:

```
package main

import (
    "fmt"
)

func functionOne(x int) {
    fmt.Println(x)
```

```
}
func main() {
   varOne := 1
   varTwo := 2
   fmt.Println("Hello there!")
   functionOne(varOne)
   functionOne(varTwo)
}
```

The nodeTreeMore.go program has two variables, named varOne and varTwo, and one additional function named functionOne. Searching the output of go tool compile -W for varOne, varTwo, and functionOne reveals the following information:

```
$ go tool compile -W nodeTreeMore.go   | grep functionOne | uniq
before functionOne
after walk functionOne
.   .       NAME-main.functionOne a(true) l(7) x(0) class(PFUNC) tc(1) used
FUNC-func(int)
$ go tool compile -W nodeTreeMore.go   | grep varTwo | uniq
.   .       NAME-main.varTwo a(true) g(2) l(13) x(0) class(PAUTO) f(1) tc(1)
used int
.   .   .       NAME-main.varTwo a(true) g(2) l(13) x(0) class(PAUTO) f(1)
tc(1) used int
$ go tool compile -W nodeTreeMore.go   | grep varOne | uniq
.   .       NAME-main.varOne a(true) g(1) l(12) x(0) class(PAUTO) f(1) tc(1)
used int
.   .   .       NAME-main.varOne a(true) g(1) l(12) x(0) class(PAUTO) f(1)
tc(1) used int
```

Thus, varOne is represented as NAME-main.varOne while varTwo is indicated by NAME-main.varTwo. The functionOne() function is referenced as NAME-main.functionOne. Consequently, the main() function is referenced as NAME-main.

Now let's examine the following code of the debug parse tree of nodeTreeMore.go:

```
before functionOne
.   AS l(8) tc(1)
.   .   NAME-main..autotmp_2 a(true) l(8) x(0) class(PAUTO) esc(N) tc(1)
assigned used int
.   .   NAME-main.x a(true) g(1) l(7) x(0) class(PPARAM) f(1) tc(1) used
int
```

This data is related to the definition of `functionOne()`. The `l(8)` string tells us that the definition of this node can be found in line 8; that is, after reading line 7. The `NAME-main..autotmp_2` integer variable is automatically generated by the compiler.

The next part of the debug parse tree output that I will explain below follows:

```
.   CALLFUNC l(15) tc(1)
.   .   NAME-main.functionOne a(true) l(7) x(0) class(PFUNC) tc(1) used FUNC-func(int)
.   CALLFUNC-list
.   .   NAME-main.varOne a(true) g(1) l(12) x(0) class(PAUTO) f(1) tc(1) used int
```

The first line says that in line 15 of the program, which is specified by `l(15)`, you will call `NAME-main.functionOne`, which is defined in line 7 of the program, as specified by `l(7)`, which is a function that requires a single integer parameter, as specified by `FUNC-func(int)`. The function list of parameters, which is specified after `CALLFUNC-list`, includes the `NAME-main.varOne` variable that is defined in line 12 of the program as `l(12)` shows.

Learning more about go build

If you want to learn more about what is happening behind the scenes when you execute a `go build` command, you should add the `-x` flag to it, as in the next example:

```
$ go build -x defer.go
WORK=/var/folders/sk/1tk8cnw501zdtr2hxcj5sv2m0000gn/T/go-build563076777
mkdir -p $WORK/command-line-arguments/_obj/
mkdir -p $WORK/command-line-arguments/_obj/exe/
cd /Users/mtsouk/Desktop/masterGo/ch/ch2/code
/usr/local/Cellar/go/1.9.2/libexec/pkg/tool/darwin_amd64/compile -o
$WORK/command-line-arguments.a -trimpath $WORK -goversion go1.9.2 -p main -
complete -buildid 594c089b099f8acfb529046cd09a886b401d57c8 -D
_/Users/mtsouk/Desktop/masterGo/ch/ch2/code -I $WORK -pack ./defer.go
cd .
/usr/local/Cellar/go/1.9.2/libexec/pkg/tool/darwin_amd64/link -o
$WORK/command-line-arguments/_obj/exe/a.out -L $WORK -extld=clang -
buildmode=exe -buildid=594c089b099f8acfb529046cd09a886b401d57c8
$WORK/command-line-arguments.a
mv $WORK/command-line-arguments/_obj/exe/a.out defer
```

Once again, there are many things happening in the background, and it is important to be aware of them. However, most of the time, you will not have to deal with the commands of the compilation process.

General Go coding advices

The following list offers practical advices that will help you write better Go code:

- If you have an error in a Go function, either log it or return it, do not do both unless you have a really good reason for doing so!
- Go **interfaces** define behaviors, not data and data structures.
- Use the `io.Reader` and `io.Writer` interfaces because they make your code more extensible.
- Make sure that you pass a pointer to a variable of a function only when needed. The rest of the time, just pass the value of the variable.
- Error variables are not `string` variables; they are `error` variables!
- Do not test your Go code on production machines!
- If you don't really know a Go feature, test it before using it for the first time, especially if you are developing an application or a utility that will be used by a large number of users.
- If you are afraid of making mistakes, you will most likely end up doing nothing really useful! So, experiment as much as you can!

Additional Resources

Have a look at the following resources:

- Learn more about the `unsafe` standard Go package by visiting its documentation page at https://golang.org/pkg/unsafe/.
- Visit the website of DTrace at http://dtrace.org/.
- You can find out more information about the functions of the `runtime` package by visiting https://golang.org/pkg/runtime/.

- Reading research papers might be difficult, but it is very rewarding. Be sure to download the *On-the-fly Garbage Collection: An Exercise in Cooperation* paper and read it. The paper can be found in many places, including https://dl.acm.org/citation.cfm?id=359655.
- Visit https://github.com/gasche/gc-latency-experiment in order to find benchmarking code for the garbage collector of various programming languages.
- Should you wish to learn more about garbage collection, definitely visit http://gchandbook.org/.
- Visit the documentation page of cgo at https://golang.org/cmd/cgo/.

Exercises

- Write an example where you use your own C code from a Go program.
- Use strace(1) on your Linux machine to inspect the operation of some standard Unix utilities such as cp(1) and ls(1). What do you see?
- If you are using a macOS machine, use dtruss(1) to see how the sync(8) utility works.
- Write a Go function, and use it in a C program.

Summary

This chapter covered many interesting Go topics including theoretical and practical information about the Go **garbage collector**; how to call C code from your Go programs; the handy, yet sometimes tricky, defer keyword; the panic() and recover() functions; the strace(1), dtrace(1), and dtruss(1) Unix tools; and the use of the unsafe standard Go package. The chapter also presented some assembly language generated by Go.

Toward the end of the chapter, it told you how to find out information about your Go environment using the runtime package and how to reveal and explain the node tree of a Go program before giving you some handy Go coding advice.

At the end of the day, what you should remember from this chapter is that tools such as the `unsafe` Go package and the ability to call C code from Go are usually used for three occasions: first, when you want the best performance and you are willing to sacrifice some Go safety for it; second, when you want to communicate with another programming language; and third, when you want to implement something that cannot be implemented in Go.

In the next chapter, you will start to learn about the basic data types that come with Go including **arrays**, **slices**, and **maps**. Despite their simplicity, those data types are the building blocks of almost every Go application because they are the basis of more complex data structures that allow you to store your data and move information inside your Go projects. Additionally, you will learn about **pointers**, which are also found in other programming languages, Go **loops**, and the unique way that Go works with dates and times.

3
Working with Basic Go Data Types

In the previous chapter, we covered many fascinating topics including the way the Go garbage collector works, the `panic()` and `recover()` functions, the `unsafe` package, how to call C code from a Go program, and how to call Go code from a C program. We also addressed the node tree created by the Go compiler when compiling a Go program.

The core subject matter of this chapter is the basic data types of Go. This list includes **arrays**, **slices**, and **maps**. Despite their simplicity, these data types can help you store, retrieve, and alter the data in your programs in a very convenient and quick way. Moreover, this chapter will talk about **pointers**, **constants**, **loops**, and working with dates and times, which is a very interesting subject.

In this chapter, you will learn the following topics:

- Go arrays
- Go slices and why slices are much better than arrays
- Go maps
- Pointers in Go
- Looping in Go
- Constants in Go
- Working with times
- Operating with dates

Go loops

Every programming language has a way of looping, and Go is no exception. Go offers the `for` loop, which allows you to iterate over many kinds of data types.

 Go does not offer support for the `while` keyword. However, the `for` loops in Go can replace `while` loops!

The for loop

The most common programming loop type is the `for` loop, which allows you to iterate for a predefined number of times or for as long as a condition is valid or according to a value that is calculated at the beginning of the `for` loop. Such values include the size of a slice or an array and the number of the keys on a map. This means that one of the most common ways for accessing all of the elements of an array, a slice, or a map is the `for` loop. The other way is with the use of the `range` keyword.

The following code shows the simplest form of a `for` loop, where a given variable takes a range of predefined values:

```
for i := 0; i < 100; i++ {
}
```

In the preceding loop, the values that `i` will take are from 0 to 99. As soon as `i` reaches 100, the execution of the `for` loop will stop. In this case, `i` is a local and temporary variable, which means that after the termination of the `for` loop, `i` will be garbage collected at some point and disappear. However, if `i` was defined outside the `for` loop, it would have kept its value after the termination of the `for` loop. In that case, the value of `i` after the termination of the `for` loop would have been 100.

You can completely exit a `for` loop using the `break` keyword. The `break` keyword also allows you to create a `for` loop without an exit condition, such as `i < 100` used in the preceding example, because the exit condition can be included in the code block of the `for` loop. You are also allowed to have multiple exit conditions in a `for` loop.

Additionally, you can skip a single iteration of a `for` loop using the `continue` keyword. The `continue` keyword stops running the block, jumps back up to the top of the `for` loop, and carries on for the next item.

The while loop

As discussed earlier, Go does not offer the `while` keyword for writing `while` loops, but it allows you to use a `for` loop instead of a `while` loop. This section will present two examples where a `for` loop does the job of a `while` loop.

The following is the typical case where you want to write something like `while(condition)`:

```
for {
}
```

It is the job of the developer to use the `break` keyword to exit this `for` loop!

However, the `for` loop can also emulate a `do...while` loop that are found in other programming languages. As an example, the following Go code is equivalent to a `do...while(anExpression)` loop:

```
for ok := true; ok; ok = anExpression {
}
```

As soon as the `ok` variable reaches the `false` value, the `for` loop will terminate.

The range keyword

Go also offers the `range` keyword, which is used in `for` loops and allows you to write easy to understand code for iterating over Go data types.

The main advantage of the `range` keyword is that you do not need to know the **cardinality** of a slice or a map in order to process its elements one by one. You will see `range` in action in a while.

Examples of Go for loops

This section will display multiple examples of `for` loops. The name of the file is `loops.go`, and it will be presented in four parts. The first code segment of `loops.go` is as follows:

```
package main

import (
    "fmt"
)
```

```go
func main() {
    for i := 0; i < 100; i++ {
        if i%20 == 0 {
            continue
        }
        if i == 95 {
            break
        }
        fmt.Print(i, " ")
    }
```

The preceding code shows a typical `for` loop as well as the use of the `continue` and `break` keywords.

The next code segment is as follows:

```go
fmt.Println()
i := 10
for {
    if i < 0 {
        break
    }
    fmt.Print(i, " ")
    i--
}
fmt.Println()
```

The preceding code emulates a typical `while` loop. Note the use of the `break` keyword to exit the `for` loop.

The third part of `loops.go` follows next:

```go
i = 0
anExpression := true
for ok := true; ok; ok = anExpression {
    if i > 10 {
        anExpression = false
    }
    fmt.Print(i, " ")
    i++
}
fmt.Println()
```

In this part, you see the use of a `for` loop that does the job of a `do...while` loop as discussed earlier in this chapter.

The last part of `loops.go` is shown in the following Go code:

```
        anArray := [5]int{0, 1, -1, 2, -2}
        for i, value := range anArray {
            fmt.Println("index:", i, "value: ", value)
        }
    }
```

Applying the `range` keyword to an array variable returns two values: an array index and the value of the element at that index, respectively. You can use both of them, one of them, or none of them; that is, in case you just want to count the elements of the array.

Executing `loops.go` will produce the following output:

```
$ go run loops.go
1 2 3 4 5 6 7 8 9 10 11 12 13 14 15 16 17 18 19 21 22 23 24 25 26 27 28 29
30 31 32 33 34 35 36 37 38 39 41 42 43 44 45 46 47 48 49 50 51 52 53 54 55
56 57 58 59 61 62 63 64 65 66 67 68 69 70 71 72 73 74 75 76 77 78 79 81 82
83 84 85 86 87 88 89 90 91 92 93 94
10 9 8 7 6 5 4 3 2 1 0
0 1 2 3 4 5 6 7 8 9 10 11
index: 0 value:  0
index: 1 value:  1
index: 2 value:  -1
index: 3 value:  2
index: 4 value:  -2
```

Go arrays

Arrays are one of the most popular data structures for two reasons: arrays are simple and easy to understand and they are very versatile and can store many different kinds of data.

You can declare an array that stores four integers as follows:

```
    anArray := [4]int{1, 2, 4, -4}
```

The size of the array is stated before its type, which is defined before its elements. You can find the length of an array with the help of the `len()` function: `len(anArray)`.

Working with Basic Go Data Types

The index of the first element of any dimension of an array is 0, the index of the second element of any array dimension is 1, and so on. This means that for an array with one dimension named a, the valid indexes are from 0 to len(a)-1.

Although you might be familiar with accessing the elements of an array in other programming languages using a for loop and a numeric variable, there are cooler ways to visit all of the elements of an array in Go that involve the use of the range keyword and that allow you to bypass the use of the len() function in the for loop. Look to the Go code of loops.go for such an example.

Multi-dimensional arrays

Arrays can have more than one dimension. However, using more than three dimensions without a serious reason can make your program difficult to read and might create bugs.

Arrays can store all the types of elements. We are just using integers here because they are easier to understand and type.

The following Go code shows how you can create an array with two dimensions (twoD) and another one with three dimensions (threeD):

```
twoD := [4][4]int{{1, 2, 3, 4}, {5, 6, 7, 8}, {9, 10, 11, 12},
{13, 14, 15, 16}}
threeD := [2][2][2]int{{{1, 0}, {-2, 4}}, {{5, -1}, {7, 0}}}
```

Accessing, assigning, or printing a single element from one of the previous two arrays can be done easily. As an example, the first element of the twoD array is twoD[0][0] and its value is 1.

Therefore, accessing all of the elements of the twoD array with the help of multiple for loops can be done as follows:

```
        for i := 0; i < len(threeD); i++ {
            for j := 0; j < len(v); j++ {
                for k := 0; k < len(m); k++ {
                }
            }
        }
```

[94]

As you can see, you need as many `for` loops as the dimensions of the array in order to access all of its elements. The same rules apply to slices, which will be presented in the next section.

The code of `usingArrays.go`, which will be presented in three parts, presents a complete example of how to deal with arrays in Go.

The first part of the code is as follows:

```
package main

import (
    "fmt"
)

func main() {
    anArray := [4]int{1, 2, 4, -4}
    twoD := [4][4]int{{1, 2, 3, 4}, {5, 6, 7, 8}, {9, 10, 11, 12}, {13, 14, 15, 16}}
    threeD := [2][2][2]int{{{1, 0}, {-2, 4}}, {{5, -1}, {7, 0}}}
```

Here, you define three array variables, named `anArray`, `twoD`, and `threeD`, respectively.

The second part of `usingArrays.go` is shown here:

```
    fmt.Println("The length of", anArray, "is", len(anArray))
    fmt.Println("The first element of", twoD, "is", twoD[0][0])
    fmt.Println("The length of", threeD, "is", len(threeD))

    for i := 0; i < len(threeD); i++ {
        v := threeD[i]
        for j := 0; j < len(v); j++ {
            m := v[j]
            for k := 0; k < len(m); k++ {
                fmt.Print(m[k], " ")
            }
        }
        fmt.Println()
    }
}
```

What you get from the first `for` loop is a two-dimensional array (`threeD[i]`), whereas what you get from the second `for` loop is an array with one dimension (`v[j]`). The last `for` loop iterates over the elements of the array with one dimension.

The last code part is shown in the following Go code:

```
for _, v := range threeD {
    for _, m := range v {
        for _, s := range m {
            fmt.Print(s, " ")
        }
    }
    fmt.Println()
}
```

The range keyword does exactly the same job as the iteration variables used in the for loops of the preceding code segment, but it does so in a more elegant and clearer way.

> The range keyword also works with Go maps, which makes it pretty handy and my preferred way of iteration.

Executing usingArrays.go generates the following output:

```
$ go run usingArrays.go
The length of [1 2 4 -4] is 4
The first element of [[1 2 3 4] [5 6 7 8] [9 10 11 12] [13 14 15 16]] is 1
The length of [[[1 0] [-2 4]] [[5 -1] [7 0]]] is 2
1 0 -2 4
5 -1 7 0
1 0 -2 4
5 -1 7 0
```

One of the biggest problems with arrays is out-of-bounds errors, which means trying to access an element that does not exist. It's like trying to access the sixth element of an array with only five elements. The Go compiler considers issues that can be detected as compiler errors because this helps the development workflow. Therefore, the Go compiler can detect out-of-bounds array access errors as follows:

```
./a.go:10: invalid array index -1 (index must be non-negative)
./a.go:10: invalid array index 20 (out of bounds for 2-element array)
```

The shortcomings of Go arrays

Go arrays have many disadvantages that will make you reconsider using them in your Go projects. First of all, once you define an array, you cannot change its size, which means that Go arrays are not dynamic. Putting it simply, if you need to add an element to an existing array that has no space left, you will need to create a bigger array and copy all of the elements of the old array to the new one. Second, when you pass an array to a function as a parameter, you actually pass a copy of the array, which means that any changes you make to an array inside a function will be lost after the function exits. Last, passing a large array to a function can be pretty slow, mostly because Go has to create a copy of the array. The solution to all of these problems is to use Go slices, which will be presented in the next section.

WARNING: Because of their disadvantages, arrays are rarely used in Go!

Go slices

Go **slices** are very powerful, and it would not be an exaggeration to say that slices could totally replace the use of arrays in Go. There are only a few occasions where you will need to use an array instead of a slice. The most obvious one is when you are absolutely sure that you will need to store a fixed number of elements.

Slices are implemented using arrays internally, which means that Go uses an underlying array for each slice.

As slices are **passed by reference** to functions, which means that what is actually passed is the memory address of the slice variable, any modifications that you make to a slice inside a function will not get lost after the function exits. Additionally, passing a big slice to a function is significantly faster than passing an array with the same number of elements because Go will not have to make a copy of the slice—it will just pass the memory address of the slice variable.

Performing basic operations on slices

You can create a new slice literal as follows:

```
aSliceLiteral := []int{1, 2, 3, 4, 5}
```

This means that **slice literals** are defined just like arrays but without the element count. If you put an element count in a definition, you will get an array instead!

However, there is also the `make()` function that allows you to create empty slices with the desired **length** and **capacity** as the parameters passed to `make()`. The capacity parameter can be omitted. In that case, the capacity of the slice will be the same as its length.

You can define a new empty slice with 20 places that can be automatically expanded when needed as follows:

```
integer := make([]int, 20)
```

Please note that Go automatically initializes the elements of an empty slice to the zero value of its type, which means that the value of the initialization depends on the type of the object stored in the slice. Although it is good to know that Go initializes the elements of every slice created with `make`, you should not make any assumptions based on this fact because that behavior might change in the future.

After that, you can access all of the elements of a slice in the following way:

```
for i := 0; i < len(integer); i++ {
    fmt.Println(integer[i])
}
```

If you want to empty an existing slice, the zero value for a slice variable is `nil`.

```
aSliceLiteral = nil
```

You can add an element to the slice, which will automatically increase its size, using the `append()` function:

```
integer = append(integer, -5000)
```

You can access the first element of the `integer` slice as `integer[0]`, whereas you can access the last element of the `integer` slice as `integer[len(integer)-1]`.

Also, you can access multiple continuous slice elements using the [:] notation. The next statement selects the second and the third elements of a slice:

```
integer[1:3]
```

Additionally, you can use [:] notation for creating a new slice from an existing slice or array:

```
s2 := integer[1:3]
```

Please note that this process is called **re-slicing**, and it can cause problems in some cases. Examine the following program:

```
package main

import "fmt"

func main() {
    s1 := make([]int, 5)
    reSlice := s1[1:3]
    fmt.Println(s1)
    fmt.Println(reSlice)

    reSlice[0] = -100
    reSlice[1] = 123456
    fmt.Println(s1)
    fmt.Println(reSlice)

}
```

First, note that in order to select the second and third elements of a slice using the [:] notation, you should use [1:3], which means starting with index number 1 and going up to index number 3, without including index number 3.

Given an array a1, you can create a slice s1 that references that array by executing s1 := a1[:].

Executing the preceding code, which is saved as `reslice.go`, will create the next output:

```
$ go run reslice.go
[0 0 0 0 0]
[0 0]
[0 -100 123456 0 0]
[-100 123456]
```

At the end of the program, the contents of the `s1` slice will be `[0 -100 123456 0 0]`, even though we did not change them directly. This means that altering the elements of a re-slice modifies the element of the original slice because both slices point to the same memory address! Put simply, the re-slice process does not make a copy of the original slice.

The second problem of re-slicing is that, even if you re-slice a slice in order to use a small part of the original slice, the underlying array from the original slice will be kept in memory for as long as the smaller re-slice exists because the original slice is being referenced by the smaller re-slice. Although this is not truly important for small slices, it can cause problems when you are reading big files into slices and you only want to use a small part of them.

Slices are being expanded automatically

Slices have two main properties: **capacity** and **length**. The tricky part is that usually these two properties have different values. The length of a slice is the same as the length of an array with the same number of elements and can be found using the `len()` function. The capacity of a slice is the current room that has been allocated for this particular slice, and it can be found with the `cap()` function. As slices are dynamic in size, if a slice runs out of room, Go automatically doubles its current length to make room for more elements.

Put simply, if the length and the capacity of a slice have the same values and you try to add another element to the slice, the capacity of the slice will be doubled, whereas its length will be increased by one. Although this might work well for small slices, adding a single element to a really huge slice might take more memory than expected.

The code of `lenCap.go` illustrates the concepts of capacity and length in more detail, and it will be presented in three parts. The first part of the program follows:

```
package main

import (
    "fmt"
)

func printSlice(x []int) {
```

```go
    for _, number := range x {
        fmt.Print(number, " ")
    }
    fmt.Println()
}
```

The `printSlice()` function helps you print a one-dimensional slice without having to repeat the same Go code all of the time.

The second part of `lenCap.go` contains the next piece of Go code:

```go
func main() {
    aSlice := []int{-1, 0, 4}
    fmt.Printf("aSlice: ")
    printSlice(aSlice)

    fmt.Printf("Cap: %d, Length: %d\n", cap(aSlice), len(aSlice))
    aSlice = append(aSlice, -100)
    fmt.Printf("aSlice: ")
    printSlice(aSlice)
    fmt.Printf("Cap: %d, Length: %d\n", cap(aSlice), len(aSlice))
```

In this part, as well as the next one, you will add some elements to the `aSlice` slice to alter its length and its capacity.

The last portion of Go code is as follows:

```go
    aSlice = append(aSlice, -2)
    aSlice = append(aSlice, -3)
    aSlice = append(aSlice, -4)
    printSlice(aSlice)
    fmt.Printf("Cap: %d, Length: %d\n", cap(aSlice), len(aSlice))
}
```

The execution of `lenCap.go` will create the following output:

```
$ go run lenCap.go
aSlice: -1 0 4
Cap: 3, Length: 3
aSlice: -1 0 4 -100
Cap: 6, Length: 4
-1 0 4 -100 -2 -3 -4
Cap: 12, Length: 7
```

As you can see, the initial size of the slice was 3. As a result, the initial value of its capacity was also 3. After adding one element to the slice, its size became 4, whereas its capacity became 6. After adding three more elements to the slice, its size became 7, whereas its capacity was doubled one more time and became 12.

Byte slices

A **byte slice** is a slice where its type is `byte`. You can create a new byte slice named s as follows:

```
s := make([]byte, 5)
```

There is nothing special in the way that you can access a byte slice compared to the other types of slices. It is just that byte slices are used in file input and output operations. You will see byte slices in action in Chapter 8, *Telling a Unix System What to Do*.

The copy() function

You can create a slice from the elements of an existing array, and you can copy an existing slice to another one using the `copy()` function. However, as the use of `copy()` can be very tricky, this subsection will try to clarify its usage with the help of the Go code of `copySlice.go`, which will be presented in four parts.

> You should be very careful when using the `copy()` function on slices because the built-in `copy(dst, src)` copies the minimum of the `len(dst)` and `len(src)` elements.

The first part of the program is shown in the following Go code:

```
package main

import (
    "fmt"
)

func main() {
    a6 := []int{-10, 1, 2, 3, 4, 5}
    a4 := []int{-1, -2, -3, -4}
    fmt.Println("a6:", a6)
    fmt.Println("a4:", a4)
```

```
copy(a6, a4)
fmt.Println("a6:", a6)
fmt.Println("a4:", a4)
fmt.Println()
```

In the preceding code, we define two slices named a6 and a4, print them, and then try to copy a4 to a6. As a6 has more elements than a4, all of the elements of a4 will be copied to a6. However, as a4 has only four elements and a6 has six elements, the last two elements of a6 will remain the same.

The second part of copySlice.go is as follows:

```
b6 := []int{-10, 1, 2, 3, 4, 5}
b4 := []int{-1, -2, -3, -4}
fmt.Println("b6:", b6)
fmt.Println("b4:", b4)
copy(b4, b6)
fmt.Println("b6:", b6)
fmt.Println("b4:", b4)
```

In this case, only the first four elements of b6 will be copied to b4 because b4 has only four elements.

The third code segment of copySlice.go is shown in the following Go code:

```
fmt.Println()
array4 := [4]int{4, -4, 4, -4}
s6 := []int{1, 1, -1, -1, 5, -5}
copy(s6, array4[0:])
fmt.Println("array4:", array4[0:])
fmt.Println("s6:", s6)
fmt.Println()
```

Here you try to copy an array with four elements to a slice with six elements. Note that the array is converted to a slice with the help of the [:] notation (array4[0:]).

Working with Basic Go Data Types

The last code portion of `copySlice.go` is as follows:

```go
    array5 := [5]int{5, -5, 5, -5, 5}
    s7 := []int{7, 7, -7, -7, 7, -7, 7}
    copy(array5[0:], s7)
    fmt.Println("array5:", array5)
    fmt.Println("s7:", s7)
}
```

Here you can see how to copy a slice to an array that has places for five elements. As `copy()` only accepts slice arguments, you should also use the `[:]` notation to convert the array into a slice.

If you try to copy an array into a slice or vice versa without using the `[:]` notation, the program will fail to compile and will display one of the following error messages:

```
# command-line-arguments
./a.go:42:6: first argument to copy should be slice; have [5]int
./a.go:43:6: second argument to copy should be slice or string; have [5]int
./a.go:44:6: arguments to copy must be slices; have [5]int, [5]int
```

Executing `copySlice.go` will create the following output:

```
$ go run copySlice.go
a6: [-10 1 2 3 4 5]
a4: [-1 -2 -3 -4]
a6: [-1 -2 -3 -4 4 5]
a4: [-1 -2 -3 -4]

b6: [-10 1 2 3 4 5]
b4: [-1 -2 -3 -4]
b6: [-10 1 2 3 4 5]
b4: [-10 1 2 3]
array4: [4 -4 4 -4]
s6: [4 -4 4 -4 5 -5]
array5: [7 7 -7 -7 7]
s7: [7 7 -7 -7 7 -7 7]
```

Multidimensional slices

Slices can have many dimensions as is also the case with arrays. The next statement creates a slice with two dimensions:

```
s1 := make([][]int, 4)
```

 If you find yourselves using slices with many dimensions all of the time, you might need to reconsider your approach and choose a simpler design that does not require multidimensional slices.

You will find a code example with a **multidimensional slice** in the next section.

Another example of slices

The Go code of the `slices.go` program will hopefully clarify many things about slices for you, and it will be presented in five parts.

The first part of the program contains the expected preamble as well as the definition of two slices:

```
package main

import (
    "fmt"
)

func main() {
    aSlice := []int{1, 2, 3, 4, 5}
    fmt.Println(aSlice)
    integer := make([]int, 2)
    fmt.Println(integer)
    integer = nil
    fmt.Println(integer)
```

Working with Basic Go Data Types

The second part of the program shows how to use the `[:]` notation to create a new slice that references an existing array. Remember that you are not creating a copy of the array, just a reference to it, which will be verified in the output of the program:

```go
anArray := [5]int{-1, -2, -3, -4, -5}
refAnArray := anArray[:]

fmt.Println(anArray)
fmt.Println(refAnArray)
anArray[4] = -100
fmt.Println(refAnArray)
```

The third code segment defines a slice with one dimension and another one with two dimensions using the `make()` function:

```go
s := make([]byte, 5)
fmt.Println(s)
twoD := make([][]int, 3)
fmt.Println(twoD)
fmt.Println()
```

As slices are automatically initialized by Go, all of the elements of the two preceding slices will have the zero value of the slice type, which for integers is 0 and for slices is `nil`. Keep in mind that the elements of a multidimensional slice are slices!

In the fourth part of `slices.go` that is shown in the next piece of Go code, you will learn how to initialize all the elements of a slice with two dimensions manually:

```go
for i := 0; i < len(twoD); i++ {
    for j := 0; j < 2; j++ {
        twoD[i] = append(twoD[i], i*j)
    }
}
```

The preceding Go code shows that in order to expand an existing slice and make it grow, you will need to use the `append()` function and not reference an index that does not exist! The latter would create a `panic: runtime error: index out of range` error message! Note that the values of the slice elements have been chosen arbitrarily.

The last part of the program shows you how to use the `range` keyword to visit and print all of the elements of a slice with two dimensions:

```
    for _, x := range twoD {
        for i, y := range x {
            fmt.Println("i:", i, "value:", y)
        }
        fmt.Println()
    }
}
```

If you execute `slices.go`, you will get the following output:

```
$ go run slices.go
[1 2 3 4 5]
[0 0]
[]
[-1 -2 -3 -4 -5]
[-1 -2 -3 -4 -5]
[-1 -2 -3 -4 -100]
[0 0 0 0 0]
[[] [] []]
i: 0 value: 0
i: 1 value: 0
i: 0 value: 0
i: 1 value: 1
i: 0 value: 0
i: 1 value: 2
```

It should not come as a surprise to you that the objects of the slice with the two dimensions are initialized to `nil` and therefore printed as empty. This happens because the zero value for the slice type is `nil`.

Sorting slices using sort.slice()

This subsection will illustrate the use of the `sort.Slice()` function that was first introduced in Go version 1.8. This means that the code presented, which is saved in `sortSlice.go`, will not run on older Go versions. The program will be presented in three parts. The first part of the program is as follows:

```
package main

import (
    "fmt"
    "sort"
)

type aStructure struct {
    person string
    height int
    weight int
}
```

Apart from the expected preamble, you can also see the definition of a **Go structure** for the first time in this book. Chapter 4, *The Uses of Composite Types*, will totally explore Go structures. For now, however, keep in mind that structures are data types with multiple variables of various types.

The second part of `sortSlice.go` is shown in the following Go code:

```
func main() {

    mySlice := make([]aStructure, 0)
    mySlice = append(mySlice, aStructure{"Mihalis", 180, 90})
    mySlice = append(mySlice, aStructure{"Bill", 134, 45})
    mySlice = append(mySlice, aStructure{"Marietta", 155, 45})
    mySlice = append(mySlice, aStructure{"Epifanios", 144, 50})
    mySlice = append(mySlice, aStructure{"Athina", 134, 40})

    fmt.Println("0:", mySlice)
```

Here you create a new slice named `mySlice` with elements from the `aStructure` structure created earlier.

The final part of the program is as follows:

```
    sort.Slice(mySlice, func(i, j int) bool {
        return mySlice[i].height < mySlice[j].height
    })
```

```
        fmt.Println("<:", mySlice)
        sort.Slice(mySlice, func(i, j int) bool {
            return mySlice[i].height > mySlice[j].height
        })
        fmt.Println(">:", mySlice)
}
```

Here you sort `mySlice` two times using `sort.Slice()` and two anonymous functions, one anonymous function at a time, using the `height` field of `aStructure`.

The `sort.Slice()` function changes the order of the elements in the slice according to the sorting function.

Executing `sortSlice.go` will create the following output:

```
$ go run sortSlice.go
0: [{Mihalis 180 90} {Bill 134 45} {Marietta 155 45} {Epifanios 144 50} {Athina 134 40}]
<: [{Bill 134 45} {Athina 134 40} {Epifanios 144 50} {Marietta 155 45} {Mihalis 180 90}]
>: [{Mihalis 180 90} {Marietta 155 45} {Epifanios 144 50} {Bill 134 45} {Athina 134 40}]
```

If you try to execute `sortSlice.go` on a Unix machine with an older than 1.8 Go version, you will get the next error message:

```
$ go version
go version go1.3.3 linux/amd64
$ go run sortSlice.go
# command-line-arguments
./sortSlice.go:24: undefined: sort.Slice
./sortSlice.go:28: undefined: sort.Slice
```

Go maps

A Go **map** is equivalent to the well-known **hash table** found in many other programming languages. The main advantage of maps is that they can use any data type as their index, which in this case is called a **map key** or just a **key**. Although Go maps do not exclude any data types from being used as keys, for a data type to be used as a key, it must be **comparable**, which means that the Go compiler must be able to differentiate one key from another or, put simply, that the keys of a map must support the == operator. The good news is that almost all data types are comparable. However, as you can understand, using the `bool` data type as the key to a map will definitely limit your options! Additionally, using floating-point numbers as keys might present problems caused by the precision used among different machines and operating systems.

A Go map is a reference to a hash table! The good thing is that Go hides the implementation of the hash table and therefore its complexity. You will learn more about implementing a hash table on your own in Go in `Chapter 5`, *Enhancing Go Code with Data Structures*.

You can create a new empty map with `string` keys and `int` values with the help of the `make()` function:

```
iMap = make(map[string]int)
```

Alternatively, you can use the next **map literal** in order to create a new map that will be populated with data:

```
anotherMap := map[string]int {
    "k1": 12
    "k2": 13
}
```

You can access the two objects of `anotherMap` as `anotherMap["k1"]` and `anotherMap["k1"]`. You can delete an object of a map using the `delete()` function:

```
delete(anotherMap, "k1")
```

You can iterate over all of the elements of a map using the following technique:

```
        for key, value := range iMap {
            fmt.Println(key, value)
        }
```

The Go code of usingMaps.go will illustrate the use of maps in more detail. The program will be presented in three parts. The first part is shown in the following Go code:

```go
package main

import (
    "fmt"
)

func main() {
    iMap := make(map[string]int)
    iMap["k1"] = 12
    iMap["k2"] = 13
    fmt.Println("iMap:", iMap)

    anotherMap := map[string]int{
        "k1": 12,
        "k2": 13,
    }
```

The second part of usingMaps.go contains the following code:

```go
    fmt.Println("anotherMap:", anotherMap)
    delete(anotherMap, "k1")
    delete(anotherMap, "k1")
    delete(anotherMap, "k1")
    fmt.Println("anotherMap:", anotherMap)

    _, ok := iMap["doesItExist"]
    if ok {
        fmt.Println("Exists!")
    } else {
        fmt.Println("Does NOT exist")
    }
```

Here you see a technique that allows you to determine whether a given key is in the map or not. This is a vital technique because without it you would not know whether a given map has the required information or not.

The bad thing is that if you try to get the value of a map key that does not exist in the map, you will end up getting zero, which gives you no way of determining whether the result was zero because the key you requested was not there, or because the element with the corresponding key actually had the zero value.

Working with Basic Go Data Types

Additionally, you can see the `delete()` function in action-calling the same `delete()` statement multiple times does not make any difference and does not generate any warning messages.

The last part of the program is as follows:

```
    for key, value := range iMap {
        fmt.Println(key, value)
    }
}
```

Here you see the use of the `range` keyword on a map, which is pretty elegant and handy.

If you execute `usingMaps.go`, you will get the following output:

```
$ go run usingMaps.go
iMap: map[k1:12 k2:13]
anotherMap: map[k1:12 k2:13]
anotherMap: map[k2:13]
Does NOT exist
k1 12
k2 13
```

You cannot and should not make any assumptions about the order in which the map pairs will be displayed because that order is totally random!

Storing to a nil map

The following Go code will work:

```
aMap := map[string]int{}
aMap["test"] = 1
```

However, the following Go code will *not* work because you have assigned the `nil` value to the map you are trying to use:

```
aMap := map[string]int{}
aMap = nil
fmt.Println(aMap)
aMap["test"] = 1
```

Saving the preceding code to `failMap.go` and trying to compile it will generate the following error message:

```
$ go run failMap.go
map[]
panic: assignment to entry in nil map
```

This means that trying to insert data into a `nil` map will fail, however, looking up, deleting, finding the length, and using `range` loops on `nil` maps will not crash your code!

When you should use a map?

Maps are more versatile than both slices and arrays, but this flexibility comes at a cost; that is, the extra processing power required for the implementation of a Go map. Nevertheless, built-in Go structures are very fast, so do not hesitate using a Go map when you need to use one.

What you should remember is that Go maps are very convenient and can store many different kinds of data, while being both easy to understand and fast when you are working with them.

Go constants

Go supports **constants**, which are variables that cannot change their values. Constants in Go are defined with the help of the `const` keyword.

Generally speaking, constants are usually global variables. Thus, you might rethink your approach if you find yourself defining too many constant variables with a local scope!

The main benefit that you get from using constants in your programs is the guarantee that their value will not change during program execution!

Strictly speaking, the value of a constant variable is defined at compile time—*not* at run time. Behind the scenes, Go uses Boolean, string, or number as the type for storing a constant variable because this gives Go more flexibility when dealing with constants!

Working with Basic Go Data Types

You can define a new constant as follows:

```
const HEIGHT = 200
```

Additionally, if you want to declare many constants at once, mainly because they are related to each other, you can use the following notation:

```
const (
    C1 = "C1C1C1"
    C2 = "C2C2C2"
    C3 = "C3C3C3"
)
```

Please note that the Go compiler considers the results of all operations applied to constants as constants. However, if a constant is part of a larger expression, this will not be the case.

And now for something completely different! The following three variable declarations mean exactly the same thing in Go:

```
s1 := "My String"
var s2 = "My String"
var s3 string = "My String"
```

However, as none of these variable declarations contains the `const` keyword in its declaration, none of them is a constant. This does not mean that you cannot define two constants in a similar way as follows:

```
const s1 = "My String"
const s2 string = "My String"
```

Although both `s1` and `s2` are constants, `s2` comes with a type declaration (`string`), which makes its declaration more restrictive than the declaration of `s1`. This is because a typed Go constant must follow all of the strict rules of a typed Go variable. On the other hand, a constant without a type need not follow all of the strict rules of a typed variable, which means that it can be mixed with expressions more liberally. Additionally, even constants without a type have a default type that is used when, and only when, no other type information is available. The main reason for this behavior is that as you do not know how a constant will be used. In that case, you do not want to use all of the available Go rules. A simple example is the definition of a numeric constant such as `const value = 123`. As you might use the `value` constant in many expressions, declaring a type would make your job much more difficult. Look at the following Go code:

```
const s1 = 123
const s2 float64 = 123
```

```
        var v1 float32 = s1 * 12
        var v2 float32 = s2 * 12
```

Although the compiler will not have a problem with the definition of v1, the code used for the definition of v2 will not compile because s2 and v2 have different types:

```
$ go run a.go
# command-line-arguments
./a.go:12:6: cannot use s2 * 12 (type float64) as type float32 in
assignment
```

 As general advice, if you are using lots of constants in your programs, it might be a good idea to gather all of them in a Go package.

The constant generator iota

The **constant generator iota** is used for declaring a sequence of related values that uses incrementing numbers without the need to type each one of them explicitly.

Most of the concepts related to the const keyword, including the constant generator iota, will be illustrated in the constants.go file, which will be presented in four parts.

The first code segment of constants.go is as follows:

```
package main

import (
    "fmt"
)

type Digit int
type Power2 int

const PI = 3.1415926

const (
    C1 = "C1C1C1"
    C2 = "C2C2C2"
    C3 = "C3C3C3"
)
```

Working with Basic Go Data Types

In this part, we declare two new types, named `Digit` and `Power2`, and four new constants, named `PI`, `C1`, `C2` and `C3`.

 A Go **type** is a way of defining a new **named type** that uses the same underlying type as an existing one. This is mainly used for differentiating between different types that might use the same kind of data.

The second part of `constants.go` is shown in the following Go code:

```
func main() {
    const s1 = 123
    var v1 float32 = s1 * 12
    fmt.Println(v1)
    fmt.Println(PI)
```

In this part of the program, you define another constant (`s1`) that is used in an expression (`v1`).

The third part of the program is as follows:

```
    const (
        Zero Digit = iota
        One
        Two
        Three
        Four
    )
    fmt.Println(One)
    fmt.Println(Two)
```

Here you see the definition of a **constant generator iota** based on `Digit`, which is equivalent to the following declaration of four constants:

```
    const (
        Zero = 0
        One = 1
        Two = 2
        Three = 3
        Four = 4
    )
```

The last part of the program of constants.go is as follows:

```
const (
    p2_0 Power2 = 1 << iota
    _
    p2_2
    _
    p2_4
    _
    p2_6
)

fmt.Println("2^0:", p2_0)
fmt.Println("2^2:", p2_2)
fmt.Println("2^4:", p2_4)
fmt.Println("2^6:", p2_6)
}
```

There is another constant generator iota here, which is a little different than the previous one. First, notice the use of the underscore character in a const block with a constant generator iota. This allows you to skip unwanted values. Second, the value of iota always increments, and it can be used in expressions, which is what occurred in this case.

Now let's see what really happens inside the const block. For p2_0, iota has the value of 0 and p2_0 is defined as 1. For p2_2, iota has the value of 2 and p2_2 is defined as the result of the expression 1 << 2, which is 00000100 in binary representation-the decimal value of 00000100 is 4, which is the result and the value of p2_2. Analogously, the value of p2_4 is 16, and the value of p2_6 is 32.

As you can see, the use of iota can save your time when it fits your needs!

Executing constants.go will generate the following output:

```
$ go run constants.go
1476
3.1415926
1
2
2^0: 1
2^2: 4
2^4: 16
2^6: 64
```

Go pointers

Go supports pointers! **Pointers** are memory addresses that offer improved speed in exchange for difficult-to-debug code and nasty bugs-C programmers should know more about this. You have already seen pointers in action in Chapter 2, *Understanding Go Internals*, when we talked about unsafe code and the unsafe package, but this section will try to shed more light on this difficult and tricky subject. Additionally, native Go pointers are safe provided that you know what you are doing.

When working with pointers, you need * to get the value of a pointer, which is called **dereferencing the pointer**, and & to get the memory address of a non-pointer variable.

Generally speaking, amateur developers should use pointers only when the libraries that they use require it, because pointers can be the cause of horrible and difficult to discover bugs when used carelessly.

You can make a function accept a pointer parameter as follows:

```
func getPointer(n *int) {
}
```

Similarly, a function can return a pointer as follows:

```
func returnPointer(n int) *int {
}
```

The use of safe Go pointers is illustrated in pointers.go, which will be presented in four parts. The first code segment of pointers.go follows next:

```
package main

import (
    "fmt"
)

func getPointer(n *int) {
    *n = *n * *n

}

func returnPointer(n int) *int {
    v := n * n
    return &v
}
```

The good thing with `getPointer()` is that it allows you to update the variable passed to it without needing to return anything to the caller function. This happens because the pointer passed as a parameter contains the memory address of the variable.

On the other hand, `returnPointer()` gets an integer parameter and returns a pointer to an integer, which is denoted by `return &v`. Although this might not appear to be that useful, you will really appreciate this capability in Chapter 4, *The Uses of Composite Types*, when we talk about pointers to Go structures, as well as in later chapters where more complex data structures will be involved.

Both the `getPointer()` and `returnPointer()` functions find the square of an integer. However, they use a totally different approach, as `getPointer()` stores the result to the provided parameter whereas `returnPointer()` returns the result and requires a different variable for storing it.

The second part of the program contains the following Go code:

```
func main() {
    i := -10
    j := 25

    pI := &i
    pJ := &j

    fmt.Println("pI memory:", pI)
    fmt.Println("pJ memory:", pJ)
    fmt.Println("pI value:", *pI)
    fmt.Println("pJ value:", *pJ)
```

Both `i` and `j` are normal integer variables. However, `pI` and `pJ` are both pointers pointing to `i` and `j`, respectively. The `pI` is the memory address of the pointer, whereas `*pI` is the value stored to that memory address.

The third part of `pointers.go` is as follows:

```
*pI = 123456
*pI--
fmt.Println("i:", i)
```

Here you can see how you can change the `i` variable through the `pI` pointer that points to `i` in two different ways. First, it does so by directly assigning a new value to it and second, using the `--` operator.

Working with Basic Go Data Types

The last code portion of `pointers.go` is shown in the following Go code:

```
    getPointer(pJ)
    fmt.Println("j:", j)
    k := returnPointer(12)
    fmt.Println(*k)
    fmt.Println(k)
}
```

Here you call the `getPointer()` function using `pJ` as its parameter. As we discussed before, any changes made to the `pJ` variable inside `getPointer()` will have an effect on the value of the `j` variable because the `pJ` variable points to the `j` variable, which can be verified by the output of the `fmt.Println("j:", j)` statement. The call to `returnPointer()` returns a pointer that is assigned to the `k` pointer variable.

Running `pointers.go` will create the following output:

```
$ go run pointers.go
pI memory: 0xc420012088
pJ memory: 0xc420012090
pI value: -10
pJ value: 25
i: 123455
j: 625
144
0xc4200120c8
```

I realize that you may have trouble understanding the Go code of `pointers.go` because we have not yet discussed functions and function definitions, so feel free to look at *Chapter 6, What You Might Not Know About Go Packages,* where topics related to functions are explained in more detail.

Strings in Go are value types, not pointers as in C.

Dealing with times and dates

In this section, you will learn how to parse date and time strings in Go, how to convert between different date and time formats, and how to print dates and times in the format you desire. Although this task might look insignificant at first, it can truly be critical when you want to synchronize multiple tasks, or when your application needs to read the date from one or more text files or directly from the user.

The time package is the star of working with times and dates in Go. You will see some of its functions in action in this section.

Before learning how to parse a string and convert it into a time or a date, you will review a simple program named usingTime.go that will introduce the time package. The program will be presented in three parts. The first part of this program follows:

```
package main

import (
    "fmt"
    "time"
)
```

The second code segment of usingTime.go is shown in the following Go code:

```
func main() {
    fmt.Println("Epoch time:", time.Now().Unix())
    t := time.Now()
    fmt.Println(t, t.Format(time.RFC3339))
    fmt.Println(t.Weekday(), t.Day(), t.Month(), t.Year())

    time.Sleep(time.Second)
    t1 := time.Now()
    fmt.Println("Time difference:", t1.Sub(t))
```

The time.Now().Unix() function returns the **Unix epoch time**, which is the number of seconds that have elapsed since 00:00:00 UTC, 1 January 1970. The Format() function allows you to convert a time variable to another format-in this case, the RFC3339 format.

You will see the `time.Sleep()` function many times in this book as an easy way of emulating the delay from the execution of a true function. The `time.Second` constant allows you to use a one second duration in Go. If you want to define a duration of 10 seconds, you will need to multiply `time.Second` by 10! Other similar constants include `time.Nanosecond`, `time.Microsecond`, `time.Millisecond`, `time.Minute`, and `time.Hour`. Thus, the smallest amount of time that can be defined with the `time` package is the nanosecond. Last, the `time.Sub()` function allows you to find the time difference between two times.

The last part of the program is as follows:

```
    formatT := t.Format("01 January 2006")
    fmt.Println(formatT)
    loc, _ := time.LoadLocation("Europe/Paris")
    londonTime := t.In(loc)
    fmt.Println("Paris:", londonTime)
}
```

Here you define a new date format using `time.Format()` in order to use it for printing out a `time` variable.

Executing `usingTime.go` generates the following output:

```
$ go run usingTime.go
Epoch time: 1510427100
2017-11-11 21:05:00.226749 +0200 EET m=+0.000268522
2017-11-11T21:05:00+02:00
Saturday 11 November 2017
Time difference: 1.004278426s
11 November 2017
Paris: 2017-11-11 20:05:00.226749 +0100 CET
```

Now that you know the basics of the `time` package, it is "time" to dig deeper into its functionality starting by working with times.

Working with times

When you have a `time` variable, it is easy to convert it into anything that is related to time or date. However, the main problem occurs when you have a string and you want to see whether it is a valid time or not. The function used for parsing time and date strings is `time.Parse()`, and it accepts two parameters. The first parameter denotes the expected format of the string that you will parse, while the second parameter is the actual string that will be parsed. The first parameter is composed of elements derived from a list of Go constants related to date and time parsing.

The list of constants that can be used for creating your own parse format can be found at `https://golang.org/src/time/format.go`. Go does not define the format of a date or a time in a form like `DDYYYYMM` or `%D%Y%M` as do the rest of the programming languages, rather it uses its own approach. Although you might find this approach strange at first, you will certainly appreciate it as it prevents the developer from making silly mistakes.

The Go constants for working with times are `15` for parsing the hour, `04` for parsing the minutes, and `05` for parsing the seconds. You can easily guess that all of these numeric values must be unique! You can use `PM` for parsing the `PM` string in uppercase and `pm` for parsing the lowercase version.

You are not obligated to use every available Go constant. The main task of the developer is putting the various Go constants in the desired order in order to match the kind of strings that the program will have to process.

You can consider the final version of the string that is passed as the first parameter to the `time.Parse()` function as a **regular expression**.

Parsing times

This section will tell you how to parse a string, which is given as a command-line argument to the `parseTime.go` utility in order to convert it into a `time` variable. This is not always possible, however, because the given string might not be in the correct format or might contain invalid characters. The `parseTime.go` utility will be presented in three parts.

The first code segment of `parseTime.go` is as follows:

```
package main

import (
    "fmt"
    "os"
    "path/filepath"
    "time"
)
```

The second part is shown in the following Go code:

```
func main() {
    var myTime string
    if len(os.Args) != 2 {
        fmt.Printf("usage: %s string\n",
 filepath.Base(os.Args[0]))
        os.Exit(1)
    }

    myTime = os.Args[1]
```

The last portion of `parseTime.go`, which is where the magic happens, is as follows:

```
    d, err := time.Parse("15:04", myTime)
    if err == nil {
        fmt.Println("Full:", d)
        fmt.Println("Time:", d.Hour(), d.Minute())
    } else {
        fmt.Println(err)
    }
}
```

Thus, in order to parse an hour and minute string, you will need to use `"15:04"`. The value of the `err` variable tells you whether the parsing was successful or not.

Executing `parseTime.go` will create the following output:

```
$ go run parseTime.go
usage: parseTime string
exit status 1
$ go run parseTime.go 12:10
Full: 0000-01-01 12:10:00 +0000 UTC
Time: 12 10
```

As you can see here, Go prints a full date time string because this is what is stored in a `time` variable. If you are only interested in the time and not in the date, you should print only the parts of a `time` variable that you want.

If you use a wrong Go constant like `22:04` when trying to parse a string and convert it into a time, you will get the following error message:

```
$ go run parseTime.go 12:10
parsing time "12:10" as "22:04": cannot parse ":10" as "2"
```

However, if you use a Go constant like `11`, which is used for parsing months when the month is given as a number, the error message will be slightly different:

```
$ go run parseTime.go 12:10
parsing time "12:10": month out of range
```

Working with dates

In this subsection, you will learn how to parse strings that denote dates in Go. Doing this still requires the use of the `time.Parse()` function.

The Go constants for working with dates are `Jan` for parsing the three-letter abbreviation used for describing a month, `2006` for parsing the year, and `02` for parsing the day of the month. If you use `January` instead of `Jan`, you will get the long name of the month instead of its three-letter abbreviation, which makes perfect sense.

Additionally, you can use `Monday` for parsing strings that contain a long weekday string and `Mon` for the abbreviated version of the weekday.

Parsing dates

The name of the Go program that will be developed in this subsection is `parseDate.go`, and it will be presented in two parts.

The first part of `parseDate.go` is as follows:

```go
package main

import (
    "fmt"
    "os"
    "path/filepath"
```

```go
    "time"
)

func main() {

    var myDate string
    if len(os.Args) != 2 {
        fmt.Printf("usage: %s string\n",
filepath.Base(os.Args[0]))
        return
    }

    myDate = os.Args[1]
```

The second part of `parseDate.go` contains the following Go code:

```go
    d, err := time.Parse("02 January 2006", myDate)
    if err == nil {
        fmt.Println("Full:", d)
        fmt.Println("Time:", d.Day(), d.Month(), d.Year())
    } else {
        fmt.Println(err)
    }
}
```

If there is a character such as – between the name of the month and the year, you can use `"02 January-2006"` instead of `"02 January 2006"` as the first parameter to `time.Parse()`.

Executing `parseDate.go` will generate the following output:

```
$ go run parseDate.go
usage: parseDate string
$ go run parseDate.go "20 July 2000"
Full: 2000-07-20 00:00:00 +0000 UTC
Time: 20 July 2000
```

As `parseDate.go` does not expect data about the time, the `00:00:00 +0000 UTC` string is automatically added to the end of the full date and time string.

Changing date and time formats

In this section, you will learn how to change the format of a string that contains both a date and a time. A very common place for finding such strings is the log files of web servers such as **Apache** and **Nginx**. As we do not know how to read a text file line by line, the text will be hard coded in the program, This fact, however, does not change the functionality of the program.

The Go code of `timeDate.go` will be presented in four parts. The first part of this program is the following expected preamble:

```
package main

import (
    "fmt"
    "regexp"
    "time"
)
```

You need to import the `regexp` standard Go package for supporting regular expressions.

The second code portion of `timeDate.go` is as follows:

```
func main() {

    logs := []string{"127.0.0.1 - - [16/Nov/2017:10:49:46 +0200] 325504",
        "127.0.0.1 - - [16/Nov/2017:10:16:41 +0200] \"GET /CVEN HTTP/1.1\" 200 12531 \"-\" \"Mozilla/5.0 AppleWebKit/537.36",
        "127.0.0.1 200 9412 - - [12/Nov/2017:06:26:05 +0200] \"GET \"http://www.mtsoukalos.eu/taxonomy/term/47\" 1507",
        "[12/Nov/2017:16:27:21 +0300]",
        "[12/Nov/2017:20:88:21 +0200]",
        "[12/Nov/2017:20:21 +0200]",
    }
```

As you cannot be sure about your data and its format, the sample data used for this program tries to cover many different cases, including incomplete data like `[12/Nov/2017:20:21 +0200]`, where there are no seconds in the time part, and erroneous data such as `[12/Nov/2017:20:88:21 +0200]`, where the value of the minutes is 88.

Working with Basic Go Data Types

The third part of `timeDate.go` contains the following Go code:

```
for _, logEntry := range logs {
    r :=
regexp.MustCompile('.*\[(\d\d/\w+/\d\d\d\d:\d\d:\d\d:\d\d.*)\].*')
    if r.MatchString(logEntry) {
        match := r.FindStringSubmatch(logEntry)
```

The main benefit that you receive from such a difficult-to-read regular expression in this particular program is that it allows your code to find out whether you have a date and time string somewhere in your line or not. After you get that string, you will feed `time.Parse()` with it and let that do the rest of the job!

The last part of the program is shown in the following Go code:

```
            dt, err := time.Parse("02/Jan/2006:15:04:05 -0700",
match[1])
            if err == nil {
                newFormat := dt.Format(time.RFC850)
                fmt.Println(newFormat)
            } else {
                fmt.Println("Not a valid date time format!")
            }
        } else {
            fmt.Println("Not a match!")
        }
    }
}
```

Once you find a string that matches the regular expression, you parse it using `time.Parse()` to make sure that it is a valid date/time string. If yes, `timeDate.go` will print the date and time according to the **RFC850 format**.

If you execute `timeDate.go`, you will get the following output:

```
$ go run timeDate.go
Thursday, 16-Nov-17 10:49:46 EET
Thursday, 16-Nov-17 10:16:41 EET
Sunday, 12-Nov-17 06:26:05 EET
Sunday, 12-Nov-17 16:27:21 +0300
Not a valid date time format!
Not a match!
```

Additional resources

Consider looking at the following resources:

- Visit the documentation page of the `time` package that can be found at https://golang.org/pkg/time/
- Visit the documentation page of the `regexp` standard Go package at https://golang.org/pkg/regexp/

Exercises

- Write a constant generator iota for the powers of number 4.
- Write a constant generator iota for the days of the week.
- Write a Go program that converts an existing array into a map.
- Write your own version of `parseTime.go`. Do not forget to test your program.
- Can you create a version of `timeDate.go` that can process two date and time formats?
- Write your own version of `parseDate.go`.

Summary

In this chapter, you learned about many interesting Go topics including maps, arrays, and slices, as well as Go pointers, Go constants and loops, and how Go allows you to work with dates and times. By now, you should understand why slices are superior to arrays.

The next chapter will cover building and using composite types in Go, which mainly includes types that are created with the `struct` keyword and are called **structures**. After that, we will talk about `string` variables and **tuples**.

Additionally, the next chapter will address **Regular expressions** and **Pattern matching**, which are tricky subjects, not only in Go but also in every other programming language as well. When used properly and carefully, however, regular expressions and pattern matching can make the life of a developer so much easier that it is totally worth learning more about them.

Also, you will learn about the `switch` keyword and the `strings` package that allow you to manipulate **UTF-8** strings.

4
The Uses of Composite Types

In the previous chapter, we talked about many core Go topics including arrays, slices, maps, pointers, constants, the `for` loop, the `range` keyword, and how to work with times and dates. This chapter will investigate more advanced Go features such as tuples and strings, the `strings` standard Go package, the `switch` statement, but most importantly structures that are created using the `struct` keyword. Another important topic of this chapter will be defining regular expressions and performing Pattern matching in Go. After all these topics, the last section of this chapter will implement a simple *key-value store* in Go.

In this chapter, you will learn the following topics:

- Go **structures** and the `struct` keyword
- Go **tuples**
- Go strings, runes, byte slices, and string literals
- Regular expressions in Go
- Pattern matching in Go
- The `switch` statement
- The functionality that the `strings` package offers
- Calculating **Pi** with great accuracy
- Developing a **key-value store**

About composite types

Although standard Go types are pretty handy, fast, and flexible, they most likely cannot cover every kind of data that you want to support in your Go code. Go solves this problem by supporting **structures**, which are custom types defined by the developer. Additionally, Go has its own way of supporting **tuples**, which mainly allows functions to return multiple values without needing to group them in structures as is the case in C.

Structures

Although arrays, slices, and maps are all very useful, they cannot group and hold multiple values in the same place. When you need to group various types of variables and create a new handy type, you can use a structure. The various elements of a structure are called **fields of the structure** or just fields.

We will start this section by explaining a simple structure that was first defined in the `sortSlice.go` source file of the previous chapter:

```
type aStructure struct {
    person string
    height int
    weight int
}
```

For reasons that will become evident in Chapter 6, *What You Might Not Know About Go Packages*, the fields of a structure usually begin with an uppercase letter. This convention will be used throughout the rest of the book.

This structure has three fields named `person`, `height`, and `weight`, respectively. You can create a new variable of the `aStructure` type as follows:

```
var s1 aStructure
```

Additionally, you can access a specific field of a structure by its name. So, in order to get the value of the `person` field of the `s1` variable, you should type `s1.person`.

A **structure literal** can be defined as follows:

```
p1 := aStructure{"fmt", 12, -2}
```

However, since remembering the order of the fields of a structure can be pretty hard, Go allows you to use another form for defining a structure literal:

```
p1 := aStructure{weight: 12, height: -2}
```

In this case, you do not need to define an initial value for every field of the structure.

Now that you know the basics of structures, it is time to show you a more practical example, which is named `structures.go` and will be presented in four parts.

The first part of `structures.go` contains the following code:

```
package main

import (
    "fmt"
)
```

 Structures in particular and Go types in general are usually defined outside the `main()` function in order to have a global scope and be available to the entire Go package, unless you want to clarify that a type is only useful within the current scope and is not expected to be used elsewhere.

The second code segment of `structures.go` is shown in the following Go code:

```
func main() {

    type XYZ struct {
        X int
        Y int
        Z int
    }

    var s1 XYZ
    fmt.Println(s1.Y, s1.Z)
```

As you can see, there is nothing that prevents you from defining a new structure type inside a function, but you should have a reason for doing so.

The third portion of structures.go follows next:

```
p1 := XYZ{23, 12, -2}
p2 := XYZ{Z: 12, Y: 13}
fmt.Println(p1)
fmt.Println(p2)
```

Here you define two structure literals named p1 and p2 that you print afterwards.

The last part of structures.go contains the following Go code:

```
pSlice := [4]XYZ{}
pSlice[2] = p1
pSlice[0] = p2
fmt.Println(pSlice)
p2 = XYZ{1, 2, 3}
fmt.Println(pSlice)
}
```

In this last part, we created an array of structures named pSlice. As you will understand from the output of structures.go, when you assign a structure to an array of structures, the structure is copied into the array, so changing the value of the original structure will have no effect on the objects of the array.

Executing structures.go will generate the following output:

```
$ go run structures.go
0 0
{23 12 -2}
{0 13 12}
[{0 13 12} {0 0 0} {23 12 -2} {0 0 0}]
[{0 13 12} {0 0 0} {23 12 -2} {0 0 0}]
```

> Note that the order in which you put the fields in the definition of a structure type is significant for the **type identity** of the defined structure. Put simply, two structures with the same fields will not be considered identical in Go if their fields are not in exactly the same order.

The output of structures.go illustrates that the zero value of a struct variable is constructed by zeroing all of the fields of the struct variable according to their types.

Pointers to structures

In the previous chapter, we talked about pointers. In this section, we will continue our discussion of pointers and present an example that is related to **pointers to structures**. The name of the program is pointerStruct.go and will be presented in four parts.

The first part of the program contains the following Go code:

```
package main

import (
    "fmt"
)

type myStructure struct {
    Name     string
    Surname  string
    Height   int32
}
```

The second code segment from pointerStruct.go follows next:

```
func createStruct(n, s string, h int32) *myStructure {
    if h > 300 {
        h = 0
    }
    return &myStructure{n, s, h}
}
```

The approach used in createStruct() for creating a new structure variable has many advantages over initializing structure variables on your own, including the fact that you are allowed to check whether the information provided is both correct and valid. Additionally, this approach is cleaner-there is a central point where structure variables are initialized so that if there is something wrong with your struct variables, you know where to look and who to blame! Note that some people might prefer to name the createStruct() function as NewStruct().

For those with a C or C++ background, it is perfectly legal for a Go function to return the memory address of a local variable. Nothing gets lost, so everybody is happy!

The third portion of `pointerStruct.go` is as follows:

```go
func retStructure(n, s string, h int32) myStructure {
    if h > 300 {
        h = 0
    }
    return myStructure{n, s, h}
}
```

This part presents the no pointer version of the `createStruct()` function named `retStructure()`. Both functions work fine, so choosing between the implementation of `createStruct()` and `retStructure()` is just a matter of personal preference. A more appropriate name for these functions might have been `NewStructurePointer` and `NewStructure`, respectively.

The last part of `pointerStruct.go` is shown in the following Go code:

```go
func main() {
    s1 := createStruct("Mihalis", "Tsoukalos", 123)
    s2 := retStructure("Mihalis", "Tsoukalos", 123)
    fmt.Println((*s1).Name)
    fmt.Println(s2.Name)
    fmt.Println(s1)
    fmt.Println(s2)
}
```

If you execute `pointerStruct.go`, you will get the following output:

```
$ go run pointerStruct.go
Mihalis
Mihalis
&{Mihalis Tsoukalos 123}
{Mihalis Tsoukalos 123}
```

Here you can see one more time that the main difference between `createStruct()` and `retStructure()` is that the former returns a pointer to a structure, which means that you will need to dereference that pointer in order to use the object to which it points. This can make your code look a little uglier.

Chapter 4

 Structures are very important in Go, and they are being used extensively in real-world programs because they allow you to group as many values as you want and treat those values as a single entity.

Using the new keyword

Go supports the `new` keyword that allows you to allocate new objects. However, there is a very important detail that you need to remember about `new`: `new` returns the memory address of the allocated object. Put simply, `new` returns a pointer!

You can create a fresh `aStructure` variable as follows:

```
pS := new(aStructure)
```

After executing the new statement, you are ready to work with your fresh variable that has its allocated memory zeroed, but not initialized.

 The main difference between `new` and `make` is that variables created with `make` are properly initialized without just zeroing the allocated memory space. Additionally, `make` can only be applied to maps, channels, and slices, and it does not return a memory address, which means that `make` does not return a pointer.

The next statement will create a slice with `new` that points to `nil`:

```
sP := new([]aStructure)
```

Tuples

Strictly speaking, a **tuple** is a finite ordered list with multiple parts. The most important thing about tuples is that Go has no support for the tuple type, which means that Go does not officially care about tuples despite the fact that it has support for certain uses of tuples.

[137]

The Uses of Composite Types

One interesting point here is that we have been using *Go tuples* in this book since Chapter 1, *Go and the Operating System*. They are used in statements like the next one, where a function returns two values that you get in a single statement:

```
min, _ := strconv.ParseFloat(arguments[1], 64)
```

The name of the Go program that will illustrate Go tuples is tuples.go, and it will be presented in three code segments. Please note that the code presented uses a function that returns three values as a tuple. You will learn more about functions in Chapter 6, *What You Might Not Know About Go Packages*.

The first part of tuples.go is as follows:

```
package main

import (
    "fmt"
)

func retThree(x int) (int, int, int) {
    return 2 * x, x * x, -x
}
```

Here you can see the implementation of a function named retThree() that returns a tuple containing three integer values. This capability permits the function to return multiple values without needing to group the various return values into a structure and to return a structure variable instead.

In Chapter 6, *What You Might Not Know About Go Packages*, you will learn how to assign names to the return values of a Go function, which is a very handy feature that can save you from various types of bugs.

The second part of tuples.go is as follows:

```
func main() {
    fmt.Println(retThree(10))
    n1, n2, n3 := retThree(20)
    fmt.Println(n1, n2, n3)
}
```

Here, we use the retThree() function twice. First, we do this without saving its return values. Second, by saving the three return values of retThree() in three different variables using a single statement, which in Go terminology is called a tuple assignment, hence the confusion about Go supporting tuples! If you do not care about one or more return values of a function, you can use an underscore character (_) in its place. Note that it is a compile time error in Go if you declare a variable and do not use it afterwards.

The third part of the tuples.go program is shown in the following Go code:

```
    n1, n2 = n2, n1
    fmt.Println(n1, n2, n3)

    x1, x2, x3 := n1*2, n1*n1, -n1
    fmt.Println(x1, x2, x3)
}
```

As you can see, tuples can do many intelligent things like swapping values without the need for a temporary variable as well as evaluating expressions.

Executing tuples.go will create the following output:

```
$ go run tuples.go
20 100 -10
40 400 -20
400 40 -20
800 160000 -400
```

Regular expressions and pattern matching

Pattern matching, which plays a key role in Go, is a technique used for searching a string for some set of characters based on a specific search pattern that is based on **regular expressions** and **grammars**. If pattern matching is successful, it allows you to extract the desired data from the string, replace it, or delete it.

The Go package responsible for defining regular expressions and performing pattern matching is called regexp. You will see it in action in a while.

When using a regular expression in your code, you should consider the definition of the regular expression as the most important part of your program.

Now for some theory

Every regular expression is compiled into a recognizer by building a generalized transition diagram called a **finite automaton**. A finite automaton can be either deterministic or nondeterministic. Nondeterministic means that more than one transition out of a state can be possible for the same input. A **recognizer** is a program that takes a string x as input and is able to tell if x is a sentence of a given language.

A **grammar** is a set of production rules for strings in a formal language. The production rules describe how to create strings from the alphabet of the language that are valid according to the syntax of the language. A grammar does not describe the meaning of a string or what can be done with it in whatever context-it only describes its form. What is important here is to realize that grammars are the heart of regular expressions because, without a grammar, you cannot define or use a regular expression.

Although regular expressions allow you to solve problems that would be extremely difficult to solve otherwise, do not try to solve every problem you face with a regular expression. Always use the right tool for the job!

The rest of this section will present three examples of regular expressions and pattern matching.

A simple example

In this subsection, you will learn how to select a particular column from a line of text. To make things more interesting, you will also learn how to read a text file line by line. However, file I/O is the subject of Chapter 8, *Telling a Unix System What to Do*, so you refer to that chapter in order to get more information about the relevant Go code.

The name of the Go source file for this example is `selectColumn.go`. It will be presented in five segments. The utility needs at least two command-line arguments to operate; the first one is the column number you want, and the second one is the path of the text file to process. Nonetheless, you can put as many text files as you want—selectColumn.go will process all of them one by one!

The first part of `selectColumn.go` is follows next:

```go
package main

import (
    "bufio"
    "fmt"
    "io"
    "os"
    "strconv"
    "strings"
)
```

The second code portion of `selectColumn.go` contains the following Go code:

```go
func main() {
    arguments := os.Args
    if len(arguments) < 2 {
        fmt.Printf("usage: selectColumn column <file1> [<file2> [... <fileN]]\n")
        os.Exit(1)
    }

    temp, err := strconv.Atoi(arguments[1])
    if err != nil {
        fmt.Println("Column value is not an integer:", temp)
        return
    }

    column := temp
    if column < 0 {
        fmt.Println("Invalid Column number!")
        os.Exit(1)
    }
```

The first test that the program performs is to make sure that it has an adequate number of command-line arguments (`len(arguments) < 2`). Additionally, you need two more tests to make sure that the provided column value is actually a number and that it is bigger than 0.

The third part of `selectColumn.go` follows:

```go
    for _, filename := range arguments[2:] {
        fmt.Println("\t\t", filename)
        f, err := os.Open(filename)
        if err != nil {
            fmt.Printf("error opening file %s\n", err)
```

The Uses of Composite Types

```
        continue
    }
    defer f.Close()
```

The program performs various tests to make sure that the text file does exist and that you can read it. The `os.Open()` function is used for opening the text file. Remember that the Unix file permissions of a text file might not allow you to read it.

The fourth code piece of `selectColumn.go` is as follows:

```
    r := bufio.NewReader(f)
    for {
        line, err := r.ReadString('\n')

        if err == io.EOF {
            break
        } else if err != nil {
            fmt.Printf("error reading file %s", err)
        }
```

As you will learn in Chapter 8, *Telling a Unix System What to Do,* the `bufio.ReadString()` function reads a file until the first occurrence of its parameter. As a result, `bufio.ReadString('\n')` tells Go to read a file line by line because \n is the Unix newline character! The `bufio.ReadString()` function returns a byte slice.

The last code segment of `selectColumn.go` is as follows:

```
        data := strings.Fields(line)
        if len(data) >= column {
            fmt.Println((data[column-1]))
        }
    }
}
```

The logic behind the program is pretty simple: you split each line of text and select the desired column. However, as you cannot be sure that the current line has the required number of fields, you check that before printing any output. This is the simplest form of pattern matching because each line is split using space characters as word separators.

If you want to have more information about the splitting of the lines—and I am sure that you do—then you will find very useful the fact that the strings.Fields() function splits a string based on the whitespace characters that are defined in the unicode.IsSpace() function and returns a slice of strings.

Executing selectColumn.go will generate the next type of output:

```
$ go run selectColumn.go 15 /tmp/swtag.log /tmp/adobegc.log   | head
/tmp/swtag.log              /tmp/adobegc.log
AdobeGCData
Successfully
Initializing
Stream
**********AdobeGC
Perform
Perform
Trying
```

The selectColumn.go utility prints the name of each processed file, even if you get no output from that file.

The important thing to remember here is that you should never trust your data, especially when it comes from nontechnical users. Put simply, always verify that the data you expect to grab is there!

A more advanced example

In this section, you will learn how to match a date and time string as found in the log files of an Apache web server. To make things even more interesting, you will also learn how to change the date and time format of the log file into a different format. Once again, this requires reading the Apache log file, which is a plain text file, line by line.

The name of the command-line utility is changeDT.go, and it will be presented in five parts. Note that changeDT.go is an improved version of the timeDate.go utility presented in Chapter 3, *Working with Basic Go Data Types*, not only because it gets its data from an external file, but also because changeDT.go uses two regular expressions and therefore is able to match strings in two different time and date formats.

The Uses of Composite Types

 The very important point to note here is not try to implement every possible feature in the first version of your utilities! A better approach is to build a working version with fewer features and improve that version in small steps.

The first chunk of code of `changeDT.go` follows:

```
package main

import (
    "bufio"
    "fmt"
    "io"
    "os"
    "regexp"
    "strings"
    "time"
)
```

Lots of packages are needed because `changeDT.go` does so many fascinating things.

The second piece for code of `changeDT.go` is the following:

```
func main() {

    arguments := os.Args
    if len(arguments) == 1 {
        fmt.Println("Please provide one text file to process!")
        os.Exit(1)
    }

    filename := arguments[1]
    f, err := os.Open(filename)
    if err != nil {
        fmt.Printf("error opening file %s", err)
        os.Exit(1)
    }
    defer f.Close()

    notAMatch := 0
    r := bufio.NewReader(f)
    for {
        line, err := r.ReadString('\n')
        if err == io.EOF {
            break
```

```
} else if err != nil {
    fmt.Printf("error reading file %s", err)
}
```

In this part, you just try to open your input file for reading in order to read it line by line. The `notAMatch` variable holds the number of lines in the input file that did not match any one of the two regular expressions of the program.

The third code segment of `changeDT.go` is shown in the following Go code:

```
            r1 := regexp.MustCompile
('.*\[(\d\d/\w+/\d\d\d\d:\d\d:\d\d:\d\d.*)\] .*')
            if r1.MatchString(line) {
                match := r1.FindStringSubmatch(line)
                d1, err := time.Parse("02/Jan/2006:15:04:05 -0700", match[1])
                if err == nil {
                    newFormat := d1.Format(time.Stamp)
                    fmt.Print(strings.Replace(line, match[1], newFormat, 1))
                else {
                    notAMatch++
                }
                continue
            }
```

Here you can see that if the first date and time format is not a match, the program will continue its execution. However, if it enters the `if` block, the `continue` statement will get executed, which means that it will skip the remaining code of the surrounding `for` loop. Thus, in the first supported format, the time and date string has the `21/Nov/2017:19:28:09 +0200` format.

The `regexp.MustCompile()` function is like `regexp.Compile()`, but it panics if the expression cannot be parsed. The parenthesis around the regular expression allows you to use the matches afterwards. In this case, you can only have one match, which you will get using the `regexp.FindStringSubmatch()` function.

The fourth part of `changeDT.go` follows next:

```
            r2 := regexp.MustCompile('.*\[(\w+\-\d\d-\d\d:\d\d:\d\d:\d\d.*)\] .*')
            if r2.MatchString(line) {
                match := r2.FindStringSubmatch(line)
                d1, err := time.Parse("Jan-02-06:15:04:05 -0700", match[1])
                if err == nil {
                    newFormat := d1.Format(time.Stamp)
                    fmt.Print(strings.Replace(line, match[1], newFormat, 1))
                else {
                    notAMatch++
```

The Uses of Composite Types

```
        }
        continue
}
```

The second supported time and date format is `Jun-21-17:19:28:09 +0200`. As you can understand, there are not that many differences between the two formats. Note that although the program uses just two date and time formats, you can have as many of these types of formats as you desire.

The last code portion of `changeDT.go` contains the following Go code:

```
    }
    fmt.Println(notAMatch, "lines did not match!")
}
```

Here you just print the number of lines that did not match any one of the two formats.

The text file that will be used for testing `changeDT.go` contains the following lines:

```
$ cat logEntries.txt
- - [21/Nov/2017:19:28:09 +0200] "GET /AMEv2.tif.zip HTTP/1.1" 200 2188249 "-"
- - [21/Jun/2017:19:28:09 +0200] "GET /AMEv2.tif.zip HTTP/1.1" 200
- - [25/Lun/2017:20:05:34 +0200] "GET /MongoDjango.zip HTTP/1.1" 200 118362
- - [Jun-21-17:19:28:09 +0200] "GET /AMEv2.tif.zip HTTP/1.1" 200
- - [20/Nov/2017:20:05:34 +0200] "GET /MongoDjango.zip HTTP/1.1" 200 118362
- - [35/Nov/2017:20:05:34 +0200] "GET MongoDjango.zip HTTP/1.1" 200 118362
```

Executing `changDT.go` will generate the following output:

```
$ go run changeDT.go logEntries.txt
- - [Nov 21 19:28:09] "GET /AMEv2.tif.zip HTTP/1.1" 200 2188249 "-"
- - [Jun 21 19:28:09] "GET /AMEv2.tif.zip HTTP/1.1" 200
- - [Jun 21 19:28:09] "GET /AMEv2.tif.zip HTTP/1.1" 200
- - [Nov 20 20:05:34] "GET /MongoDjango.zip HTTP/1.1" 200 118362
2 lines did not match!
```

Matching IPv4 addresses

An **IPv4 address**, or simply an **IP address**, has four discrete parts. As an IPv4 address is stored using 8-bit binary numbers. Each part can have values from 0, which is 00000000 in the binary format, to 255, which is equal to 11111111 in the binary format.

The format of an IPv6 address is much more complicated than the format of an IPv4 address so the program presented here will not work with IPv6 addresses.

The name of the program is `findIPv4.go`, and it will be presented in five parts. The first part of `findIPv4.go` is shown here:

```
package main

import (
    "bufio"
    "fmt"
    "io"
    "net"
    "os"
    "path/filepath"
    "regexp"
)
```

As `findIPv4.go` is a pretty sophisticated utility, it needs many standard Go packages.

The second part is shown in the following Go code:

```
func findIP(input string) string {
    partIP := "(25[0-5]|2[0-4][0-9]|1[0-9][0-9]|[1-9]?[0-9])"
    grammar := partIP + "\\." + partIP + "\\." + partIP + "\\." + partIP
    matchMe := regexp.MustCompile(grammar)
    return matchMe.FindString(input)
}
```

The preceding code contains the definition of the regular expression that will help you discover an IPv4 address inside a function. This is the most critical part of the program, because if you define the regular expression incorrectly, you will never be able to catch any IPv4 addresses!

I will explain the regular expression a little further shortly, but before I do that, it is important to understand that prior to defining one or more regular expressions, you should be aware of the problem that you are trying to solve. In other words, if you are not aware of the fact that the decimal values of an IPv4 address cannot be larger than 255, no regular expression can save you!

Now that we are on the same page, let's talk about the next two statements:

```
partIP := "(25[0-5]|2[0-4][0-9]|1[0-9][0-9]|[1-9]?[0-9])"
grammar := partIP + "\\." + partIP + "\\." + partIP + "\\." + partIP
```

The Uses of Composite Types

The regular expression defined in `partIP` matches each one of the four parts of an IP address. A valid IPv4 address can begin with 25 and end with 0, 1, 2, 3, 4, or 5 because that is the biggest 8-bit binary number (`25[0-5]`), or it can begin with 2 followed by 0, 1, 2, 3, or 4 and end with 0, 1, 2, 3, 4, 5, 6, 7, 8, or 9 (`2[0-4][0-9]`). Alternatively, it can begin with 1 followed by two more digits from 0 to 9 (`1[0-9][0-9]`). The last alternative would be a natural number that has one or two digits. The first digit, which is optional, can be from 1 to 9, and the second, which is mandatory, can be from 0 to 9 (`[1-9]?[0-9]`).

The `grammar` variable tells us that what we are looking for has four distinct parts, and each one of them must match `partIP`. This `grammar` variable is what matches the complete IPv4 address that we seek!

As `findIPv4.go` works with regular expressions to find an IPv4 address in a file, it can also process any kind of text file that contains valid IPv4 addresses!

One last thing: if you have any special requirements, such as excluding certain IPv4 addresses or watching for specific addresses or networks, you can easily change the Go code of `findIPv4.go` and add the extra functionality you desire. This is the kind of flexibility achieved when you develop your own tools.

The third part of the `findIPv4.go` utility contains the following Go code:

```
func main() {
    arguments := os.Args
    if len(arguments) < 2 {
        fmt.Printf("usage: %s logFile\n", filepath.Base(arguments[0]))
        os.Exit(1)
    }

    for _, filename := range arguments[1:] {
        f, err := os.Open(filename)
        if err != nil {
            fmt.Printf("error opening file %s\n", err)
            os.Exit(-1)
        }
        defer f.Close()
```

First, you make sure that you have a sufficient number of command-line parameters by checking the length of `os.Args`. Then, you can use a `for` loop to iterate over all of the command-line arguments.

The fourth code portion of `findIPv4.go` is as follows:

```
r := bufio.NewReader(f)
for {
    line, err := r.ReadString('\n')
    if err == io.EOF {
        break
    } else if err != nil {
        fmt.Printf("error reading file %s", err)
        break
    }
```

As happened in `selectColumn.go`, you use `bufio.ReadString()` to read your input line by line.

The last part of `findIPv4.go` contains the following Go code:

```
        ip := findIP(line)
        trial := net.ParseIP(ip)
        if trial.To4() == nil {
            continue
        } else {
            fmt.Println(ip)
        }
    }
  }
}
```

For each line of the input text file, you call the `findIP()` function. The `net.ParseIP()` function double checks that you are dealing with a valid IPv4 address-it is never a bad thing to double check! If the call to `net.ParseIP()` is successful, you print the IPv4 that you just found. After that, the program will deal with the next line of input.

If you want, you can remove the `else` block from the preceding code and put the `fmt.Println(ip)` statement after the `if` block since `continue` will just jump away.

Executing `findIPv4.go` will generate the following type of output:

```
$ go run findIPv4.go /tmp/auth.log
116.168.10.9
192.31.20.9
10.10.16.9
10.31.160.9
192.168.68.194
```

The Uses of Composite Types

Thus the output of `findIPv4.go` can have lines that are displayed multiple times. Apart from that detail, the output of the utility is pretty straightforward.

Processing the preceding output with some traditional Unix command line utilities might help you reveal more information about your data:

```
$ go run findIPv4.go /tmp/auth.log.1 /tmp/auth.log | sort -rn | uniq -c |
sort -rn
    38 xxx.zz.116.9
    33 x.zz.2.190
    25 xx.zzz.1.41
    20 178.132.1.18
    18 x.zzz.63.53
    17 178.zzz.y.9
    15 103.yyy.xxx.179
    10 213.z.yy.194
    10 yyy.zzz.110.4
     9 yy.xx.65.113
```

What we did here is to find out the top-10 IPv4 addresses located in the processed text files using the `sort(1)` and `uniq(1)` Unix command-line utilities. The logic behind this pretty long `bash(1)` shell command is simple: the output of the `findIPv4.go` utility will become the input of the first `sort -rn` command in order to be sorted numerically in reverse order. Then the `uniq -c` command removes the lines that appear multiple times by replacing them with a single line that is preceded with the count of the number of times the line occurred in the input. The output then is sorted once again so that the IPv4 addresses with the higher number of occurrences will appear first!

Once again, it is important to realize that the core functionality of `findIPv4.go` is implemented through the regular expression. If the regular expression is defined incorrectly, or if it does not match all of the cases (**false negative**), or if it matches things that should not be matched (**false positive**), then your program will not work correctly!

[150]

Strings

Strictly speaking, a **string** is not a composite type, but there are so many Go functions that support strings that I will describe strings in this chapter. As discussed in Chapter 3, *Working with Basic Go Data Types*, strings in Go are value types not pointers, as is the case with **C strings**. Additionally, Go supports UTF-8 strings by default, which means that you do not need to load any special packages or do anything tricky in order to print Unicode characters. However, there are subtle differences between a character, a rune, and a byte, as well as differences between a string and a string literal, which will be clarified here.

A Go string is a read-only **byte slice** that can hold any type of bytes, and it can have an arbitrary length.

You can define a new **string literal** as follows:

```
const sLiteral = "\x99\x42\x32\x55\x50\x35\x23\x50\x29\x9c"
```

You might be surprised by the look of a string literal in Go. You can define a string variable as follows:

```
s2 := "€£³"
```

You can find the length of a string variable or a string literal using the `len()` function.

The `strings.go` file will illustrate many standard operations related to strings. It will be presented in five parts. The first part is shown in the following Go code:

```
package main

import (
    "fmt"
)
```

The second portion of Go code is as follows:

```
func main() {
    const sLiteral = "\x99\x42\x32\x55\x50\x35\x23\x50\x29\x9c"
    fmt.Println(sLiteral)
    fmt.Printf("x: %x\n", sLiteral)
    fmt.Printf("sLiteral length: %d\n", len(sLiteral))
```

Each `\xAB` sequence represents a single character of `sLiteral`. As a result, calling `len(sLiteral)` will return the number of characters of `sLiteral`. Using `%x` in `fmt.Printf()` will return the `AB` part of a `\xAB` sequence.

The third code segment of strings.go is shown in the following Go code:

```go
for i := 0; i < len(sLiteral); i++ {
    fmt.Printf("%x ", sLiteral[i])
}
fmt.Println()

fmt.Printf("q: %q\n", sLiteral)
fmt.Printf("+q: %+q\n", sLiteral)
fmt.Printf(" x: % x\n", sLiteral)

fmt.Printf("s: As a string: %s\n", sLiteral)
```

Here you see that you can access a string literal as if it were a slice. Using %q in fmt.Printf() with a string argument will print a double-quoted string safely escaped with Go syntax. Using %+q in fmt.Printf() with a string argument will guarantee ASCII-only output.

Last, using % x (note the space between the % character and the x character) in fmt.Printf() will put spaces between the printed bytes. In order to print a string literal as a string, you will need to call fmt.Printf() with %s.

The fourth portion of the code of strings.go is as follows:

```go
s2 := "€£³"
for x, y := range s2 {
    fmt.Printf("%#U starts at byte position %d\n", y, x)
}

fmt.Printf("s2 length: %d\n", len(s2))
```

Here you define a string named s2 with three Unicode characters. Using fmt.Printf() with %#U will print the characters in the U+0058 format. Using the range keyword on a string that contains Unicode characters allows you to process its Unicode characters one by one.

The output of len(s2) might surprise you a little. As the s2 variable contains Unicode characters, its byte size is larger than the number of characters in it.

The last part of strings.go is as follows:

```
    const s3 = "ab12AB"
    fmt.Println("s3:", s3)
    fmt.Printf("x: % x\n", s3)

    fmt.Printf("s3 length: %d\n", len(s3))

    for i := 0; i < len(s3); i++ {
        fmt.Printf("%x ", s3[i])
    }
    fmt.Println()
}
```

Running strings.go will generate the following output:

```
$ go run strings.go
�B2UP5#P)�
x: 9942325550352350299c
sLiteral length: 10
99 42 32 55 50 35 23 50 29 9c
q: "\x99B2UP5#P)\x9c"
+q: "\x99B2UP5#P)\x9c"
 x: 99 42 32 55 50 35 23 50 29 9c
s: As a string: �B2UP5#P)�
U+20AC '€' starts at byte position 0
U+00A3 '£' starts at byte position 3
U+00B3 '³' starts at byte position 5
s2 length: 7
s3: ab12AB
x: 61 62 31 32 41 42
s3 length: 6
61 62 31 32 41 42
```

It's no surprise if you find the information presented in this section pretty strange and complex, especially if you are unfamiliar with Unicode and UTF-8 representations of characters and symbols. The good thing is that you will not need most of them in your everyday life as a Go developer, and that you will most likely get away with using simple fmt.Println() and fmt.Printf() commands in your programs to print your output. However, if you are living outside of Europe or the USA, you might find some of the information in this section pretty handy.

What is a rune?

A **rune** is an `int32` value, and therefore it is a Go type that is used for representing a **Unicode code point**. A Unicode code point or code position is a numerical value that is usually used for representing single Unicode characters; however, it can also have alternative meanings, such as providing formatting information.

NOTE: You can consider a string as a collection of runes.

A **rune literal** is a character in single quotes. You may also consider a rune literal as a **rune constant**. Behind the scenes, a rune literal is associated with a Unicode code point.

Runes will be illustrated in `runes.go`, which is will be presented in three parts. The first part of `runes.go` follows next:

```
package main

import (
    "fmt"
)
```

The second part of `runes.go` contains the following code:

```
func main() {
    const r1 = '€'
    fmt.Println("(int32) r1:", r1)
    fmt.Printf("(HEX) r1: %x\n", r1)
    fmt.Printf("(as a String) r1: %s\n", r1)
    fmt.Printf("(as a character) r1: %c\n", r1)
```

So, first you define a rune literal named `r1`. (Please note that the Euro sign does not belong to the ASCII table of characters.) Then you print `r1` using various statements. First you print its `int32` value and then its hexadecimal value. After that, you try printing it as a string. Finally, you print it as a character, which is what gives you the same output as the one used in the definition of `r1`.

The third and last code segment of `runes.go` is shown in the following Go code:

```
    fmt.Println("A string is a collection of runes:", []byte("Mihalis"))
    aString := []byte("Mihalis")
    for x, y := range aString {
        fmt.Println(x, y)
```

```
                fmt.Printf("Char: %c\n", aString[x])
        }
        fmt.Printf("%s\n", aString)
}
```

Here you see that a **byte slice** is a collection of runes, and that printing a byte slice with `fmt.Println()` might not return what you may have expected. In order to convert a rune into a character, you should use `%c` in a `fmt.Printf()` statement, and in order to print a byte slice as a string, you will need to use `fmt.Printf()` with `%s`.

Executing `runes.go` will create the following output:

```
$ go run runes.go
(int32) r1: 8364
(HEX) r1: 20ac
(as a String) r1: %!s(int32=8364)
(as a character) r1: €
A string is a collection of runes: [77 105 104 97 108 105 115]
0 77
Char: M
1 105
Char: i
2 104
Char: h
3 97
Char: a
4 108
Char: l
5 105
Char: i
6 115
Char: s
Mihalis
```

Finally, the easiest way to get an `illegal rune literal` error message is by using single quotes instead of double quotes when importing a package:

```
$ cat a.go
package main
import (
    'fmt'
)
func main(){
}
$ go run a.go
package main:a.go:4:2: illegal rune literal
```

The Unicode package

The `unicode` standard Go package contains various handy functions. One of them, which is called `unicode.IsPrint()`, can help you identify the parts of a string that are printable using runes. This technique will be illustrated in the Go code of `unicode.go`, which will be presented in two parts.

The first part of `unicode.go` is as follows:

```
package main

import (
    "fmt"
    "unicode"
)

func main() {
    const sL = "\x99\x00ab\x50\x00\x23\x50\x29\x9c"
```

The second code segment of `unicode.go` is shown in the following Go code:

```
    for i := 0; i < len(sL); i++ {
        if unicode.IsPrint(rune(sL[i])) {
            fmt.Printf("%c\n", sL[i])
        } else {
            fmt.Println("Not printable!")
        }
    }
}
```

As stated before, all of the dirty work is done by the `unicode.IsPrint()` function that returns `true` when a rune is printable and `false` otherwise. If you are really into Unicode characters, you should definitely check the documentation page of the `unicode` package.

Executing `unicode.go` will generate the following output:

```
$ go run unicode.go
Not printable!
Not printable!
a
b
P
Not printable!
#
```

```
P
)
Not printable!
```

The strings package

The `strings` standard Go package allows you to manipulate UTF-8 strings in Go and includes many powerful functions. Most of these functions will be illustrated in the `useStrings.go` source file, which will be presented in five parts. Note that the functions of the `strings` package that are related to file input and output will be demonstrated in Chapter 8, *Telling a Unix System What to Do*.

The first part of `useStrings.go` follows next:

```
package main

import (
    "fmt"
    s "strings"
    "unicode"
)

var f = fmt.Printf
```

You can apply this great tip to all Go packages. There is a difference in the way that the `strings` package is imported. This kind of import statement makes Go create an alias for that package. So, instead of writing `strings.FunctionName()`, you can now write `s.FunctionName()`, which is a bit shorter. Please note that you will not be able to call a function of the `strings` package as `strings.FunctionName()` anymore!

Another handy trick is that if you find yourself using the same function all of the time, and you want to use something shorter instead, you can assign a variable name to that function and use that variable name instead. Here you see that feature applied to the `fmt.Printf()` function. Nevertheless, you should not overuse this feature because you might end up having difficult to read code!

The second part of the `useStrings.go` program contains the following Go code:

```
func main() {
    upper := s.ToUpper("Hello there!")
    f("To Upper: %s\n", upper)
    f("To Lower: %s\n", s.ToLower("Hello THERE"))
```

The Uses of Composite Types

```
f("%s\n", s.Title("tHis wiLL be A title!"))

f("EqualFold: %v\n", s.EqualFold("Mihalis", "MIHAlis"))
f("EqualFold: %v\n", s.EqualFold("Mihalis", "MIHAli"))
```

In this code segment, you see many functions that allow you to play with the case of a string. Additionally, you see that the `strings.EqualFold()` function allows you to determine whether two strings are the same in spite of their differences in their letters.

The third code portion of `useStrings.go` follows here:

```
f("Prefix: %v\n", s.HasPrefix("Mihalis", "Mi"))
f("Prefix: %v\n", s.HasPrefix("Mihalis", "mi"))
f("Suffix: %v\n", s.HasSuffix("Mihalis", "is"))
f("Suffix: %v\n", s.HasSuffix("Mihalis", "IS"))

f("Index: %v\n", s.Index("Mihalis", "ha"))
f("Index: %v\n", s.Index("Mihalis", "Ha"))
f("Count: %v\n", s.Count("Mihalis", "i"))
f("Count: %v\n", s.Count("Mihalis", "I"))
f("Repeat: %s\n", s.Repeat("ab", 5))

f("TrimSpace: %s\n", s.TrimSpace(" \tThis is a line. \n"))
f("TrimLeft: %s", s.TrimLeft(" \tThis is a\t line. \n", "\n\t "))
f("TrimRight: %s\n", s.TrimRight(" \tThis is a\t line. \n", "\n\t "))
```

The `strings.Count()` function counts the number of non-overlapping times the second parameter appears in the string that is given as the first parameter. The `strings.HasPrefix()` function returns `true` when the first parameter string begins with the second parameter string and `false` otherwise. Similarly, the `strings.HasSuffix()` function returns `true` when the first parameter, which is a string, ends with the second parameter, which is also a string, and `false` otherwise.

The fourth code segment of `useStrings.go` contains the following Go code:

```
f("Compare: %v\n", s.Compare("Mihalis", "MIHALIS"))
f("Compare: %v\n", s.Compare("Mihalis", "Mihalis"))
f("Compare: %v\n", s.Compare("MIHALIS", "MIHalis"))

f("Fields: %v\n", s.Fields("This is a string!"))
f("Fields: %v\n", s.Fields("Thisis\na\tstring!"))

f("%s\n", s.Split("abcd efg", ""))
```

This code portion contains some pretty advanced and ingenious functions. The first handy function is strings.Split(), which allows you to split the given string according to the desired separator string. The strings.Split() function returns a **string slice**.

Using "" as the second parameter of strings.Split() will allow you to process a string character by character!

The strings.Compare() function compares two strings lexicographically, and it may return 0 if the two strings are identical and -1 or +1 otherwise.

Last, the strings.Fields() function splits the string parameter using white space characters as separators. The white space characters are defined in the unicode.IsSpace() function.

strings.Split() is a powerful function that you should learn because sooner rather than later you will have to use it in your programs!

The last part of useStrings.go is shown in the following Go code:

```go
        f("%s\n", s.Replace("abcd efg", "", "_", -1))
        f("%s\n", s.Replace("abcd efg", "", "_", 4))
        f("%s\n", s.Replace("abcd efg", "", "_", 2))

        lines := []string{"Line 1", "Line 2", "Line 3"}
        f("Join: %s\n", s.Join(lines, "+++"))

        f("SplitAfter: %s\n", s.SplitAfter("123++432++", "++"))

        trimFunction := func(c rune) bool {
            return !unicode.IsLetter(c)
        }
        f("TrimFunc: %s\n", s.TrimFunc("123 abc ABC \t .", trimFunction))
}
```

As in the previous part of useStrings.go, the last code segment contains functions that implement some very intelligent functionality in a simple to understand and easy to use way.

The Uses of Composite Types

The `strings.Replace()` function takes four parameters. The first one is the string that you want to process. The second parameter contains the string that, if found, will be replaced by the third parameter of `strings.Replace()`. The last parameter is the maximum number of replacements that will happen. If that parameter has a negative value, then there is no limit to the number of replacements that can take place.

The last two statements of the program define a **trim function**, which allows you to keep the runes of a string that interest you and utilize that function as the second argument to the `strings.TrimFunc()` function.

Last, the `strings.SplitAfter()` function splits its first parameter string into substrings based on the separator string that is given as the second parameter to the function.

For the full list of functions similar to `unicode.IsLetter()`, you should visit the documentation page of the `unicode` standard Go package.

Executing `useStrings.go` will display the following output:

```
$ go run useStrings.go
To Upper: HELLO THERE!
To Lower: hello there
THis WiLL Be A Title!
EqualFold: true
EqualFold: false
Prefix: true
Prefix: false
Suffix: true
Suffix: false
Index: 2
Index: -1
Count: 2
Count: 0
Repeat: abababab
TrimSpace: This is a line.
TrimLeft: This is a      line.
TrimRight:       This is a    line.
Compare: 1
Compare: 0
Compare: -1
Fields: [This is a string!]
Fields: [Thisis a string!]
[a b c d   e f g]
_a_b_c_d_ _e_f_g
__a_b_c_d efg
```

[160]

```
_a_bcd efg
Join: Line 1+++Line 2+++Line 3
SplitAfter: [123++ 432++ ]
TrimFunc: abc ABC
```

Please note that the list of functions presented from the `strings` package is far from complete. You should examine the documentation page of the `strings` package at https://golang.org/pkg/strings/ for the complete list of available functions.

If you are working with text and text processing, you will definitely need to learn all of the gory details and functions of the `strings` package, so make sure that you experiment with all these functions a lot and create many examples that will help you clarify things.

The switch statement

The main reason for presenting the `switch` statement in this chapter is because a `switch` case can use regular expressions! First, however, take a look at this simple `switch` block:

```
switch asString {
case "1":
      fmt.Println("One!")
case "0":
      fmt.Println("Zero!")
default:
      fmt.Println("Do not care!")
}
```

The preceding `switch` block can differentiate the `asString` value among the "1" string, the "0" string, and everything else (`default`).

> Having a *match all remaining cases* case in a `switch` block is considered a very good practice. However, as the order of the cases in a `switch` block does matter, the *match all remaining cases* case should be put last. In Go, the name of the *match all remaining cases* case is `default`.

A `switch` statement, however, can be much more flexible and adaptable:

```
switch {
case number < 0:
    fmt.Println("Less than zero!")
case number > 0:
    fmt.Println("Bigger than zero!")
default:
```

The Uses of Composite Types

```
        fmt.Println("Zero!")
}
```

The preceding `switch` block does the job of identifying whether you are dealing with a positive integer, a negative integer, or zero. As you can see, the branches of a `switch` statement can have conditions. You will soon see that the branches of a `switch` statement can also have regular expressions in them!

All of these examples and some additional ones can be found in `switch.go`, which will be presented in five parts.

The first part of `switch.go` is as follows:

```
package main

import (
    "fmt"
    "os"
    "regexp"
    "strconv"
)

func main() {
    arguments := os.Args
    if len(arguments) < 2 {
        fmt.Println("usage: switch number")
        os.Exit(1)
    }
```

The `regexp` package is needed for supporting regular expressions in `switch`.

The second code segment of `switch.go` is shown in the following Go code:

```
    number, err := strconv.Atoi(arguments[1])
    if err != nil {
        fmt.Println("This value is not an integer:", number)
    } else {
        switch {
        case number < 0:
            fmt.Println("Less than zero!")
        case number > 0:
            fmt.Println("Bigger than zero!")
        default:
            fmt.Println("Zero!")
        }
    }
```

The third part of switch.go is as follows:

```
asString := arguments[1]
switch asString {
case "5":
    fmt.Println("Five!")
case "0":
    fmt.Println("Zero!")
default:
    fmt.Println("Do not care!")
}
```

In this code segment, you can see that a switch case can also contain hardcoded values. This mainly occurs when the switch keyword is followed by the name of a variable.

The fourth code portion of switch.go contains the following Go code:

```
var negative = regexp.MustCompile('-')
var floatingPoint = regexp.MustCompile('\d?\.\d')
var email = regexp.MustCompile('^[^@]+@[^@.]+\.[^@.]+')
switch {
case negative.MatchString(asString):
    fmt.Println("Negative number")
case floatingPoint.MatchString(asString):
    fmt.Println("Floating point!")
case email.MatchString(asString):
    fmt.Println("It is an email!")
    fallthrough
default:
    fmt.Println("Something else!")
}
```

Many interesting things are happening here. First, you define three regular expressions named negative, floatingPoint, and email, respectively. Then you use all three of them in the switch block with the help of the regexp.MatchString() function that does the actual matching.

Last, the fallthrough keyword also tells Go to execute the branch that follows the current one, which in this case is the default branch. This means that when the code of the email.MatchString(asString) case is the one that will get executed, the default case will also be executed.

The last part of `switch.go` is as follows:

```
    var aType error = nil
    switch aType.(type) {
    case nil:
        fmt.Println("It is nil interface!")
    default:
        fmt.Println("Not nil interface!")
    }
}
```

Here you can see that `switch` can differentiate between types. You will learn more about working with `switch` and Go **interfaces** at Chapter 7, *Reflection and Interfaces for All Seasons*.

Executing `switch.go` with various input arguments will generate the following type of output:

```
$ go run switch.go
usage: switch number.
exit status 1
$ go run switch.go mike@g.com
This value is not an integer: 0
Do not care!
It is an email!
Something else!
It is nil interface!
$ go run switch.go 5
Bigger than zero!
Five!
Something else!
It is nil interface!
$ go run switch.go 0
Zero!
Zero!
Something else!
It is nil interface!
$ go run switch.go 1.2
This value is not an integer: 0
Do not care!
Floating point!
It is nil interface!
$ go run switch.go -1.5
This value is not an integer: 0
Do not care!
Negative number
It is nil interface!
```

Calculating Pi with great accuracy

In this section, you will learn how to calculate **Pi** with great accuracy using a standard Go package named `math/big` and the special purpose types offered by that package.

 This section contains the ugliest Go code that I have even seen! Even **Java** code looks better than this.

The name of the program that uses Bellard's formula to calculate Pi is `calculatePi.go`, and it will be presented in four parts.

The first part of `calculatePi.go` follows next:

```
package main

import (
    "fmt"
    "math"
    "math/big"
    "os"
    "strconv"
)

var precision uint = 0
```

The `precision` variable holds the desired precision of the calculations, and it is made global in order to be accessible from everywhere in the program.

The second code segment of `calculatePi.go` is shown in the following Go code:

```
func Pi(accuracy uint) *big.Float {
    k := 0
    pi := new(big.Float).SetPrec(precision).SetFloat64(0)
    k1k2k3 := new(big.Float).SetPrec(precision).SetFloat64(0)
    k4k5k6 := new(big.Float).SetPrec(precision).SetFloat64(0)
    temp := new(big.Float).SetPrec(precision).SetFloat64(0)
    minusOne := new(big.Float).SetPrec(precision).SetFloat64(-1)
    total := new(big.Float).SetPrec(precision).SetFloat64(0)

    two2Six := math.Pow(2, 6)
    two2SixBig := new(big.Float).SetPrec(precision).SetFloat64(two2Six)
```

The Uses of Composite Types

The `new(big.Float)` call creates a new `big.Float` variable with the desired precision, which is set by `SetPrec()`.

The third part of `calculatePi.go` contains the remaining Go code of the `Pi()` function:

```
    for {
        if k > int(accuracy) {
            break
        }
        t1 := float64(float64(1) / float64(10*k+9))
        k1 := new(big.Float).SetPrec(precision).SetFloat64(t1)
        t2 := float64(float64(64) / float64(10*k+3))
        k2 := new(big.Float).SetPrec(precision).SetFloat64(t2)
        t3 := float64(float64(32) / float64(4*k+1))
        k3 := new(big.Float).SetPrec(precision).SetFloat64(t3)
        k1k2k3.Sub(k1, k2)
        k1k2k3.Sub(k1k2k3, k3)
        t4 := float64(float64(4) / float64(10*k+5))
        k4 := new(big.Float).SetPrec(precision).SetFloat64(t4)
        t5 := float64(float64(4) / float64(10*k+7))
        k5 := new(big.Float).SetPrec(precision).SetFloat64(t5)
        t6 := float64(float64(1) / float64(4*k+3))
        k6 := new(big.Float).SetPrec(precision).SetFloat64(t6)
        k4k5k6.Add(k4, k5)
        k4k5k6.Add(k4k5k6, k6)
        k4k5k6 = k4k5k6.Mul(k4k5k6, minusOne)
        temp.Add(k1k2k3, k4k5k6)

        k7temp := new(big.Int).Exp(big.NewInt(-1), big.NewInt(int64(k)), nil)
        k8temp := new(big.Int).Exp(big.NewInt(1024), big.NewInt(int64(k)), nil)
        k7 := new(big.Float).SetPrec(precision).SetFloat64(0)
        k7.SetInt(k7temp)
        k8 := new(big.Float).SetPrec(precision).SetFloat64(0)
        k8.SetInt(k8temp)

        t9 := float64(256) / float64(10*k+1)
        k9 := new(big.Float).SetPrec(precision).SetFloat64(t9)
        k9.Add(k9, temp)
        total.Mul(k9, k7)
        total.Quo(total, k8)
        pi.Add(pi, total)

        k = k + 1
    }
```

```
        pi.Quo(pi, two2SixBig)
        return pi
}
```

This part of the program is the Go implementation of Bellard's formula. The bad thing about math/big is that you need a special function of it for almost every type of calculation. This is the case primarily because those functions keep the precision at the desired level. Thus, without using big.Float and big.Int variables, as well as the functions of math/big all of the time, you cannot compute Pi with the desired precision.

The last part of calculatePi.go shows the implementation of the main() function:

```
func main() {
    arguments := os.Args
    if len(arguments) == 1 {
        fmt.Println("Please provide one numeric argument!")
        os.Exit(1)
    }

    temp, _ := strconv.ParseUint(arguments[1], 10, 32)
    precision = uint(temp) * 3

    PI := Pi(precision)
    fmt.Println(PI)
}
```

Executing calculatePi.go will generate the following type of output:

```
$ go run calculatePi.go
Please provide one numeric argument!
exit status 1
$ go run calculatePi.go 20
3.141592653589793258
$ go run calculatePi.go 200
3.1415926535897932569603993617387624040191831562485732434931792835710464502
4891346711851178431761535428201792941629280905081393787528343561058631336354
8602436768047706489838924381929
```

The moral of this section is that there are many different data types, and that you should use the right data type each time!

Developing a key/value store in Go

In this section, you will learn how to develop an unsophisticated version of a **key-value store** in Go, which means that you will learn how to implement the core functionality of a key-value store without any additional bells and whistles. The idea behind a key-value store is modest: answer queries fast and, generally speaking, work as fast as possible. This translates into using simple algorithms and simple data structures.

The program presented now will basically implement the four fundamental tasks of a key-value store:

1. Adding a new element
2. Deleting an existing element from the key-value store based on a key
3. Looking up for the value of a specific key in the store, and
4. Changing the value of an existing key

These four functions allow you to have full control over the key-value store. The commands for these four functions will be named ADD, DELETE, LOOKUP, and CHANGE, respectively. This means that the program will only operate when its gets one of these four commands. Additionally, the program will stop when you enter the STOP word as input, and it will print the full contents of the key-value store when you enter the PRINT command.

The name of the program in this example will be `keyValue.go`, and it will be presented in five code segments.

The first code segment of `keyValue.go` follows next:

```go
package main

import (
    "bufio"
    "fmt"
    "os"
    "strings"
)

type myElement struct {
    Name    string
    Surname string
    Id      string
}

var DATA = make(map[string]myElement)
```

The key-value store is stored in a native Go map because using a built-in Go structure is usually faster. The map variable is defined as a global variable, where its keys are `string` variables and its values are `myElement` variables. You can also see the definition of the `myElement struct` type here.

The second code segment of `keyValue.go` is as follows:

```go
func ADD(k string, n myElement) bool {
    if k == "" {
        return false
    }

    if LOOKUP(k) == nil {
        DATA[k] = n
        return true
    }
    return false
}

func DELETE(k string) bool {
    if LOOKUP(k) != nil {
        delete(DATA, k)
        return true
    }
    return false
}
```

This code contains the implementation of two functions that support the functionality of the `ADD` and `DELETE` commands. Note that if the user tries to add a new element to the store without giving enough values to populate the `myElement` struct, the `ADD` function will not fail. For this particular program, the missing fields of the `myElement` struct will be set to the empty string. However, if you try to add a key that already exists, you will get an error message instead of modifying the existing key-value.

Notice that I rarely use function names with all caps unless I want them to stand out in the code.

The third portion of `keyValue.go` contains the following code:

```go
func LOOKUP(k string) *myElement {
    _, ok := DATA[k]
    if ok {
        n := DATA[k]
        return &n
    } else {
        return nil
```

The Uses of Composite Types

```
        }
    }

    func CHANGE(k string, n myElement) bool {
        DATA[k] = n
        return true
    }

    func PRINT() {
        for k, d := range DATA {
            fmt.Printf("key: %s value: %v\n", k, d)
        }
    }
```

In this Go code segment, you see the implementation of the functions that support the functionality of the LOOKUP and CHANGE commands. If you try to change a key that does not exist, the program will add that key to the store without generating any error messages.

In this part, you can also see the implementation of the PRINT() function that prints the full contents of the key-value store.

The fourth part of keyValue.go is as follows:

```
    func main() {
        scanner := bufio.NewScanner(os.Stdin)
        for scanner.Scan() {
            text := scanner.Text()
            text = strings.TrimSpace(text)
            tokens := strings.Fields(text)

            switch len(tokens) {
            case 0:
                    continue
            case 1:
                    tokens = append(tokens, "")
                    tokens = append(tokens, "")
                    tokens = append(tokens, "")
                    tokens = append(tokens, "")
            case 2:
                    tokens = append(tokens, "")
                    tokens = append(tokens, "")
                    tokens = append(tokens, "")
            case 3:
                    tokens = append(tokens, "")
                    tokens = append(tokens, "")
```

```
        case 4:
                tokens = append(tokens, "")
        }
```

In this part of keyValue.go, you read the input from the user. First, the for loop makes sure that the program will keep running for as long as the user provides some input. Additionally, the program makes sure that the tokens slice has at least five elements, even though only the ADD command requires that number of elements. Thus, for an ADD operation to be complete and not to have any missing values, you will need an input that looks like ADD aKey Field1 Field2 Field3.

The last part of keyValue.go is shown in the following Go code:

```
                switch tokens[0] {
                case "PRINT":
                        PRINT()
                case "STOP":
                        return
                case "DELETE":
                        if !DELETE(tokens[1]) {
                                fmt.Println("Delete operation failed!")
                        }
                case "ADD":
                        n := myElement{tokens[2], tokens[3], tokens[4]}
                        if !ADD(tokens[1], n) {
                                fmt.Println("Add operation failed!")
                        }
                case "LOOKUP":
                        n := LOOKUP(tokens[1])
                        if n != nil {
                                fmt.Printf("%v\n", *n)
                        }
                case "CHANGE":
                        n := myElement{tokens[2], tokens[3], tokens[4]}
                        if !CHANGE(tokens[1], n) {
                                fmt.Println("Update operation failed!")
                        }
                default:
                        fmt.Println("Unknown command - please try again!")
                }
        }
}
```

The Uses of Composite Types

In this part of the program, you process the input from the user. The `switch` statement makes the design of the program very clean, and it saves you from having to use multiple `if...else` statements.

Executing and using `keyValue.go` will create the following output:

```
$ go run keyValue.go
UNKNOWN
Unknown command - please try again!
ADD 123 1 2 3
ADD 234 2 3 4
ADD 345
PRINT
key: 123 value: {1 2 3}
key: 234 value: {2 3 4}
key: 345 value: {  }
ADD 345 3 4 5
Add operation failed!
PRINT
key: 123 value: {1 2 3}
key: 234 value: {2 3 4}
key: 345 value: {  }
CHANGE 345 3 4 5
PRINT
key: 123 value: {1 2 3}
key: 234 value: {2 3 4}
key: 345 value: {3 4 5}
DELETE 345
PRINT
key: 123 value: {1 2 3}
key: 234 value: {2 3 4}
DELETE 345
Delete operation failed!
PRINT
key: 123 value: {1 2 3}
key: 234 value: {2 3 4}
ADD 345 3 4 5
ADD 567 -5 -6 -7
PRINT
key: 123 value: {1 2 3}
key: 234 value: {2 3 4}
key: 345 value: {3 4 5}
key: 567 value: {-5 -6 -7}
CHANGE 345
PRINT
key: 123 value: {1 2 3}
key: 234 value: {2 3 4}
```

```
key: 345 value: {  }
key: 567 value: {-5 -6 -7}
STOP
```

You will have to wait until `Chapter 8`, *Telling a Unix System What to Do*, in order to learn how to add data persistence to the key-value store.

You can also improve `keyValue.go` by adding goroutines and channels to it. However, adding goroutines and channels to a single user application has no practical meaning. Nevertheless, if you make `keyValue.go` capable of operating over TCP/IP networks, then the use of goroutines and channels will allow it to accept multiple connections and serve multiple users. You will learn more about routines and channels in `Chapter 9`, *Go Concurrency – Goroutines, Channels and Pipelines* and `Chapter 10`, *Go Concurrency – Advanced topics*. You will also learn about creating network applications in Go in `Chapter 12`, *The Foundations of Network Programming in Go*, and in `Chapter 13`, *Network Programming – Building Servers and Clients*.

Additional resources

You might find the following resources useful:

- Read the documentation of the `regexp` standard Go package that can be found at `https://golang.org/pkg/regexp/`
- Visit the man page of the `grep(1)` utility
- Develop a Go program that finds all of the IPv4 addresses of a log file that downloaded ZIP files
- You can find more information about the `math/big` Go package at `https://golang.org/pkg/math/big/`
- Review the documentation of the `unicode` standard Go package at `https://golang.org/pkg/unicode/`
- Although you might find it hard at first, start reading **The Go Programming Language Specification** at `https://golang.org/ref/spec`

Exercises

- Try to write a Go program that prints the invalid part or parts of an IPv4 address
- Describe the differences between `make` and `new` without looking at the chapter
- Describe the differences between a character, a byte, and a rune
- Using the code of `findIPv4.go`, write a Go program that prints the most popular IPv4 addresses found in a log file without processing the output with any Unix utilities
- Develop a Go program that finds the IPv4 addresses in a log file that generated a 404 HTML error message
- Using the `math/big` standard Go package, write a Go program that calculates square roots with high precision. Choose the algorithm on your own
- Write a Go utility that finds a given date and time format in its input and returns just the time part of it
- Develop a Go utility that uses a regular expression in order to match integers from 200 to 400
- Try to improve `keyValue.go` by adding logging to it
- Modify the `findIP()` function of the `findIPv4.go` utility so that the regular expression is compiled only once and not every time the `findIP()` function is called

Summary

In this chapter, you learned about many handy Go features, including creating and using structures, tuples, strings, runes, and the functionality of the `unicode` standard Go package. Additionally, you learned about pattern matching and regular expressions, the `switch` statement, the `strings` standard Go package, you developed a key-value store in Go, and learned how to use the types of the `math/big` package to calculate Pi with the required accuracy.

In the next chapter, you will learn how to arrange and manipulate data using more advanced features like binary trees, linked lists, doubly-linked lists, queues, stacks and hash tables, as well as the structures that can be found in the `container` standard Go package. The last topic of the next chapter will be **random numbers** and generating difficult-to-guess strings that can be used as passwords.

5
Enhancing Go Code with Data Structures

In the previous chapter, we discussed composite data types, which are constructed using the `struct` keyword as well as topics such as regular expressions, pattern matching, tuples, runes, strings and the `unicode` and `strings` standard Go packages before developing a simple key-value store in Go. There are times, however, that the structures offered by a programming language will not fit a particular problem. In such cases, you will need to create your own data structures to store, search, and receive your data in explicit and specialized ways. Consequently, this chapter is all about developing and using many well-known data structures in Go including binary trees, linked lists, hash tables, stacks, and queues and learning about their advantages. As nothing describes a data structure better than an image, you will see many explanatory figures in this chapter! The last part of the chapter will address random number generation, which can help you generate difficult-to-guess passwords! In this chapter of Mastering Go, you will learn about the following topics:

- Graphs and nodes
- Measuring the complexity of an algorithm
- Binary trees in Go
- Hash tables in Go
- Linked lists in Go
- Doubly linked lists in Go
- Queues in Go
- Stacks in Go
- The data structures offered by the `container` standard Go package

- Generating random numbers in Go
- Building random strings that can be used as difficult-to-crack passwords

About graphs and nodes

A **graph** G(V,E) is a finite, nonempty set of vertices V (or **nodes**) and a set of edges E. There are two main types of graphs: **cyclic graphs** and **acyclic graphs**. A cyclic graph is one where all or a number of its vertices are connected in a closed chain. In acyclic graphs, there are no closed chains. A **Directed Graph** is one whose edges have a direction associated with them. A **Directed Acyclic Graph** is a directed graph with no cycles in it.

As a node may contain any kind of information, nodes are usually implemented using Go structures.

Algorithm complexity

The efficiency of an algorithm is judged by its computational complexity, which mostly has to do with the number of times the algorithm needs to access its input data to do its job.

The **Big O notation** is used in computer science for describing the complexity of an algorithm. Thus, an $O(n)$ algorithm, which needs to access its input only once, is considered better than an $O(n^2)$ algorithm, which is better than an $O(n^3)$ algorithm, and so on. The worst algorithms, however, are the ones with an $O(n!)$ running time, which makes them almost unusable for inputs with more than 300 elements.

Last, most Go lookup operations in built-in types, such as finding the value of a map key or accessing an array element, have a constant time, which is represented by $O(1)$. This means that built-in types are generally faster than custom types, and that you should generally favor using them, unless you want full control over what is going on behind the scenes. Furthermore, not all data structures are created equal. Generally speaking, array operations are faster than map operations, which is the price you have to pay for the versatility of a map.

Although every algorithm has its drawbacks, if you do not have lots of data, the algorithm is not really important as long as it performs the desired job accurately.

Binary trees in Go

A **binary tree** is a data structure where underneath each node there are two other nodes *at most*. At most means that a node can be connected to one, two, or no other nodes. The **root of a tree** is the first node of the tree. The **depth of a tree**, which is also called the height of a tree, is defined as the longest path from the root to a node, whereas the **depth of a node** is the number of edges from the node to the root of the tree. A **leaf** is a node with no children.

A tree is considered **balanced** when the longest distance from the root to a leaf is at most one more than the shortest such distance. As the name suggests, an **Unbalanced Tree** is a tree that is not balanced. Balancing a tree might be a difficult and slow operation, so it is better to keep your tree balanced from the very beginning than trying to balance it after you have created it, especially when your tree has many nodes.

The following figure shows an unbalanced binary tree. Its root node is J, and nodes A, G, W, and D are leaves.

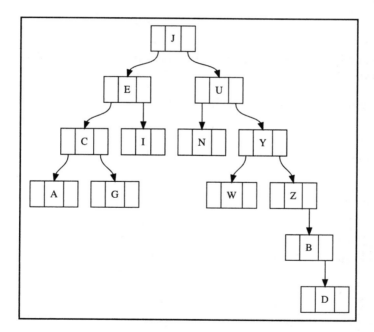

An unbalanced binary tree

Implementing a binary tree in Go

This section illustrates how to implement a binary tree in Go using the source code found in `binTree.go` as an example. The contents of `binTree.go` will be presented in five parts. The first part follows next:

```go
package main

import (
    "fmt"
    "math/rand"
    "time"
)

type Tree struct {
    Left  *Tree
    Value int
    Right *Tree
}
```

What you see here is the definition of the node of the tree using a Go structure. The `math/rand` package is used for populating the tree with random numbers, as we do not yet have any real data.

The second code part of `binTree.go` is shown in the following Go code:

```go
func traverse(t *Tree) {
    if t == nil {
        return
    }
    traverse(t.Left)
    fmt.Print(t.Value, " ")
    traverse(t.Right)
}
```

The `traverse()` function reveals how you can visit all of the nodes of a binary tree using recursion.

The third code segment of `binTree.go` is as follows:

```go
func create(n int) *Tree {
    var t *Tree
    rand.Seed(time.Now().Unix())
    for i := 0; i < 2*n; i++ {
        temp := rand.Intn(n * 2)
        t = insert(t, temp)
    }
    return t
}
```

The `create()` function is only used for populating the binary tree with random integers.

The fourth part of the program is as follows:

```go
func insert(t *Tree, v int) *Tree {
    if t == nil {
        return &Tree{nil, v, nil}
    }

    if v == t.Value {
        return t
    }

    if v < t.Value {
        t.Left = insert(t.Left, v)
        return t
    }
    t.Right = insert(t.Right, v)
    return t
}
```

The `insert()` function does many important things using `if` statements. The first `if` statement checks whether you are dealing with an empty tree or not. If it is indeed an empty tree, then the new node will be the root of the tree and it will be created as `&Tree{nil, v, nil}`. The second `if` statement determines whether the value you are trying to insert already exists in the binary tree or not. If it exists, the function returns without doing anything else. The third `if` statement determines whether the value you are trying to insert will go to the left or to the right side of the node that is currently being examined and acts accordingly.

Enhancing Go Code with Data Structures

Notice that the presented implementation creates unbalanced binary trees.

The last part of `binTree.go` contains the following Go code:

```
func main() {
    tree := create(10)
    fmt.Println("The value of the root of the tree is", tree.Value)
    traverse(tree)
    fmt.Println()
    tree = insert(tree, -10)
    tree = insert(tree, -2)
    traverse(tree)
    fmt.Println()
    fmt.Println("The value of the root of the tree is", tree.Value)
}
```

Executing `binTree.go` will generate the following type of output:

```
$ go run binTree.go
The value of the root of the tree is 18
0 3 4 5 7 8 9 10 11 14 16 17 18 19
-10 -2 0 3 4 5 7 8 9 10 11 14 16 17 18 19
The value of the root of the tree is 18
```

Advantages of binary trees

You cannot beat a tree when you need to represent hierarchical data. For that reason, trees are extensively used when the compiler of a programming language parses a computer program.

Additionally, trees are *ordered* by design, which means that you do not have to make any special effort to order them; putting an element into its correct place keeps them ordered. However, deleting an element from a tree is not always trivial because of the way that trees are constructed.

If a binary tree is balanced, its search, insert, and delete operations take about *log(n)* steps, where *n* is the total number of elements that the tree holds. Additionally, the height of a balanced binary tree is approximately *log2(n)*, which means that a balanced tree with 10,000 elements has a height of about 14, and that is remarkably small. Similarly, the height of a balanced tree with 100,000 elements will be about 17, and the height of a balanced tree with 1,000,000 elements will be about 20. In other words, putting a significantly large number of elements into a balanced binary tree does not change the speed of the tree in an extreme way. Stated differently, you can reach any node of a tree with 1,000,000 nodes in less than 20 steps!

A major disadvantage of binary trees is that the shape of the tree depends on the order in which its elements were inserted. If the keys of a tree are long and complex, then inserting or searching for an element might be slow due to the large number of comparisons required. Last, if a tree is not balanced, then the performance of the tree is unpredictable.

Although you can create a linked list or an array faster than a binary tree, the flexibility that a binary tree offers in search operations might worth the extra overhead and maintenance. When searching for an element on a binary tree, you check whether the value of the element that you are looking for is bigger or smaller than the value of the current node and use that decision to choose which part of the tree you will go down next,. Doing this saves you a lot of time.

Hash tables in Go

Strictly speaking, a **Hash table** is a data structure that stores one or more key and value pairs and uses a hash function to compute an index into an array of buckets or slots from which the correct value can be discovered. Ideally, the **hash function** should assign each key to a unique bucket, provided that you have the required number of buckets, which is usually the case.

A good hash function must be able to produce a uniform distribution of the hash values, because it is inefficient to have unused buckets or big differences in the cardinalities of the buckets. Additionally, the hash function should work consistently and output the same hash value for identical keys. Otherwise, it would be impossible to locate the information you want!

The next figure shows what a hash table with 10 buckets looks like.

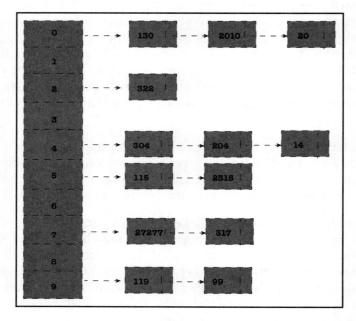

A hash table with 10 buckets

Implementing a hash table in Go

The Go code of hashTable.go, which will be presented in five parts, will help clarify many things about hash tables.

The first part of hashTable.go is as follows:

```
package main

import (
    "fmt"
)

const SIZE = 15

type Node struct {
    Value int
    Next  *Node
}
```

In this part, you can see the definition of the node of the hash table, which, as expected, is defined using a Go structure. The SIZE constant variable holds the number of buckets of the hash table.

The second segment from hashTable.go is shown in the following Go code:

```go
type HashTable struct {
    Table map[int]*Node
    Size  int
}

func hashFunction(i, size int) int {
    return (i % size)
}
```

In this code segment, you can see the implementation of the hash function used in this particular hash. The hashFunction() uses the **modulo operator**. The main reason for choosing the modulo operator is because this particular hash table has to cope with integer values. If you were dealing with strings or floating-point numbers, then you should have used a different logic in your hash function.

The actual hash is stored in a HashTable structure that has two fields. The second field is the size of the hash table, and the first field is a map that associates an integer with a linked list (*Node). As a result, this hash table will have as many linked lists as the number of its buckets. This also means that the nodes of each bucket of the hash table will be stored in linked lists. You will learn more about linked lists in a little while.

The third code portion of hashTable.go is as follows:

```go
func insert(hash *HashTable, value int) int {
    index := hashFunction(value, hash.Size)
    element := Node{Value: value, Next: hash.Table[index]}
    hash.Table[index] = &element
    return index
}
```

The insert() function is called for inserting elements into the hash table. Note that the current implementation of the insert() function does not check for duplicate values.

The fourth part of hashTable.go comes next:

```go
func traverse(hash *HashTable) {
    for k := range hash.Table {
        if hash.Table[k] != nil {
            t := hash.Table[k]
```

```
            for t != nil {
                fmt.Printf("%d -> ", t.Value)
                t = t.Next
            }
            fmt.Println()
        }
    }
}
```

The `traverse()` function is used for printing all of the values in the hash table. The function visits each one of the linked lists of the hash table and prints the stored values, linked list by linked list.

The last code portion of `hashTable.go` is as follows:

```
func main() {
    table := make(map[int]*Node, SIZE)
    hash := &HashTable{Table: table, Size: SIZE}
    fmt.Println("Number of spaces:", hash.Size)
    for i := 0; i < 120; i++ {
        insert(hash, i)
    }
    traverse(hash)
}
```

In this part of the code, you create a new hash table named `hash` using the `table` variable, which is a map that holds the buckets of the hash table. As you already know, the slots of a hash table are implemented using linked lists. The main reason for using a map to hold the linked lists of a hash table instead of a slice or an array is that the keys of a slice or an array can only be positive integers, while the keys of a map can be almost anything you need.

Executing `hashTable.go` will produce the following output:

```
$ go run hashTable.go
Number of spaces: 15
108 -> 93 -> 78 -> 63 -> 48 -> 33 -> 18 -> 3 ->
109 -> 94 -> 79 -> 64 -> 49 -> 34 -> 19 -> 4 ->
117 -> 102 -> 87 -> 72 -> 57 -> 42 -> 27 -> 12 ->
119 -> 104 -> 89 -> 74 -> 59 -> 44 -> 29 -> 14 ->
111 -> 96 -> 81 -> 66 -> 51 -> 36 -> 21 -> 6 ->
112 -> 97 -> 82 -> 67 -> 52 -> 37 -> 22 -> 7 ->
116 -> 101 -> 86 -> 71 -> 56 -> 41 -> 26 -> 11 ->
114 -> 99 -> 84 -> 69 -> 54 -> 39 -> 24 -> 9 ->
118 -> 103 -> 88 -> 73 -> 58 -> 43 -> 28 -> 13 ->
105 -> 90 -> 75 -> 60 -> 45 -> 30 -> 15 -> 0 ->
106 -> 91 -> 76 -> 61 -> 46 -> 31 -> 16 -> 1 ->
107 -> 92 -> 77 -> 62 -> 47 -> 32 -> 17 -> 2 ->
```

```
110 -> 95 -> 80 -> 65 -> 50 -> 35 -> 20 -> 5 ->
113 -> 98 -> 83 -> 68 -> 53 -> 38 -> 23 -> 8 ->
115 -> 100 -> 85 -> 70 -> 55 -> 40 -> 25 -> 10 ->
```

This particular hash table is perfectly balanced because it has to deal with continuous numbers that are placed in a slot according to the results of the **modulo operator**. Real-world problems might not generate such convenient results!

The remainder of a **Euclidean division** between two natural numbers a and b can be calculated according to the $a = bq + r$ formula, where q is the quotient and r is the remainder. The values allowed for the remainder can be between 0 and $b-1$, which are the possible results of the modulo operator.

Note that if you execute `hashTable.go` several times, you will most likely get an output where the lines are in a different order, because the way that Go outputs the key and value pairs of a map is totally random!

Implementing the lookup functionality

In this section, you will see an implementation of the `lookup()` function that allows you to determine whether a given element is already in the hash table or not. The code of the `lookup()` function is based on that of the `traverse()` function, as follows:

```
func lookup(hash *HashTable, value int) bool {
    index := hashFunction(value, hash.Size)
    if hash.Table[index] != nil {
        t := hash.Table[index]
        for t != nil {
            if t.Value == value {
                return true
            }
            t = t.Next
        }
    }
    return false
}
```

You can find the preceding code in the `hashTableLookup.go` source file. Executing `hashTableLookup.go` will create the following output:

```
$ go run hashTableLookup.go
120 is not in the hash table!
121 is not in the hash table!
```

```
122 is not in the hash table!
123 is not in the hash table!
124 is not in the hash table!
```

The preceding output means that the `lookup()` function does its job pretty well!

Advantages of hash tables

If you think that hash tables are not that useful, handy, or smart, consider the following: When a hash table has n keys and k buckets, the search speed for the n keys goes from $O(n)$ for a linear search down to $O(n/k)$! Although the improvement might look small, you must realize that for a hash array with only 20 slots, the search time will be reduced by 20 times! This makes hash tables perfect for applications such as dictionaries or any other analogous application where you have to search large amounts of data.

Linked lists in Go

A **Linked list** is a data structure with a finite set of elements where each element uses at least two memory locations, one for storing the actual data and the other for storing a pointer that links the current element to the next one, thus creating a sequence of elements that construct the linked list.

The first element of a linked list is called the *head* whereas the last element of a linked list is often called the *tail*. The first thing that you should do when defining a linked list is to keep the head of the list in a separate variable because the head is the only thing that you have to access the entire linked list. Note that if you lose the pointer to that first node of a single linked list, there is no way to find it again.

The following figure shows a linked list with five nodes:

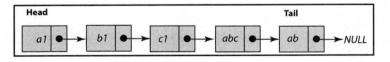

A linked list with five nodes

The next figure shows you how to remove an existing node from a linked list in order to understand better the steps that are involved in the process. The main thing that you will need to do is to arrange the pointer to the left node of the node that you are removing in order to point to the right node of the node that is being removed.

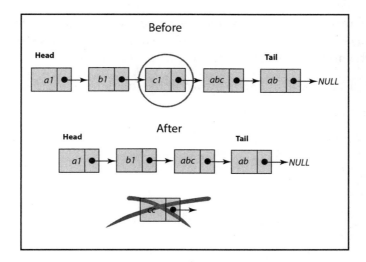

Removing a node from a linked list

The linked list implementation that follows is relatively simple and will not include the delete node functionality, which is left as an exercise for the reader to tackle.

Implementing a linked list in Go

The Go source file for the implementation of the linked list is called `linkedList.go`, and it will be presented in five parts.

The first code segment of `linkedList.go` follows next:

```
package main

import (
    "fmt"
)

type Node struct {
    Value int
    Next  *Node
}

var root = new(Node)
```

Enhancing Go Code with Data Structures

In this part of the program, you define the `Node` structure type that will be used for the nodes of the linked list as well as the `root` global variable that holds the first element of the linked list, which will be accessible from everywhere in the code.

The second part of `linkedList.go` is shown in the following Go code:

```go
func addNode(t *Node, v int) int {
    if root == nil {
        t = &Node{v, nil}
        root = t
        return 0
    }

    if v == t.Value {
        fmt.Println("Node already exists:", v)
        return -1
    }

    if t.Next == nil {
        t.Next = &Node{v, nil}
        return -2
    }

    return addNode(t.Next, v)
}
```

Due to the way linked lists work, they do not normally contain duplicate entries. Furthermore, new nodes are usually added at the end of a linked list when the linked list is not sorted.

Thus, the `addNode()` function is used for adding new nodes to the linked list. There are three distinct cases in the implementation that are examined using `if` statements. In the first case, you test whether you are dealing with an empty linked list or not. In the second case, you check whether the value that you want to add is already in the list. In the third case, you check whether you have reached the end of the linked list. In this case, you add a new node at the end of the list with the desired value using `t.Next = &Node{v, nil}`. If none of these conditions is true, you repeat the same process with the `addNode()` function for the next node of the linked list using `return addNode(t.Next, v)`.

The third code segment of the `linkedList.go` program contains the implementation of the `traverse()` function:

```go
func traverse(t *Node) {
    if t == nil {
        fmt.Println("-> Empty list!")
```

```go
        return
    }

    for t != nil {
        fmt.Printf("%d -> ", t.Value)
        t = t.Next
    }
    fmt.Println()
}
```

The fourth part of `linkedList.go` is as follows:

```go
func lookupNode(t *Node, v int) bool {
    if root == nil {
        t = &Node{v, nil}
        root = t
        return false
    }

    if v == t.Value {
        return true
    }

    if t.Next == nil {
        return false
    }

    return lookupNode(t.Next, v)
}

func size(t *Node) int {
    if t == nil {
        fmt.Println("-> Empty list!")
        return 0
    }

    i := 0
    for t != nil {
        i++
        t = t.Next
    }
    return i
}
```

Enhancing Go Code with Data Structures

In this part, you see the implementation of two very handy functions: `lookupNode()` and `size()`. The former checks whether a given element exists in the linked list, while the latter returns the size of the linked list, which is the number of nodes in the linked list.

The logic behind the implementation of the `lookupNode()` function is easy to understand: You start accessing all of the elements of the single linked list in order to search for the value you crave. If you reach the tail of the linked list without having found the desired value, then you know that the linked list does not contain that value.

The last part of `linkedList.go` contains the implementation of the `main()` function:

```go
func main() {
    fmt.Println(root)
    root = nil
    traverse(root)
    addNode(root, 1)
    addNode(root, -1)
    traverse(root)
    addNode(root, 10)
    addNode(root, 5)
    addNode(root, 45)
    addNode(root, 5)
    addNode(root, 5)
    traverse(root)
    addNode(root, 100)
    traverse(root)

    if lookupNode(root, 100) {
        fmt.Println("Node exists!")
    } else {
        fmt.Println("Node does not exist!")
    }

    if lookupNode(root, -100) {
        fmt.Println("Node exists!")
    } else {
        fmt.Println("Node does not exist!")
    }
}
```

Executing `linkedList.go` will generate the following output:

```
$ go run linkedList.go
&{0 <nil>}
-> Empty list!
1 -> -1 ->
Node already exists: 5
```

```
Node already exists: 5
1 -> -1 -> 10 -> 5 -> 45 ->
1 -> -1 -> 10 -> 5 -> 45 -> 100 ->
Node exists!
Node does not exist!
```

Advantages of linked lists

The greatest advantages of linked lists are that they are easy to understand and implement, and they are generic enough so that they can be used in many different situations. This means that they can be used to model many different kinds of data. Additionally, linked lists are really fast at sequential searching when used with pointers.

Linked lists cannot only help you sort your data, but they can also assist you in keeping your data sorted even after inserting or deleting elements. Deleting a node from a sorted linked list is the same as in an unsorted linked list; however, inserting a new node into a sorted linked list is different because the new node has to go to the right place in order to keep the list sorted. In practice, this means that if you have lots of data and you know that you will need to delete data all the time, using a linked list is a better choice than using a hash table or a binary tree.

Last, **sorted linked lists** allow you to use various optimization techniques when searching or inserting a node. The most common technique is keeping a pointer at the center node of the sorted linked list and starting your lookups from there. This simple optimization can reduce the time of the lookup operation in half!

Doubly linked lists in Go

A **doubly linked list** is one where each node keeps a pointer to the previous element on the list as well as the next element.

The following figure shows what a doubly linked list looks like:

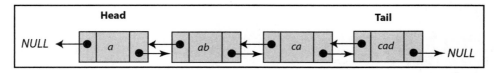

A doubly linked list

Enhancing Go Code with Data Structures

Thus, on a doubly linked list, the next link of the first node points to the second node, while its previous link points to `nil` (also called as **NULL**). Analogously, the next link of the last node points to `nil`, while its previous link points to the next-to-last node of the doubly linked list.

The last figure of this chapter illustrates the addition of a node in a doubly linked list. As you can imagine, the main task that needs to be accomplished is dealing with the pointers of three nodes: the new node, the node that will be on the left of the new node, and the node that will be on the right of the new node.

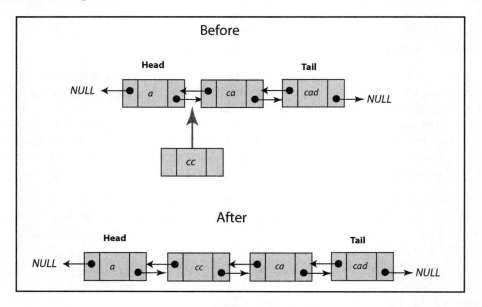

Inserting a new node in the middle of a doubly linked list

Thus in reality, the main difference between a single linked list and a doubly linked list is that the latter requires more housekeeping. This is the price that you will have to pay for being able to access your doubly linked list both ways.

Implementing a doubly linked list in Go

The name of the program with the Go implementation of a doubly linked list is `doublyLList.go`, and will be offered to you in five parts. The general idea behind a doubly linked list is that same as in a single linked list, but you just have to do more housekeeping due to the presence of two pointers in each node of the list.

The first part of `doublyLList.go` follows next:

```
package main

import (
    "fmt"
)

type Node struct {
    Value    int
    Previous *Node
    Next     *Node
}
```

In this part, you can see the definition of the node of the doubly linked list using a Go structure. However, this time, the `struct` has two pointer fields for apparent reasons.

The second code portion of `doublyLList.go` contains the following Go code:

```
func addNode(t *Node, v int) int {
    if root == nil {
        t = &Node{v, nil, nil}
        root = t
        return 0
    }

    if v == t.Value {
        fmt.Println("Node already exists:", v)
        return -1
    }

    if t.Next == nil {
        temp := t
        t.Next = &Node{v, temp, nil}
        return -2
    }

    return addNode(t.Next, v)
}
```

As happened in the case of the single linked list, each new node is placed at the end of the current doubly linked list. However, this is not mandatory as you can decide that you want to have a sorted doubly linked list.

The third part of `doublyLList.go` is as follows:

```
func traverse(t *Node) {
    if t == nil {
        fmt.Println("-> Empty list!")
        return
    }

    for t != nil {
        fmt.Printf("%d -> ", t.Value)
        t = t.Next
    }
    fmt.Println()
}

func reverse(t *Node) {
    if t == nil {
        fmt.Println("-> Empty list!")
        return
    }

    temp := t
    for t != nil {
        temp = t
        t = t.Next
    }

    for temp.Previous != nil {
        fmt.Printf("%d -> ", temp.Value)
        temp = temp.Previous
    }
    fmt.Printf("%d -> ", temp.Value)
    fmt.Println()
}
```

Here you see the Go code for the `traverse()` and `reverse()` functions. The implementation of the `traverse()` function is the same as in the `linkedList.go` program. However, the logic behind the `reverse()` function is very interesting, as we do not keep a pointer to the tail of the doubly linked list, we need to go to the end of the doubly linked list before being able to access its nodes in reverse order.

The fourth part of doublyLList.go contains the following Go code:

```go
func size(t *Node) int {
    if t == nil {
        fmt.Println("-> Empty list!")
        return 0
    }

    n := 0
    for t != nil {
        n++
        t = t.Next
    }
    return n
}

func lookupNode(t *Node, v int) bool {
    if root == nil {
        return false
    }

    if v == t.Value {
        return true
    }

    if t.Next == nil {
        return false
    }

    return lookupNode(t.Next, v)
}
```

The last code segment of doublyLList.go contains the following Go code:

```go
var root = new(Node)

func main() {
    fmt.Println(root)
    root = nil
    traverse(root)
    addNode(root, 1)
    addNode(root, 1)
    traverse(root)
    addNode(root, 10)
    addNode(root, 5)
    addNode(root, 0)
    addNode(root, 0)
    traverse(root)
```

```
        addNode(root, 100)
        fmt.Println("Size:", size(root))
        traverse(root)
        reverse(root)
}
```

If you execute `doublyLList.go`, you will get the following output:

```
$ go run doublyLList.go
&{0 <nil> <nil>}
-> Empty list!
Node already exists: 1
1 ->
Node already exists: 0
1 -> 10 -> 5 -> 0 ->
Size: 5
1 -> 10 -> 5 -> 0 -> 100 ->
100 -> 0 -> 5 -> 10 -> 1 ->
```

As you can see, the `reverse()` function works just fine!

Advantages of doubly linked lists

Doubly linked lists are more versatile than single linked lists because you can traverse them in any direction you want and also you can insert and delete elements from a doubly linked list more easily. Additionally, even if you lose the pointer to the head of a doubly linked list, you can still find the head node of that list. However, this versatility comes at a price; that is, maintaining two pointers for each node. It is up to the developer to decide whether that extra complexity is justified or not.

After all, your music player might be using a doubly linked list to represent your current list of songs and be able to go to the previous song as well as the next one!

Queues in Go

A **Queue** is a special kind of a linked list where each new element is inserted to the head and removed from the tail of the linked list. We do not need a figure to describe a queue; imagine going to a bank and waiting for the people that came before you to finish their transactions before you can talk to a bank teller. That is a queue!

The main advantage of queues is simplicity! You only need two functions to access a queue, which means that you have to worry about fewer things going wrong and you can implement a queue any way you want so long as you can offer support for those two functions!

Implementing a queue in Go

The program that will illustrate the Go implementation of a queue is called `queue.go`, and it will be presented in five parts. Note that a linked list will be used for the implementation of the queue. The `Push()` and a `Pop()` functions are used for adding and removing nodes from the queue, respectively.

The first part of the code for `queue.go` follows next:

```
package main

import (
    "fmt"
)

type Node struct {
    Value int
    Next  *Node
}

var size = 0
var queue = new(Node)
```

Having a variable (`size`) for keeping the number of nodes that you have on the queue is handy but not compulsory. However, the implementation presented here supports this functionality because it makes things simpler.

The second code portion of `queue.go` contains the following Go code:

```
func Push(t *Node, v int) bool {
    if queue == nil {
        queue = &Node{v, nil}
        size++
        return true
    }

    t = &Node{v, nil}
    t.Next = queue
    queue = t
```

```
        size++
        return true
}
```

This part displays the implementation of the `Push()` function, which is straightforward. If the queue is empty, then the new node will become the queue. If the queue is not empty, then you create a new node that is placed in front of the current queue. After that, the head of the queue becomes the node that was just created.

The third part of `queue.go` contains the following Go code:

```
func Pop(t *Node) (int, bool) {
    if size == 0 {
        return 0, false
    }

    if size == 1 {
        queue = nil
        size--
        return t.Value, true
    }

    temp := t
    for (t.Next) != nil {
        temp = t
        t = t.Next
    }

    v := (temp.Next).Value
    temp.Next = nil

    size--
    return v, true
}
```

The preceding code shows the implementation of the `Pop()` function, which removes the oldest element of the queue. If the queue is empty (`size == 0`), there is nothing to extract. If the queue has only one node, then you extract the value of that node and the queue becomes empty. Otherwise, you extract the last element of the queue, remove the last node of the queue, and fix the required pointers before returning the desired value.

The fourth part of `queue.go` contains the following Go code:

```
func traverse(t *Node) {
    if size == 0 {
```

```
            fmt.Println("Empty Queue!")
            return
    }

    for t != nil {
        fmt.Printf("%d -> ", t.Value)
        t = t.Next
    }
    fmt.Println()
}
```

Strictly speaking, the `traverse()` function is not necessary for the operation of a queue, but it gives you a practical way of looking at all of the nodes of the queue.

The last code segment of `queue.go` is shown in the following Go code:

```
func main() {
    queue = nil
    Push(queue, 10)
    fmt.Println("Size:", size)
    traverse(queue)

    v, b := Pop(queue)
    if b {
        fmt.Println("Pop:", v)
    }
    fmt.Println("Size:", size)

    for i := 0; i < 5; i++ {
        Push(queue, i)
    }
    traverse(queue)
    fmt.Println("Size:", size)

    v, b = Pop(queue)
    if b {
        fmt.Println("Pop:", v)
    }
    fmt.Println("Size:", size)

    v, b = Pop(queue)
    if b {
        fmt.Println("Pop:", v)
    }
    fmt.Println("Size:", size)
    traverse(queue)
}
```

Enhancing Go Code with Data Structures

Almost all of the Go code in `main()` is for checking the operation of the queue. The most important code in here are the two `if` statements that let you know whether the `Pop()` function returned an actual value, or if the queue was empty and there is nothing to return.

Executing `queue.go` will produce the following type of output:

```
$ go run queue.go
Size: 1
10 ->
Pop: 10
Size: 0
4 -> 3 -> 2 -> 1 -> 0 ->
Size: 5
Pop: 0
Size: 4
Pop: 1
Size: 3
4 -> 3 -> 2 ->
```

Stacks in Go

A **Stack** is a data structure that looks like a pile of plates. The last plate that goes on the top of the pile is the one that will be used first when you need to use a new plate.

The main advantage of a stack is its simplicity because you only have to worry about implementing two functions in order to be able to work with a stack-adding a new node to the stack and removing a node from the stack.

Implementing a stack in Go

It is now time to look at the implementation of a stack in Go. This will be illustrated in the `stack.go` source file. Once again, a linked list will be used for implementing the stack. As you know, you will need two functions: one function named `Push()` for putting things on the stack and another one named `Pop()` for removing things from the stack. Although it is not necessary, it is useful to keep the number of elements that you have on the stack on a separate variable in order to be able to tell whether you are dealing with an empty stack or not without having to access the linked list itself.

The source code of `stack.go` will be presented in four parts. The first part follows next:

```
package main

import (
    "fmt"
)

type Node struct {
    Value int
    Next  *Node
}

var size = 0
var stack = new(Node)
```

The second part of `stack.go` contains the implementation of the `Push()` function:

```
func Push(v int) bool {
    if stack == nil {
        stack = &Node{v, nil}
        size = 1
        return true
    }

    temp := &Node{v, nil}
    temp.Next = stack
    stack = temp
    size++
    return true
}
```

If the stack is not empty, then you create a new node (`temp`), which is placed in front of the current stack. After that, this new node becomes the head of the stack. The current version of the `Push()` function always returns `true`, but if your stack does not have unlimited space, you might want to modify it and return `false` when you are about to exceed its capacity.

The third part contains the implementation of the `Pop()` function:

```
func Pop(t *Node) (int, bool) {
    if size == 0 {
        return 0, false
    }

    if size == 1 {
        size = 0
```

```
            stack = nil
            return t.Value, true
    }

    stack = stack.Next
    size--
    return t.Value, true
}
```

The fourth code segment of stack.go is as follows:

```
func traverse(t *Node) {
    if size == 0 {
        fmt.Println("Empty Stack!")
        return
    }

    for t != nil {
        fmt.Printf("%d -> ", t.Value)
        t = t.Next
    }
    fmt.Println()
}
```

As the stack is implemented using a linked list, it is traversed as such.

The last part of stack.go is shown in the following Go code:

```
func main() {
    stack = nil
    v, b := Pop(stack)
    if b {
        fmt.Print(v, " ")
    } else {
        fmt.Println("Pop() failed!")
    }

    Push(100)
    traverse(stack)
    Push(200)
    traverse(stack)

    for i := 0; i < 10; i++ {
        Push(i)
    }

    for i := 0; i < 15; i++ {
        v, b := Pop(stack)
```

```
            if b {
                fmt.Print(v, " ")
            } else {
                break
            }
        }
        fmt.Println()
        traverse(stack)
    }
```

As you just saw, the source code of stack.go is a little shorter than the Go code of queue.go, primarily because the idea behind a stack is simpler than the idea behind a queue.

Executing stack.go will generate the following type of output:

```
$ go run stack.go
Pop() failed!
100 ->
200 -> 100 ->
9 8 7 6 5 4 3 2 1 0 200 100
Empty Stack!
```

 So far, you have seen how a linked list is used in the implementation of a hash table, queue, and a stack. These examples should help you to realize the usefulness and the importance of linked lists in programming and computer science in general!

The container package

In this section, I will explain the use of the container standard Go package. The container package supports three data structures: a heap, list, and ring. These data structures are implemented in container/heap, container/list, and container/ring, respectively.

For those of you who are unfamiliar with rings, a ring is a **circular list**, which means that the last element of a ring points to its first element. In essence, this means that all of the nodes of a ring are equivalent and that a ring does not have a beginning and an end. As a result, each element of a ring can help you traverse the entire ring.

The next three subsections will illustrate each one of the packages contained in the `container` package. The rational advice is that if the functionality of the `container` standard Go package suits your needs, use it; otherwise, you should implement and use your own data structures.

Using container/heap

In this subsection, you will see the functionality that the `container/heap` package offers. First of all, you should know that the `container/heap` package implements a heap, which is a tree where the value of each node of the tree is the smallest element in its subtree. Note that I am using the phrase *smallest element* instead of *minimum value* in order to make clear that a heap does not only support numerical values.

However, as you can guess, in order to implement a heap tree in Go, you will have to develop a way to tell which one of two elements is *smaller* than the other on your own. In such cases, Go uses interfaces because they allow you to define such a behavior.

This means that the `container/heap` package is more advanced than the other two packages found in `container`, and that you will have to define some things before being able to use the functionality of the `container/heap` package. Strictly speaking, the `container/heap` package requires that you implement the `container/heap.Interface`, which is defined as follows:

```
type Interface interface {
    sort.Interface
    Push(x interface{}) // add x as element Len()
    Pop() interface{}   // remove and return element Len() - 1.
}
```

You will learn more about **Interfaces** in Chapter 7, *Reflection and Interfaces for All Seasons*. For now, just remember that compliance with a Go interface requires the implementation of one or more functions or other interfaces, which in this case is `sort.Interface` as well as the `Push()` and `Pop()` functions. The `sort.Interface` requires that you implement the `Len()`, `Less()`, and `Swap()` functions, which makes perfect sense because you cannot perform any kind of sorting without being able to swap two elements, being able to calculate a *value* for the things that you want to sort, and being able to tell which element between two elements is bigger than the other based on the value that you calculated previously. Although you might think that this is a lot of work, keep in mind that most of the times the implementation of these functions is either trivial or rather simple.

Since the purpose of this section is to illustrate the use of container/heap and not to make your life difficult, the data type for the elements in this example will be float32.

The Go code of conHeap.go will be presented in five parts. The first part follows next:

```
package main

import (
    "container/heap"
    "fmt"
)

type heapFloat32 []float32
```

The second part of the conHeap.go is shown in the following Go code:

```
func (n *heapFloat32) Pop() interface{} {
    old := *n
    x := old[len(old)-1]
    new := old[0 : len(old)-1]
    *n = new
    return x
}

func (n *heapFloat32) Push(x interface{}) {
    *n = append(*n, x.(float32))
}
```

Although you define two functions named Pop() and Push() here, these two functions are used for interface compliance. In order to add and remove elements from the heap, you should call heap.Push() and heap.Pop() respectively.

The third code segment of conHeap.go contains the following Go code:

```
func (n heapFloat32) Len() int {
    return len(n)
}

func (n heapFloat32) Less(a, b int) bool {
    return n[a] < n[b]
}

func (n heapFloat32) Swap(a, b int) {
    n[a], n[b] = n[b], n[a]
}
```

Enhancing Go Code with Data Structures

This part implements the three functions needed by the `sort.Interface` interface.

The fourth part of `conHeap.go` is as follows:

```go
func main() {
    myHeap := &heapFloat32{1.2, 2.1, 3.1, -100.1}
    heap.Init(myHeap)
    size := len(*myHeap)
    fmt.Printf("Heap size: %d\n", size)
    fmt.Printf("%v\n", myHeap)
```

The last code portion of `conHeap.go` follows next:

```go
    myHeap.Push(float32(-100.2))
    myHeap.Push(float32(0.2))
    fmt.Printf("Heap size: %d\n", len(*myHeap))
    fmt.Printf("%v\n", myHeap)
    heap.Init(myHeap)
    fmt.Printf("%v\n", myHeap)
}
```

In this last part of `conHeap.go`, you add two new elements to `myHeap` using `heap.Push()`. However, in order for the heap to get properly re-sorted, you will need to make another call to `heap.Init()`.

Executing `conHeap.go` will generate the following output:

```
$ go run conHeap.go
Heap size: 4
&[-100.1 1.2 3.1 2.1]
Heap size: 6
&[-100.1 1.2 3.1 2.1 -100.2 0.2]
&[-100.2 -100.1 0.2 2.1 1.2 3.1]
```

If you find it strange that the `2.1 1.2 3.1` triplet in the last line of the output is not sorted in the linear logic, remember that a heap is a tree-not a linear structure like an array or a slice.

Using container/list

This subsection will illustrate the operation of the `container/list` package using the Go code of `conList.go`, which will be presented in three parts.

 The `container/list` package implements a **linked list**.

The first part of `conList.go` contains the following Go code:

```
package main

import (
    "container/list"
    "fmt"
    "strconv"
)

func printList(l *list.List) {
    for t := l.Back(); t != nil; t = t.Prev() {
        fmt.Print(t.Value, " ")
    }
    fmt.Println()

    for t := l.Front(); t != nil; t = t.Next() {
        fmt.Print(t.Value, " ")
    }

    fmt.Println()
}
```

Here, you see a function named `printList()`, which allows you to print the contents of a `list.List` variable passed as a pointer. The Go code shows you how to print the elements of a `list.List` starting from the first element and going to the last element and vice versa. Usually, you will need to use only one of the two methods in your programs. The `Prev()` and `Next()` functions allow you to iterate over the elements of a list backwards and forward.

Enhancing Go Code with Data Structures

The second code segment of `conList.go` follows next:

```
func main() {

    values := list.New()

    e1 := values.PushBack("One")
    e2 := values.PushBack("Two")
    values.PushFront("Three")
    values.InsertBefore("Four", e1)
    values.InsertAfter("Five", e2)
    values.Remove(e2)
    values.Remove(e2)
    values.InsertAfter("FiveFive", e2)
    values.PushBackList(values)

    printList(values)

    values.Init()
```

The `list.PushBack()` function allows you to insert an object at the back of a linked list, whereas the `list.PushFront()` function allows you to insert an object at the front of a list. The return value of both functions is the element inserted in the list. If you want to insert a new element after a specific element, then you should use the `list.InsertAfter()` function. Similarly, if you want to insert a new element before a specific element, you should use the `list.InsertBefore()` function. If the element does not exist, then the list will not change. The `list.PushBackList()` inserts a copy of an existing list at the end of another list, whereas the `list.PushFrontList()` function puts a copy of an existing list at the front of another list. The `list.Remove()` function removes a specific element from a list.

Note the use of the `values.Init()` function that either empties an existing list or initializes a new list.

The last portion of the code of `conList.go` is shown in the following Go code:

```
    fmt.Printf("After Init(): %v\n", values)

    for i := 0; i < 20; i++ {
        values.PushFront(strconv.Itoa(i))
    }

    printList(values)
}
```

Here, you create a new list using a `for` loop-the `strconv.Itoa()` function converts an integer value into a string.

In conclusion, the use of the functions of the `container/list` package is straightforward and comes with no surprises.

Executing `conList.go` will generate the following type of output:

```
$ go run conList.go
Five One Four Three Five One Four Three
Three Four One Five Three Four One Five
After Init(): &{{0xc420074180 0xc420074180 <nil> <nil>} 0}
0 1 2 3 4 5 6 7 8 9 10 11 12 13 14 15 16 17 18 19
19 18 17 16 15 14 13 12 11 10 9 8 7 6 5 4 3 2 1 0
```

Using container/ring

This section will illustrate the use of the `container/ring` package using the Go code of `conRing.go`, which will be presented in four parts. Note that the `container/ring` package is much simpler than both `container/list` and `container/heap`, which means that it contains fewer functions than the other two packages.

The first code segment of `conRing.go` follows next:

```
package main

import (
    "container/ring"
    "fmt"
)

var size int = 10
```

The `size` variable holds the size of the ring that will be created.

The second part of `conRing.go` contains the following Go code:

```
func main() {
    myRing := ring.New(size + 1)
    fmt.Println("Empty ring:", *myRing)

    for i := 0; i < myRing.Len()-1; i++ {
        myRing.Value = i
        myRing = myRing.Next()
```

```
    }
    myRing.Value = 2
```

Thus, a new ring is created with the help of the `ring.New()` function, which requires a single parameter-the size of the ring. The `myRing.Value = 2` statement at the end adds the value 2 to the ring. That value, however, already exists in the ring as it was added in the `for` loop. Last, the zero value of a ring is a ring with a single element whose value is `nil`.

The third part of `conRing.go` is shown in the following Go code:

```
sum := 0
myRing.Do(func(x interface{}) {
    t := x.(int)
    sum = sum + t
})
fmt.Println("Sum:", sum)
```

The `ring.Do()` function allows you to call a function for each element of a ring in forward order. However, if that function makes any changes to the ring, then the behavior of `ring.Do()` is undefined. The `x.(int)` statement is called **type assertion**. You will learn more about type assertions in Chapter 7, *Reflection and Interfaces for All Seasons*. For now, just know that it shows that x is of type `int`.

The last part of the `conRing.go` program is as follows:

```
for i := 0; i < myRing.Len()+2; i++ {
        myRing = myRing.Next()
        fmt.Print(myRing.Value, " ")
    }
    fmt.Println()
}
```

The only problem with rings is that you can keep calling `ring.Next()` indefinitely, so you will need to find a way to put a stop to that. In this case, this is accomplished with the help of the `ring.Len()` function. Personally, I prefer to use the `ring.Do()` function for iterating over all of the elements of a ring because it generates cleaner code, but using a `for` loop is just as good!

Executing `conRing.go` will generate the following type of output:

```
$ go run conRing.go
Empty ring: {0xc42000a080 0xc42000a1a0 <nil>}
Sum: 47
0 1 2 3 4 5 6 7 8 9 2 0 1
```

The output verifies that a ring can contain duplicate values, which means that unless you use the `ring.Len()` function, you have no safe way of knowing the size of a ring.

Generating random numbers

Random number generation is an art as well as a research area in Computer science. This happens because computers are purely logical machines, and it turns out that using them to generate random numbers is extremely difficult!

Go uses the `math/rand` package for generating random numbers. It needs a seed to start producing random numbers. The **seed** is used for initializing the entire process, and it is extremely important because if you always start with the same seed, you will always get the same sequence of random numbers. This means that everybody can regenerate that sequence, and that particular sequence will not be random after all!

The name of the utility that will help us generate random numbers is `randomNumbers.go`, and it will be presented in four parts. The utility takes various parameters, which are the lower and upper limits of the random numbers that will be generated as well as the amount of random numbers to generate. If you use a fourth command parameter, the program will use that as the seed of the random number generator, which will help you regenerate the same number sequence in the future that mainly happens for testing purposes.

The first part of the utility follows next:

```
package main

import (
    "fmt"
    "math/rand"
    "os"
    "strconv"
    "time"
)

func random(min, max int) int {
    return rand.Intn(max-min) + min
}
```

The `random()` function does all of the job, which is generating random numbers in the given range by calling `rand.Intn()`.

The second part of the command-line utility follows next:

```
func main() {
    MIN := 0
    MAX := 100
    TOTAL := 100
    SEED := time.Now().Unix()

    arguments := os.Args
```

In this part, you initialize the variables that will be used in the program.

The third part of `randomNumbers.go` contains the following Go code:

```
    switch len(arguments) {
    case 2:
            fmt.Println("Usage: ./randomNumbers MIN MAX TOTAL SEED")
            MIN, _ = strconv.Atoi(arguments[1])
            MAX = MIN + 100
    case 3:
            fmt.Println("Usage: ./randomNumbers MIN MAX TOTAL SEED")
            MIN, _ = strconv.Atoi(arguments[1])
            MAX, _ = strconv.Atoi(arguments[2])
    case 4:
            fmt.Println("Usage: ./randomNumbers MIN MAX TOTAL SEED")
            MIN, _ = strconv.Atoi(arguments[1])
            MAX, _ = strconv.Atoi(arguments[2])
            TOTAL, _ = strconv.Atoi(arguments[3])
    case 5:
            MIN, _ = strconv.Atoi(arguments[1])
            MAX, _ = strconv.Atoi(arguments[2])
            TOTAL, _ = strconv.Atoi(arguments[3])
            SEED, _ = strconv.ParseInt(arguments[4], 10, 64)
    default:
            fmt.Println("Using default values!")
    }
```

The logic behind this `switch` block is relatively simple: depending on the number of command-line arguments you have, you use either the initial values of the missing arguments or the values given by the user. For reasons of simplicity, the `error` variables of the `strconv.Atoi()` and `strconv.ParseInt()` functions are not being ignored using underscore characters. If this was a commercial program, the error variables of the `strconv.Atoi()` and `strconv.ParseInt()` functions would not have been ignored!

Last, the reason for using `strconv.ParseInt()` for setting a new value to the SEED variable is that the `rand.Seed()` function requires an `int64` parameter. The first parameter of `strconv.ParseInt()` is the string to parse, the second parameter is the base of the generated number, and the third parameter is the bit size of the generated number. As we want to create a decimal integer that uses 64 bits, we are using `10` as the base and `64` as the bit size. Note that should you have wished to parse an unsigned integer, you should have used the `strconv.ParseUint()` function instead.

The last part of `randomNumbers.go` is shown in the following Go code:

```
rand.Seed(SEED)
for i := 0; i < TOTAL; i++ {
        myrand := random(MIN, MAX)
        fmt.Print(myrand)
        fmt.Print(" ")
}
fmt.Println()
}
```

> **TIP**
> Instead of using the Unix epoch time as the seed for the random number generator, you can use the `/dev/random` system device. You will learn about reading from `/dev/random` in Chapter 8, *Telling a Unix system What to Do*.

Executing `randomNumbers.go` will create the following type of output:

```
$ go run randomNumbers.go
Using default values!
75 69 15 75 62 67 64 8 73 1 83 92 7 34 8 70 22 58 38 8 54 34 91 65 1 50 76
5 82 61 90 10 38 40 63 6 28 51 54 49 27 52 92 76 35 44 9 66 76 90 10 29 22
20 83 33 92 80 50 62 26 19 45 56 75 40 30 97 23 87 10 43 11 42 65 80 82 25
53 27 51 99 88 53 36 37 73 52 61 4 81 71 57 30 72 51 55 62 63 79
$ go run randomNumbers.go 1 3 2
Usage: ./randomNumbers MIN MAX TOTAL SEED
1 1
$ go run randomNumbers.go 1 3 2
Usage: ./randomNumbers MIN MAX TOTAL SEED
2 2
$ go run randomNumbers.go 1 5 10 10
3 1 4 4 1 1 4 4 4 3
$ go run randomNumbers.go 1 5 10 10
3 1 4 4 1 1 4 4 4 3
```

If you are really interested in random number generation, you should start by reading the second volume of *The Art of Computer Programming* by *Donald E. Knuth* (*Addison-Wesley Professional, 2011*).

Should you wish to generate more secure random numbers in Go, use the `crypto/rand` package. This package implements a cryptographically secure pseudorandom number generator. You can find more information about the `crypto/rand` package by visiting its documentation page at https://golang.org/pkg/crypto/rand/.

Generating random strings

Once you know how a computer represents single characters, it is not difficult to go from random numbers to random strings. This section will present a technique for creating difficult-to-guess passwords based on the Go code of `randomNumbers.go`, which was presented in the previous section. The name of the Go program for this task will be `generatePassword.go`, and it will be presented in four parts. The utility requires just one command-line parameter, which is the length of the password that you want to generate.

The first part of `generatePassword.go` contains the following Go code:

```go
package main

import (
    "fmt"
    "math/rand"
    "os"
    "strconv"
    "time"
)

func random(min, max int) int {
    return rand.Intn(max-min) + min
}
```

The second code portion of generatePassword.go contains the following Go code:

```go
func main() {
    MIN := 0
    MAX := 94
    SEED := time.Now().Unix()
    var LENGTH int64 = 8

    arguments := os.Args
```

As we only want to get printable ASCII characters, we limit the range of random numbers that can be generated. The number of printable characters in the ASCII table is 94. This means that the range of the random numbers that the program can generate should be from 0 to 94, not including 94.

The third code segment of generatePassword.go is shown in the following Go code:

```go
switch len(arguments) {
case 2:
        LENGTH, _ = strconv.ParseInt(os.Args[1], 10, 64)
default:
        fmt.Println("Using default values!")
}

rand.Seed(SEED)
```

The last part of generatePassword.go is as follows:

```go
startChar := "!"
var i int64 = 1
for {
        myRand := random(MIN, MAX)
        newChar := string(startChar[0] + byte(myRand))
        fmt.Print(newChar)
        if i == LENGTH {
                break
        }
        i++
}
fmt.Println()
}
```

The `startChar` variable holds the first ASCII character that can be generated by the utility, which in this case is the exclamation mark that has a decimal ASCII value of 33. Given that the program can generate random numbers up to 94, the maximum ASCII value that can be generated is 93 + 33, which is equal to 126, which is the ASCII value of ~. The following output shows the ASCII table with the corresponding decimal values for each character:

```
The decimal set:

    0 nul     1 soh     2 st      3 etx     4 eot     5 enq     6 ack     7 bel
    8 bs      9 ht     10 nl     11 vt     12 np     13 cr     14 so     15 si
   16 dle    17 dc1    18 dc     19 dc3    20 dc4    21 nak    22 syn    23 etb
   24 can    25 em     26 su     27 esc    28 fs     29 gs     30 rs     31 us
   32 sp     33 !      34 "      35 #      36 $      37 %      38 &      39 '
   40 (      41 )      42 *      43 +      44 ,      45 -      46 .      47 /
   48 0      49 1      50 2      51 3      52 4      53 5      54 6      55 7
   56 8      57 9      58 :      59 ;      60 <      61 =      62 >      63 ?
   64 @      65 A      66 B      67 C      68 D      69 E      70 F      71 G
   72 H      73 I      74 J      75 K      76 L      77 M      78 N      79 O
   80 P      81 Q      82 R      83 S      84 T      85 U      86 V      87 W
   88 X      89 Y      90 Z      91 [      92 \      93 ]      94 ^      95 _
   96 '      97 a      98 b      99 c     100 d     101 e     102 f     103 g
  104 h     105 i     106 j     107 k     108 l     109 m     110 n     111 o
  112 p     113 q     114 r     115 s     116 t     117 u     118 v     119 w
  120 x     121 y     122 z     123 {     124 |     125 }     126 ~     127 del
```

Typing `man ascii` on your favorite Unix shell will also generate the ASCII table in a readable form.

Executing `generatePassword.go` with the appropriate command-line parameters will create the following type of output:

```
$ go run generatePassword.go
Using default values!
ugs$5mvl
$ go run generatePassword.go
Using default values!
PA/8hA@?
$ go run generatePassword.go 20
HBR+=3\UA'B@ExT4QG|o
$ go run generatePassword.go 20
XLcr|R{*pX/::'t2u^T'
```

Additional Resources

You should find the following resources very useful:

- Examine the **Graphviz** utility website. This utility lets you draw graphs using its own language: http://graphviz.org/.
- Read the documentation page of the sub packages of the `container` standard Go package by visiting https://golang.org/pkg/container/.
- Should you wish to learn more about data structures, you should read *The Design and Analysis of Computer Algorithms* by *Alfred V. Aho*, *John E. Hopcroft*, and *Jeffrey D. Ullman* (*Addison-Wesley, 1974*). It is an excellent book!
- Other really interesting books about algorithms and data structures are *Programming Pearls* by *Jon Bentley* (*Addison-Wesley Professional, 1999*) and *More Programming Pearls: Confessions of a Coder* also by *John Bentley* (*Addison-Wesley Professional, 1988*). Reading both books will make you a better programmer!

Exercises

- Try to change the logic behind `generatePassword.go` by picking the password from a list of passwords found in a Go slice combined with the current system time or date.
- Make the necessary changes to the code of `queue.go` in order to store floating-point numbers instead of integers.
- Change the Go code of `stack.go` so that its nodes have three data fields of integer type, named `Value`, `Number`, and `Seed`. Apart from the apparent changes to the definition of the `Nodestruct`, what is the main change that you will need to make to the rest of the program?
- Can you change the code in `linkedList.go` in order to keep the nodes of the linked list sorted?
- Similarly, can you change the Go code of `doublyLList.go` in order to keep the nodes of the list sorted? Can you develop a function for deleting existing nodes?
- Change the code of `hashTableLookup.go` so that you do not have duplicate values in your hash table. Use the `lookup()` function for that!

- Can you modify the Go code of `generatePassword.go` in order to generate passwords that only contain uppercase letters?
- Try to change the code of `conHeap.go` in order to support a custom and more complex structure instead of just `float32` elements.
- Implement the delete node functionality that is missing from `linkedList.go`.
- Do you think that a doubly linked list would make the code of the `queue.go` program better? Try to implement a queue using a doubly linked list instead of a single linked list.

Summary

In this chapter, we discussed many interesting and useful topics related to Go data structures, including implementing linked lists, doubly linked lists, hash tables, queues, and stacks, as well as using the functionality of the `container` standard Go package and generating random numbers and difficult-to-guess passwords in Go. What you should remember from this chapter is that the foundation of every data structure is the definition and the implementation of its node.

I am sure that you will find the next chapter as one of the most interesting and valuable chapters of this book. The main topic of the next chapter is Go packages. It also includes information about how to define and use the various types of Go functions in your programs. So don't waste any more time—start reading it now!

6
What You Might Not Know About Go Packages

In the previous chapter, we talked about developing and using custom data structures such as linked lists, binary trees, and hash tables as well as generating random numbers and difficult-to-guess passwords in Go.

The main focus of this chapter is Go packages, which is the Go way of organizing, delivering, and using code. The most common components of a Go package are **functions**, which are pretty flexible in Go. In the last part of this chapter you will see some advanced packages that belong to the Go standard library in order to understand better that not all Go packages are created equal.

In this chapter of *Mastering Go*, you will learn the following topics:

- Developing functions in Go
- Anonymous functions
- Functions that return multiple values
- Naming the return values of a function
- Functions that return other functions
- Functions that get other functions as parameters
- Developing Go packages
- Private and Public package objects
- Using the `init()` function in packages
- The sophisticated `html/template` standard Go package

- The `text/template` standard package, another truly sophisticated Go package that has its own language
- The `syscall` standard Go package, which is a low-level package that, although you might not use directly, it is used extensively by other Go packages

About Go packages

Everything in Go is delivered in the form of packages. A Go **package** is a Go source file that begins with the `package` keyword followed by the name of the package. Some packages have a structure. For example, the `net` package has several subdirectories, named `http`, `mail`, `rpc`, `smtp`, `textproto`, and `url`, which should be imported as `net/http`, `net/mail`, `net/rpc`, `net/smtp`, `net/textproto`, and `net/url`, respectively.

Packages are mainly used for grouping related functions, variables, and constants so that you can transfer them easily and use them in your own Go programs. Note that apart from the `main` package, Go packages are not autonomous programs and cannot be compiled into executable files. This means that they need to be called directly or indirectly from a `main` package in order to be used. As a result, if you try to execute a Go package as if it was an autonomous program, you will be disappointed:

```
$ go run aPackage.go
go run: cannot run non-main package
```

About Go functions

Functions are an important element of every programming language because they allow you to break big programs into smaller and more manageable parts. Functions must be as independent from each other as possible and must do one job and only one job well. So, if you find yourselves writing functions that do multiple things, you might consider replacing them by multiple functions instead! The single most popular Go function is `main()`, which is used in every independent Go program. You should already know that function definitions begin with the `func` keyword.

Anonymous functions

Anonymous functions can be defined inline without the need for a name, and they are usually used for implementing things that require a small amount of code. In Go, a function can return an anonymous function or take an anonymous function as one of its arguments. Additionally, anonymous functions can be attached to Go variables.

Note that anonymous functions are also called **closures**, especially in functional programming terminology.

It is considered a good practice for anonymous functions to have a small implementation and a local focus. If an anonymous function does not have a local focus, then you might need to consider making it a regular function.

When an anonymous function is suitable for a job, then it is extremely convenient and makes your life easier; just do not use too many anonymous functions in your programs without having a good reason. You will see anonymous functions in action in a little while.

Functions that return multiple values

As you already know from functions such as `strconv.Atoi()`, Go functions can return multiple distinct values, which saves you from having to create a dedicated structure in order to be able to receive multiple values at once from a function. You can declare a function that returns four values, two `int` values, a `float64` value, and a `string`, as follows:

```
func aFunction() (int, int, float64, string) {
}
```

It is now time to illustrate functions; that is, anonymous functions and functions that return multiple values in more detail using the Go code of `functions.go` as an example. The relevant code will be presented in five parts.

The first code portion of `functions.go` is as follows:

```
package main

import (
    "fmt"
    "os"
    "strconv"
)
```

The second code segment from `functions.go` is shown in the following Go code:

```go
func doubleSquare(x int) (int, int) {
    return x * 2, x * x
}
```

Here you can see the definition and implementation of a function named `doubleSquare()`, which requires a single `int` parameter and returns two `int` values.

The third part of the `functions.go` program is as follows:

```go
func main() {
    arguments := os.Args
    if len(arguments) != 2 {
        fmt.Println("The program needs 1 argument!")
        return
    }

    y, err := strconv.Atoi(arguments[1])
    if err != nil {
        fmt.Println(err)
        return
    }
```

The preceding code deals with the command line arguments of the program.

The fourth portion of the `functions.go` program contains the following Go code:

```go
    square := func(s int) int {
        return s * s
    }
    fmt.Println("The square of", y, "is", square(y))

    double := func(s int) int {
        return s + s
    }
    fmt.Println("The double of", y, "is", double(y))
```

Each one of the `square` and `double` variables holds an anonymous function. The bad part is that you are allowed to change the value of `square`, `double`, or any other variable that holds an anonymous function afterwards, which means that the meaning of those variables can change and calculate something else instead.

It is not considered a good programming practice to alter the code of variables that hold anonymous functions because this might be the root cause of nasty bugs!

The last part of functions.go is as follows:

```
    fmt.Println(doubleSquare(y))
    d, s := doubleSquare(y)
    fmt.Println(d, s)
}
```

So, you can either print the return values of a function, such as doubleSquare(), or assign them to distinct variables.

Executing functions.go will generate the following output:

```
$ go run functions.go 1 21
The program needs 1 argument!
rMacBook:code mtsouk
$ go run functions.go 10
The square of 10 is 100
The double of 10 is 20
20 100
20 100
```

The return values of a function can be named!

Unlike C, Go allows you to name the return values of a Go function. Additionally, when such a function has a return statement without any arguments, then the function automatically returns the current value of each named return value in the order in which they were declared in the definition of the function!

Named returned values are a very handy Go feature that can save you from various types of bugs, so use them! My personal advice is to name the return values of your functions unless there is a very good reason not to do so.

The source code that illustrates Go functions that have named return values is returnNames.go, and it will be presented in three parts.

The first part of the `returnNames.go` program is as follows:

```
package main

import (
    "fmt"
    "os"
    "strconv"
)

func namedMinMax(x, y int) (min, max int) {
    if x > y {
        min = y
        max = x
    } else {
        min = x
        max = y
    }
    return
}
```

In this code segment, you can see the implementation of the `namedMinMax()` function, which uses named return parameters. However, there is a tricky point here: the `namedMinMax()` function does not explicitly return any variables in its `return` statement. Nevertheless, as this function has named return values in its signature, the `min` and `max` parameters are automatically returned in the order in which they were put into the function definition.

The second code segment from `returnNames.go` follows next:

```
func minMax(x, y int) (min, max int) {
    if x > y {
        min = y
        max = x
    } else {
        min = x
        max = y
    }
    return min, max
}
```

The `minMax()` function also uses named return values, but its `return` statement specifically defines the order and the variables that are going to be returned.

The last code portion from `returnNames.go` is shown in the following Go code:

```go
func main() {
    arguments := os.Args
    if len(arguments) < 3 {
            fmt.Println("The program needs at least 2 arguments!")
            return
    }

    a1, _ := strconv.Atoi(arguments[1])
    a2, _ := strconv.Atoi(arguments[2])

    fmt.Println(minMax(a1, a2))
    min, max := minMax(a1, a2)
    fmt.Println(min, max)

    fmt.Println(namedMinMax(a1, a2))
    min, max = namedMinMax(a1, a2)
    fmt.Println(min, max)
}
```

The purpose of the Go code in the `main()` function is to verify that all methods generate the same results.

Executing `returnNames.go` will produce the following output:

```
$ go run returnNames.go -20 1
-20 1
-20 1
-20 1
-20 1
```

Functions with pointer parameters

Function can take pointer parameters provided that their signature allows it. The Go code of `ptrFun.go` will illustrate the use of pointers as function parameters.

The first part of `ptrFun.go` follows next:

```go
package main

import (
    "fmt"
)
```

```
func getPtr(v *float64) float64 {
    return *v * *v
}
```

So, the `getPtr()` function accepts a pointer parameter that points to a `float64` value.

The second part of the program is shown in the following Go code:

```
func main() {
    x := 12.2
    fmt.Println(getPtr(&x))
    x = 12
    fmt.Println(getPtr(&x))
}
```

The tricky part here is that you need to pass the address of the variable to the `getPtr()` function because it requires a pointer parameter, which can be done by putting an ampersand in front of a variable (`&x`).

Executing `ptrFun.go` will generate the following kind of output:

```
$ go run ptrFun.go
148.83999999999997
144
```

If you try to pass a plain value such as `12.12` to `getPtr()` and call it like `getPtr(12.12)`, the compilation of the program will fail, as shown in the following the next error message:

```
$ go run ptrFun.go
# command-line-arguments
./ptrFun.go:15:21: cannot use 12.12 (type float64) as type *float64 in argument to getPtr
```

Functions that return pointers

As you learned in Chapter 4, *The Uses of Composite Types,* as illustrated in the `pointerStruct.go` program, it is considered a good practice to create new structure variables using a separate function and return a pointer to them from that function. So, the scenario of functions returning pointers is very common. Generally speaking, such a function simplifies the structure of a program and allows the developer to concentrate on more important things instead of copying the same Go code all the time. This section will use a much simpler example, as found in the Go code of `returnPtr.go`.

The first part of `returnPtr.go` contains the following Go code:

```
package main

import (
    "fmt"
)

func returnPtr(x int) *int {
    y := x * x
    return &y
}
```

Apart from the expected preamble, this portion defines a new function that returns a pointer to an `int` variable. The only thing to remember is to use `&y` in the `return` statement in order to return the memory address of the `y` variable.

The second part of `returnPtr.go` is as follows:

```
func main() {
    sq := returnPtr(10)
    fmt.Println("sq:", *sq)
```

As you already know, the `*` character **dereferences a pointer variable**, which means that it returns the actual value stored at the memory address instead of the memory address itself.

The last code segment from `returnPtr.go` is shown in the following Go code:

```
    fmt.Println("sq:", sq)
}
```

The preceding code will return the memory address of the `sq` variable, not the `int` value stored in it!

If you execute `returnPtr.go`, you will see the following output (the memory address will differ):

```
$ go run returnPtr.go
sq: 100
sq: 0xc420014088
```

Functions that return other functions

In this section, you are going to learn how to implement a Go function that returns another function using the Go code of `returnFunction.go`, which will be presented in three segments.

The first code segment of `returnFunction.go` is as follows:

```
package main

import (
    "fmt"
)

func funReturnFun() func() int {
    i := 0
    return func() int {
        i++
        return i * i
    }
}
```

As you can see from the implementation of `funReturnFun()`, its return value is an anonymous function!

The second code segment from `returnFunction.go` contains the following code:

```
func main() {
    i := funReturnFun()
    j := funReturnFun()
```

In this code, you call `funReturnFun()` two times and assign its return value, which is a function, to two separate variables named `i` and `j`. As you will see in the output of the program, the two variables are totally unrelated to each other.

The last code section of `returnFunction.go` follows next:

```
    fmt.Println("1:", i())
    fmt.Println("2:", i())
    fmt.Println("j1:", j())
    fmt.Println("j2:", j())
    fmt.Println("3:", i())
}
```

So, in this Go code, you use the i variable three times as i() and the j variable two times as j(). The important thing here is that although both i and j were created by calling funReturnFun(), they are totally independent of each other and share nothing. As a result, although their return values come from the same sequence, they do not interfere with each other in any way.

Executing returnFunction.go will produce the following output:

```
$ go run returnFunction.go
1: 1
2: 4
j1: 1
j2: 4
3: 9
```

As you can see from the output of returnFunction.go, the value of i in funReturnFun() keeps increasing and does not become 0 after each call either to i() or j().

Functions that accept other functions as parameters

Go functions can accept other Go functions as parameters, which is a feature that adds versatility to what you can do with a Go function. The single most common use of this functionality is for sorting elements. However, in the example presented here, which is named funFun.go, we will implement a much simpler case that deals with integer values. The relevant code will be presented in three parts.

The first code segment of funFun.go is shown in the following Go code:

```
package main

import "fmt"

func function1(i int) int {
    return i + i
}

func function2(i int) int {
    return i * i
}
```

What we have here is two functions that both accept an int and return an int. These functions will be used as parameters to another function in a short while.

The second code segment of funFun.go contains the following code:

```
func funFun(f func(int) int, v int) int {
    return f(v)
}
```

The funFun() function accepts two parameters, a function parameter named f and an int value. The f parameter should be a function that takes one int argument and returns an int value.

The last code segment of funFun.go follows next:

```
func main() {
    fmt.Println("function1:", funFun(function1, 123))
    fmt.Println("function2:", funFun(function2, 123))
    fmt.Println("Inline:", funFun(func(i int) int {return i * i *i}, 123))
}
```

The first fmt.Println() call uses funFun() with function1 without any parentheses as its first parameter, whereas the second fmt.Println() call uses funFun() with function2 as its first parameter.

In the last fmt.Println() statement, something magical happens: the implementation of the function parameter is defined inside the call to funFun()! Although this method works fine for simple and small function parameters, it might not work that well for functions with many lines of Go code.

Executing funFun.go will produce the following output:

```
$ go run funFun.go
function1: 246
function2: 15129
Inline: 1860867
```

Developing your own Go packages

The source code of a Go package, which can contain multiple files and multiple directories, can be found within a single directory that is named after the package name with the obvious exception of the `main` package that can be located anywhere.

For the purposes of this section, a simple Go package named `aPackage` will be developed. The source file of the package is called `aPackage.go`, and its source code will be presented in two parts.

The first part of `aPackage.go` is shown in the following Go code:

```
package aPackage

import (
    "fmt"
)

func A() {
    fmt.Println("This is function A!")
}
```

The second code segment of `aPackage.go` follows next:

```
func B() {
    fmt.Println("privateConstant:", privateConstant)
}

const MyConstant = 123
const privateConstant = 21
```

As you can see, developing a new Go package is pretty easy! Right now you cannot use that package on its own, and you need to create a package named `main` with a `main()` function in it in order to create an executable file. In this case, the name of program that will use `aPackage` is `useAPackage.go`, and it's included in the following Go code:

```
package main

import (
    "aPackage"
    "fmt"
)

func main() {
    fmt.Println("Using aPackage!")
    aPackage.A()
```

```
        aPackage.B()
        fmt.Println(aPackage.MyConstant)
}
```

If you try to execute `useAPackage.go` right now, however, you will get an error message, which means that you are not done yet:

```
$ go run useAPackage.go
useAPackage.go:4:2: cannot find package "aPackage" in any of:
        /usr/local/Cellar/go/1.9.2/libexec/src/aPackage (from $GOROOT)
        /Users/mtsouk/go/src/aPackage (from $GOPATH)
```

There is another thing that you will also need to handle. As you already know from Chapter 1, *Go and the Operating System*, Go requires to execute specific commands from the Unix shell in order to install all external packages, which also includes packages that you have developed. Therefore, you will need to put the preceding package in the appropriate directory and make it available to the current Unix user. Thus, installing one of your own packages involves the execution of the following commands from your favorite Unix shell:

```
$ mkdir ~/go/src/aPackage
$ cp aPackage.go ~/go/src/aPackage/
$ go install aPackage
$ cd ~/go/pkg/darwin_amd64/
$ ls -l aPackage.a
-rw-r--r--  1 mtsouk  staff  4980 Dec 22 06:12 aPackage.a
```

If the `~/go` directory does not already exist, you will need to create it with the help of the `mkdir(1)` command. In that case, you will also need to do the same for the `~/go/src` directory.

Executing `useAPackage.go` will create the following output:

```
$ go run useAPackage.go
Using aPackage!
This is function A!
privateConstant: 21
123
```

Compiling a Go package

Although you cannot execute a Go package if it does not include a `main()` function, you are still allowed to compile it and create an object file as follows:

```
$ go tool compile aPackage.go
$ ls -l aPackage.*
-rw-r--r--@ 1 mtsouk  staff   201 Dec 22 06:06 aPackage.go
-rw-r--r--  1 mtsouk  staff  4677 Dec 22 06:20 aPackage.o
```

If you successfully compile a Go package that you are developing using the preceding method, you must also make sure that you do not have any syntax errors in your code.

Private variables and functions

What differentiates private variables and functions from public ones is that private ones can be strictly used and called internally in a package. Controlling which functions, constants, and variables are public or not is also known as **encapsulation**.

Go follows a simple rule which states that functions, variables, types, and so forth that begin with an uppercase letter are public, whereas functions, variables, types, and so on that begin with a lowercase letter are private. This is the reason that `fmt.Println()` is named `Println()` instead of `println()`. However, this rule does not affect package names that are allowed to begin with uppercase and lowercase letters.

The init() function

Every Go package can optionally have a function named `init()` that is automatically executed at the beginning of the execution time.

The `init()` function is a private function by design, which means that it cannot be called from outside the package in which it is contained. Additionally, as the user of a package has no control over the `init()` function, you should think carefully before using an `init()` function in public packages.

I will now present a code example with multiple `init()` functions from multiple Go packages. Examine the code of the following basic Go package, which is simply called a:

```go
package a

import (
    "fmt"
)

func init() {
    fmt.Println("init() a")
}

func FromA() {
    fmt.Println("fromA()")
}
```

The a package implements an `init()` function and a public one named `FromA()`.

After that, you will need to execute the following commands from your Unix shell so that the package becomes available to the current Unix user:

```
$ mkdir ~/go/src/a
$ cp a.go ~/go/src/a/
$ go install a
```

Now, look at the code of the next Go code package, which is named b:

```go
package b

import (
    "a"
    "fmt"
)

func init() {
    fmt.Println("init() b")
}

func FromB() {
    fmt.Println("fromB()")
    a.FromA()
}
```

What is happening here? Package a uses the fmt standard Go package. However, package b needs to import package a as it uses a.FromA(). Both a and b have an init() function.

As before, you will need to install that package and make it available to the current Unix user by executing the following commands from your Unix shell:

```
$ mkdir ~/go/src/b
$ cp b.go ~/go/src/b
$ go install b
```

Thus, we currently have two Go packages that both have an init() function. Now try to guess the output that you will get from executing manyInit.go, which has the following code:

```
package main

import (
    "a"
    "b"
    "fmt"
)

func init() {
    fmt.Println("init() manyInit")
}

func main() {
    a.FromA()
    b.FromB()
}
```

The actual question could have been: How many times is the init() function of package a going to be executed? Executing manyInit.go will generate the following output and shed some light on this question:

```
$ go run manyInit.go
init() a
init() b
init() manyInit
fromA()
fromB()
fromA()
```

The preceding output shows that the `init()` function of a was executed only once, despite the fact that the a package is imported two times by two different packages. Additionally, as the `import` block from `manyInit.go` is executed first, the `init()` function of package a and package b are executed before the `init()` function of `manyInit.go`, which makes perfect sense. The main reason for this is that the `init()` function of `manyInit.go` is allowed to use an element from either a or b.

Reading the Go code of a standard Go package

In this section, you will learn some techniques for reading the Go code of existing Go packages. Generally speaking, regularly studying Go code developed by experienced programmers is a very good habit that will increase the quality of your own Go code, so do not underestimate its value.

Now we are going to have a quick look at the Go code of `net/url` and `log/syslog`.

Exploring the code of the net/url package

On a macOS machine that uses Homebrew, you can find the Go code of the `net/url` package at `/usr/local/Cellar/go/1.9.2/libexec/src/net/url/url.go`. If for some reason you cannot locate the source code of the `net/url` package on your Unix machine, you can execute the following command and wait for its output:

```
$ find / -name url.go 2>/dev/null
```

The output of the preceding command on a Debian Linux machine is as follows:

```
/usr/share/go/src/pkg/net/url/url.go
/usr/share/go/src/pkg/html/template/url.go
```

Apart from the `net/url` package, there is also an `url.go` file that belongs to the `html/template` standard Go package. However, what interests us is the `/usr/share/go/src/pkg/net/url/url.go` file. Now that you have located the `url.go` file, we can start playing with it.

The preamble of the net/url package is as follows:

```
// Copyright 2009 The Go Authors. All rights reserved.
// Use of this source code is governed by a BSD-style
// license that can be found in the LICENSE file.

// Package url parses URLs and implements query escaping.
package url
```

More or less, all packages in the standard Go library have a similar preamble. Now, imagine that you want to get a list of all the functions in the net/url package. Standard Unix command line utilities such as grep(1) are here to help you:

```
$ grep '^func' /usr/local/Cellar/go/1.9.2/libexec/src/net/url/url.go
func (e *Error) Error() string { return e.Op + " " + e.URL + ": " +
e.Err.Error() }
func (e *Error) Timeout() bool {
func (e *Error) Temporary() bool {
func ishex(c byte) bool {
func unhex(c byte) byte {
func (e EscapeError) Error() string {
...
func (u *URL) MarshalBinary() (text []byte, err error) {
func (u *URL) UnmarshalBinary(text []byte) error {
```

Now let's say that you want to look at the implementation of one of the functions found in net/url, and the function that interests you is, say, User(). Its implementation is as follows:

```
// User returns a Userinfo containing the provided username
// and no password set.
func User(username string) *Userinfo {
        return &Userinfo{username, "", false}
}
```

The documentation of the User() function states that it returns a Userinfo, which is a structure defined in net/url:

```
// The Userinfo type is an immutable encapsulation of username and
// password details for a URL. An existing Userinfo value is
// guaranteed to have a username set (potentially empty,
// as allowed by RFC 2396), and optionally a password.
type Userinfo struct {
        username    string
        password    string
        passwordSet bool
}
```

Now things are becoming clearer. If you continue delving into the code of net/url, you will find more in-depth information about it.

 The general advice here is that if you have doubts about the inner workings of a package in which you are interested, and its documentation does not help you very much, you can always look at its implementation to clarify things.

The next subsection will look into the Go code of a more interesting package.

Looking at the Go code of the log/syslog package

This subsection will examine the contents of the log/syslog standard Go package. On a Homebrew Go 1.9.2 installation on a macOS machine, the syslog.go file can be found inside the /usr/local/Cellar/go/1.9.2/libexec/src/log/syslog directory. The full contents of that directory are as follows:

```
$ ls -l /usr/local/Cellar/go/1.9.2/libexec/src/log/syslog/
total 64
-rw-r--r--  1 mtsouk  admin  1010 Oct 25 21:30 doc.go
-rw-r--r--  1 mtsouk  admin   543 Oct 25 21:30 example_test.go
-rw-r--r--  1 mtsouk  admin  7567 Oct 25 21:30 syslog.go
-rw-r--r--  1 mtsouk  admin  8589 Oct 25 21:30 syslog_test.go
-rw-r--r--  1 mtsouk  admin   785 Oct 25 21:30 syslog_Unix.go
```

Now let us try to discover the constants found in the syslog.go file:

```
$ grep -n const syslog.go
26:const severityMask = 0x07
27:const facilityMask = 0xf8
29:const (
44:const (
```

As the third and the fourth const definitions have more than one line, you should go at lines 29 and 44 to have a better look at them. So, the first const definition contains the following constants:

```
const (
        // Severity.

        // From /usr/include/sys/syslog.h.
        // These are the same on Linux, BSD, and OS X.
        LOG_EMERG Priority = iota
        LOG_ALERT
```

```
            LOG_CRIT
            LOG_ERR
            LOG_WARNING
            LOG_NOTICE
            LOG_INFO
            LOG_DEBUG
)
```

Similarly, you can search for new Go `type` declarations as follows:

```
$ grep -n type syslog.go
24:type Priority int
76:type Writer struct {
91:// return a type that satisfies this interface and simply calls the C
93:type serverConn interface {
98:type netConn struct {
```

Now that you know where the `type` declarations are located, it should be easy to go there and have a more detailed look. So, the definition of the `netConn struct`, which allows you to send logging data over the network, is as follows:

```
type netConn struct {
        local bool
        conn  net.Conn
}
```

Should you wish to go deeper and have a look at the implementation of the functions contained in the `log/syslog` standard Go package, feel free to continue searching on your own.

Creating good Go packages

This section will provide some handy advice that will help you generate better Go packages. Up until now, you know that Go packages are organized in directories and can contain public and private elements. Public elements can be use both internally and externally from other packages, whereas private elements can only be used internally in a package.

Here are several good rules to follow to create superior Go packages:

- The first unofficial rule of a successful package is that its elements must be related in some way. Thus, you can create a package for supporting cars, but it would not be a good idea to create a single package for supporting both cars and bicycles. Put simply, it is better to split the functionality of a package unnecessarily into many packages than to add too much functionality to a single Go package. Additionally, packages should be made simple and stylish-but not too simplistic and fragile.
- A second practical rule is that you should use your own packages first for a reasonable amount of time before giving them to the public. This will help you discover silly bugs and make sure that your packages operate as expected. After that, give them to some fellow developers for additional testing before making them publicly available.
- Next, try to imagine the kinds of users who will use your packages, and make sure that these types of users will use your packages happily and that your packages will not create more problems for them than they can solve.
- Unless there is a very good reason, your packages should not export an endless list of functions. Packages that export a short list of functions are understood better and used more easily. After that, try to title your functions using descriptive yet not very long names.
- **Interfaces** can improve the usefulness of your functions, so when you think it is appropriate, use an interface instead of a single type as a function parameter or return type.
- When updating one of your packages, try not to break things and create incompatibilities with older versions unless it is absolutely necessary. This is a really important rule that you should always try to follow.
- When developing a new Go package, try to use multiple files in order to group similar tasks or concepts.
- Additionally, try to follow the rules that exist in the Go packages of the standard library. Reading the code of a Go package that belongs to the standard library will help you on this.
- Do not create a package that already exists from scratch! Make changes to the existing package and maybe create your own version of it.
- Nobody wants a Go package that prints logging information on the screen. It would be more professional to have a flag for turning on logging when needed.

- The Go code of your package should be in harmony with the other Go code of a program. This means that if you look at a program that uses your packages and your function names stand out in the code in a bad way, it would be better to change the names of your functions. As the name of a package is used almost everywhere, try to use a concise and expressive package name.
- It is more convenient if you put new Go-type definitions near the place that they will be used for the first time because nobody, including you, wants to search source files for finding definitions of new data types!
- Try to create test files for your packages, because packages with test files are considered more professional than ones without them-small details make all the difference and give people confidence that you are a serious developer! You will learn more about testing in Chapter 11, *Code Testing, Optimization, and Profiling*.
- Finally, do not write a Go package because you do not have anything better to do—in that case, find something better to do and do not waste your time!

Always remember that apart from the fact that the actual Go code in a package should be bug-free, the next most important element of a successful package is its **documentation**, as well as some code examples that clarify its use and showcase the idiosyncrasies of the functions of the package.

The syscall package

This section will present a small portion of the functionality of the `syscall` standard Go package. Note that the `syscall` package offers a plethora of functions and types related to low-level operating system primitives. Additionally, the `syscall` package is extensively used by other Go packages such as `os`, `net`, and `time`, which all provide a portable interface to the operating system. This means that the `syscall` package is not the most portable package in the Go library-that's not its job. Although Unix systems have many similarities, they also exhibit various differences, especially when we talk about their system internals. The job of the `syscall` package is to deal with all of these incompatibilities as gently as possible. The fact that this is not a secret and is well-documented makes `syscall` a successful package.

Strictly speaking, a **system call** is a programmatic way for an application to request something from the kernel of an operating system. As a consequence, system calls are responsible for accessing and working with most Unix low-level elements such as processes, storage devices, printing data, network interfaces and all kinds of files. Put simply, you cannot work on a Unix system without using system calls! You can inspect the system calls of a Unix process using utilities such as `strace(1)` and `dtrace(1)`, which were presented in `Chapter 2`, *Understanding Go Internals*.

The use of the `syscall` package will be illustrated in the `useSyscall.go` program, which will be presented in four parts.

You might not directly need to use the `syscall` package unless you are working on pretty low-level stuff. Not all Go packages are for everyone!

The first code portion of `useSyscall.go` is as follows:

```
package main

import (
    "fmt"
    "os"
    "syscall"
)
```

This is the easy part of the program, where you just import the required Go packages.

The second part of `useSyscall.go` is shown in the following Go code:

```
func main() {
    pid, _, _ := syscall.Syscall(39, 0, 0, 0)
    fmt.Println("My pid is", pid)
    uid, _, _ := syscall.Syscall(24, 0, 0, 0)
    fmt.Println("User ID:", uid)
```

In this part, you find out information about the process id and the user id using two `syscall.Syscall()` calls. The first parameter of the `syscall.Syscall()` call determines the information that you request.

The third code segment of `useSyscall.go` contains the following Go code:

```
    message := []byte{'H', 'e', 'l', 'l', 'o', '!', '\n'}
    fd := 1
    syscall.Write(fd, message)
```

In this part, you print a message on the screen using `syscall.Write()`. The first parameter is the file descriptor to which you will write and the second parameter is a byte slice that holds the actual message. The `syscall.Write()` function is portable.

The last part of the `useSyscall.go` program is as follows:

```
    fmt.Println("Using syscall.Exec()")
    command := "/bin/ls"
    env := os.Environ()
    syscall.Exec(command, []string{"ls", "-a", "-x"}, env)
}
```

In the last part of the program, you see how to use the `syscall.Exec()` function for executing an external command. However, you have no control over the output of the command, which is automatically printed on the screen.

Executing `useSyscall.go` on macOS High Sierra will generate the following output:

```
$ go run useSyscall.go
My pid is 14602
User ID: 501
Hello!
Using syscall.Exec()
.                   ..                  a.go
funFun.go           functions.go        html.gohtml
htmlT.db            htmlT.go            manyInit.go
ptrFun.go           returnFunction.go   returnNames.go
returnPtr.go        text.gotext         textT.go
useAPackage.go      useSyscall.go
```

Executing the same program on a Debian Linux machine will generate the following output:

```
$ go run useSyscall.go
My pid is 20853
User ID: 0
Hello!
Using syscall.Exec()
.                   ..                  a.go
funFun.go           functions.go        html.gohtml
htmlT.db            htmlT.go            manyInit.go
ptrFun.go           returnFunction.go   returnNames.go
returnPtr.go        text.gotext         textT.go
useAPackage.go      useSyscall.go
```

So, although most of the output is the same as before, the `syscall.Syscall(39, 0, 0, 0)` call does not work on Linux because the user id of the Linux user is not 0, which means that this command is not portable!

If you want to discover which standard Go packages use the `syscall` package, you can execute the following command from your Unix shell:

```
$ grep \"syscall\" `find /usr/local/Cellar/go/1.9.2/libexec/src -name "*.go"`
```

Please replace `/usr/local/Cellar/go/1.9.2/libexec/src` with the appropriate directory path.

Finding out how fmt.Println() really works

If you really want to grasp the usefulness of the `syscall` package, start by reading this subsection. The implementation of the `fmt.Println()` function, as found in `https://golang.org/src/fmt/print.go`, is as follows:

```
func Println(a ...interface{}) (n int, err error) {
    return Fprintln(os.Stdout, a...)
}
```

This means that, at the end of the day, the `fmt.Println()` function calls `fmt.Fprintln()` to do its job. The implementation of `fmt.Fprintln()`, as found in the same file, follows next:

```
func Fprintln(w io.Writer, a ...interface{}) (n int, err error) {
    p := newPrinter()
    p.doPrintln(a)
    n, err = w.Write(p.buf)
    p.free()
    return
}
```

This means that the actual writing in `fmt.Fprintln()` is done by the `Write()` function of the `io.Writer` interface. In this case, the `io.Writer` is `os.Stdout`, which is defined as follows in `https://golang.org/src/os/file.go`:

```
var (
    Stdin  = NewFile(uintptr(syscall.Stdin), "/dev/stdin")
    Stdout = NewFile(uintptr(syscall.Stdout), "/dev/stdout")
    Stderr = NewFile(uintptr(syscall.Stderr), "/dev/stderr")
)
```

Now look at the implementation of `NewFile()`, which can be found inside https://golang.org/src/os/file_plan9.go:

```go
func NewFile(fd uintptr, name string) *File {
    fdi := int(fd)
    if fdi < 0 {
        return nil
    }
    f := &File{&file{fd: fdi, name: name}}
    runtime.SetFinalizer(f.file, (*file).close)
    return f
}
```

When you see a Go source file named `file_plan9.go`, you should suspect that it contains commands specific to a Unix variant, which means that it contains code that is not portable!

What we have here is the `file` structure type that is embedded in the `File` type, which is the one that is being exported due to its name. So, start looking for functions inside https://golang.org/src/os/file_plan9.go that are applied to a `File` structure or to a pointer to a `File` structure and that allow you to write data. As the function we are seeking is named `Write()`-look at the implementation of `Fprintln()`-we will have to search all of the source files of the `os` package to find it:

```
$ grep "func (f \*File) Write(" *.go
file.go:func (f *File) Write(b []byte) (n int, err error) {
```

The implementation of `Write()` as found in https://golang.org/src/os/file.go follows next:

```go
func (f *File) Write(b []byte) (n int, err error) {
    if err := f.checkValid("write"); err != nil {
        return 0, err
    }
    n, e := f.write(b)
    if n < 0 {
        n = 0
    }
    if n != len(b) {
        err = io.ErrShortWrite
    }

    epipecheck(f, e)

    if e != nil {
        err = f.wrapErr("write", e)
    }
```

```
        return n, err
}
```

This means that we now have to search for the `write()` function. Searching for the `write` string in `https://golang.org/src/os/file_plan9.go` reveals the following function inside `https://golang.org/src/os/file_plan9.go`:

```
func (f *File) write(b []byte) (n int, err error) {
    if len(b) == 0 {
        return 0, nil
    }
    return fixCount(syscall.Write(f.fd, b))
}
```

This tells us that, at the end of the day, a call to the `fmt.Println()` function is implemented using a call to `syscall.Write()`. This underscores how useful and necessary the `syscall` package is!

Text and HTML templates

The subject of this section will surely surprise you in a good way, because both packages presented give you so much flexibility that I am sure you will find many creative ways to use them.

Templates are mainly used for separating the formatting part and the data part of the output. Please note that a Go template can be either a file or a string-the general idea is to use inline strings for smaller templates and external files for bigger ones.

 You cannot import both `text/template` and `html/template` on the same Go program because these two packages share the same package name (`template`). If absolutely necessary, you should define an alias for one of them. See the `useStrings.go` code in Chapter 4, *The Uses of Composite Types*.

Text output is usually presented on your screen, whereas HTML output is seen with the help of a web browser. However, as text output is usually cleaner than HTML output, if you think that you will need to process the output of a Go utility using other Unix command line utilities, you should use `text/template` instead of `html/template`.

Note that both the `text/template` and `html/template` packages are good examples of how sophisticated a Go package can be! As you will see shortly, both packages support their own kind of programming language-good software makes complex things look simple!

Generating text output

If you need to create plain text output, then using the `text/template` package is a good choice. The use of the `text/template` package will be illustrated in `textT.go`, which will be presented in five parts.

As templates are usually stored in external files, the example presented will use the `text.gotext` template file, which will be analyzed in three parts. Data is typically read from text files or from the Internet. However, for reasons of simplicity, the data for `textT.go` will be hardcoded in the program using a slice.

We will start by looking at the Go code of `textT.go`. The first code portion of `textT.go` follows next:

```
package main

import (
    "fmt"
    "os"
    "text/template"
)
```

The second code segment from `textT.go` is as follows:

```
type Entry struct {
    Number int
    Square int
}
```

You will need to define a new data type for storing your data unless you are dealing with very simplistic data.

The third part of `textT.go` is shown in the following Go code:

```
func main() {
    arguments := os.Args
    if len(arguments) != 2 {
        fmt.Println("Need the template file!")
        return
```

```
        }

        tFile := arguments[1]
        DATA := [][]int{{-1, 1}, {-2, 4}, {-3, 9}, {-4, 16}}
```

The `DATA` variable, which is a slice with two dimensions, holds the initial version of your data.

The fourth part of `textT.go` contains the following Go code:

```
        var Entries []Entry

        for _, i := range DATA {
            if len(i) == 2 {
                temp := Entry{Number: i[0], Square: i[1]}
                Entries = append(Entries, temp)
            }
        }
```

The preceding code creates a **slice of structures** from the `DATA` variable.

The last code segment from `textT.go` is the following:

```
        t := template.Must(template.ParseGlob(tFile))
        t.Execute(os.Stdout, Entries)
    }
```

The `template.Must()` function is used for making the required initializations. Its return data type is `Template`, which is a structure that holds the representation of a parsed template. The `template.ParseGlob()` reads the external template file. Note that I like to use the `gohtml` extension for the template files, but you can use any extension that you want-just be consistent. Also, the `template.Execute()` function does all the work, which includes processing the data and printing the output to the desired file, which in this case is `os.Stdout`.

Now it is time to look at the code of the template file. The first part of the text template file is as follows:

```
    Calculating the squares of some integers
```

Note that empty lines in a text template file are significant and will be shown as empty lines in the final output.

The second part of the template is the following:

```
{{ range . }} The square of {{ printf "%d" .Number}} is {{ printf "%d" .Square}}
```

There are many interesting things happening here. The `range` keyword allows you to iterate over the lines of the input, which is given as a slice of structures. Plain text is printed as such, whereas variables and dynamic text must begin with `{{` and end with `}}`. The fields of the structure are accessed as `.Number` and `.Square`. Note the dot character in front of the field name of the `Entry` data type. Also, the `printf` command is used for formatting the final output.

The third part of the `text.gotext` file follows next:

```
{{ end }}
```

So, a `{{ range }}` command is ended with `{{ end }}`. Accidentally putting `{{ end }}` in the wrong place will affect your output. Once again, keep in mind that empty lines in text template files are significant and will be shown in the final output.

Executing `textT.go` will generate the following type of output:

```
$ go run textT.go text.gotext
Calculating the squares of some integers
  The square of -1 is 1
  The square of -2 is 4
  The square of -3 is 9
  The square of -4 is 16
```

Constructing HTML output

This section illustrates the use of the `html/template` package with an example named `htmlT.go`. It will be presented in six parts. The philosophy of the `html/template` package is the same with the `text/template` package. The main difference between these two packages is that the `html/template` package generates HTML output that is safe against code injection.

 Although you can create HTML output with the `text/template` package-after all, HTML is just plain text-if you want to create HTML output, then you should use the `html/template` package instead.

For reasons of simplicity, the example presented below will read data from an SQLite database, but you can use any database that you want, provided that you have or you can write the appropriate Go drivers. To make things even easier, the example will populate a database table before reading from it!

The first code portion from `htmlT.go` follows next:

```go
package main

import (
    "database/sql"
    "fmt"
    _ "github.com/mattn/go-sqlite3"
    "html/template"
    "net/http"
    "os"
)

type Entry struct {
    Number int
    Double int
    Square int
}

var DATA []Entry
var tFile string
```

You can see a new package named `net/http` in the `import` block, which is used for creating HTTP servers and clients in Go. You will learn more about network programming in Go and the use of the `net` and `net/http` standard Go packages in Chapter 12, *The foundations of Network Programming in Go* and Chapter 13, *Network Programming – Building Servers and Clients*.

Apart from `net/http`, you can also see the definition of the `Entry` data type that will hold the records read from the SQLite3 table as well as two global variables named `DATA` and `tFile`, which hold the data that is going to be passed to the template file and the filename of the template file, respectively.

Also, you can see the use of the `https://github.com/mattn/go-sqlite3` package for communicating with the SQLite3 database with the help of the `database/sql` interface.

The second part of htmlT.go is the following:

```
func myHandler(w http.ResponseWriter, r *http.Request) {
    fmt.Printf("Host: %s Path: %s\n", r.Host, r.URL.Path)
    myT := template.Must(template.ParseGlob(tFile))
    myT.ExecuteTemplate(w, tFile, DATA)
}
```

The simplicity and the effectiveness of the `myHandler()` function is phenomenal, especially if you consider the size of the function! The `template.ExecuteTemplate()` function does all the work for you. Its first parameter is the variable that holds the connection with the HTTP client, its second parameter is the template file that will be used for formatting the data, and its third parameter is the slice of structures with the data that will be processed.

The third code segment from htmlT.go is shown in the following Go code:

```
func main() {
    arguments := os.Args
    if len(arguments) != 3 {
        fmt.Println("Need Database File + Template File!")
        return
    }

    database := arguments[1]
    tFile = arguments[2]
```

The fourth code portion from htmlT.go is where you start dealing with the database:

```
    db, err := sql.Open("sqlite3", database)
    if err != nil {
        fmt.Println(nil)
        return
    }

    fmt.Println("Emptying database table.")
    _, err = db.Exec("DELETE FROM data")
    if err != nil {
        fmt.Println(nil)
        return
    }

    fmt.Println("Populating", database)
    stmt, _ := db.Prepare("INSERT INTO data(number, double, square) values(?,?,?)")
    for i := 20; i < 50; i++ {
        _, _ = stmt.Exec(i, 2*i, i*i)
    }
```

The `sql.Open()` function opens the connection with the desired database. With `db.Exec()`, you can execute database commands without expecting any feedback from them. Also, the `db.Prepare()` function allows you to execute a database command multiple times by changing only its parameters and calling `Exec()` afterwards.

The fifth part of `htmlT.go` contains the following Go code:

```
rows, err := db.Query("SELECT * FROM data")
if err != nil {
        fmt.Println(nil)
        return
}

var n int
var d int
var s int
for rows.Next() {
        err = rows.Scan(&n, &d, &s)
        temp := Entry{Number: n, Double: d, Square: s}
        DATA = append(DATA, temp)
}
```

In this part of the program, we read the data from the desired table using `db.Query()` and multiple calls to `Next()` and `Scan()`. While reading the data, you put it into a slice of structures and you are done dealing with the database.

The last part of the program is all about setting up the web server, and it contains the following Go code:

```
    http.HandleFunc("/", myHandler)
    err = http.ListenAndServe(":8080", nil)
    if err != nil {
            fmt.Println(err)
            return
    }
}
```

Here, the `http.HandleFunc()` function tells the web server embedded in the program which URLs will be supported and by which handler function (`myHandler()`). The current handler supports the / URL, which in Go matches all URLs. This saves you from having to create any extra static or dynamic pages.

The code of the `htmlT.go` program is divided in two virtual parts. The first part is about getting the data from the database and putting it into a slice of structures, whereas the second part, which is similar to `textT.go`, is about displaying your data in a web browser.

The two biggest advantages of SQLite are that you do not need to run a server process for the database server and that SQLite databases are stored in self-contained files, which means that single files hold entire SQLite databases.

Note that in order to reduce the Go code and be able to run the `htmlT.go` program multiple times, you will need to create the database table and the SQLite3 database manually, which is as simple as executing the following commands:

```
$ sqlite3 htmlT.db
SQLite version 3.19.3 2017-06-27 16:48:08
Enter ".help" for usage hints.
sqlite> CREATE TABLE data (
   ...> number INTEGER PRIMARY KEY,
   ...> double INTEGER,
   ...> square INTEGER );
sqlite> ^D
$ ls -l htmlT.db
-rw-r--r--  1 mtsouk  staff  8192 Dec 26 22:46 htmlT.db
```

The first command is executed from the Unix shell, and it is needed for creating the database file. The second command is executed from the SQLite3 shell and has to do with creating a database table named `data`, which has three fields named `number`, `double`, and `square`, respectively.

Additionally, you are going to need an external template file, which will be named `html.gohtml`. It is going to be used for the generation of the output of the program.

The first part of `html.gohtml` is the following:

```
<!doctype html>
<html lang="en">
<head>
    <meta charset="UTF-8">
        <title>Doing Maths in Go!</title>
        <style>
            html {
                font-size: 14px;
            }
            table, th, td {
                border: 2px solid blue;
            }
        </style>
</head>
<body>
```

The HTML code that a web browser will read is going to be based on the contents of `html.gohtml`. This means that you will need to create proper HTML output, hence the preceding HTML code, which also includes some in-line CSS code for formatting the generated HTML table.

The second part of `html.gohtml` contains the following code:

```
<table>
   <thead>
        <tr>
             <th>Number</th>
             <th>Double</th>
             <th>Square</th>
        </tr>
   </thead>
   <tbody>
{{ range . }}
   <tr>
        <td> {{ .Number }} </td>
        <td> {{ .Double }} </td>
        <td> {{ .Square }} </td>
   </tr>
{{ end }}
   </tbody>
</table>
```

As you can see from the preceding code, you still have to use `{{ range }}` and `{{ end }}` in order to iterate over the elements of the slice of structures that was passed to `template.ExecuteTemplate()`. However, this time the `html.gohtml` template file contains lots of HTML code in order to format the data in the slice of structures better.

The last part of the HTML template file is the following:

```
</body>
</html>
```

The last part of `html.gohtml` is mainly used for properly ending the generated HTML code according to the HTML standards.

Before being able to compile and execute `htmlT.go`, you will need to download the package that will help the Go programming language to communicate with SQLite3. You can do this by executing the following command:

```
$ go get github.com/mattn/go-sqlite3
```

As you already know, you can find the source code of the downloaded package inside `~/go/src` and its compiled version inside `~/go/pkg/darwin_amd64` if you are on a macOS machine. Otherwise, check the contents of `~/go/pkg` to find out your own architecture. Note that the `~` character denotes the home directory of the current user.

Keep in mind that additional Go packages exist that can help you communicate with an SQLite3 database. However, the one used here is the only one that currently supports the `database/sql` interface.

Executing `htmlT.go` will produce the kind of output on a web browser that you can see in the following screenshot:

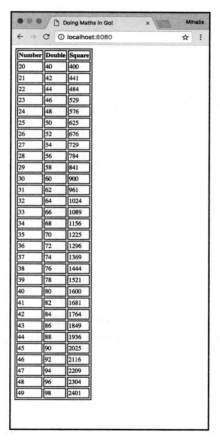

The output of the htmlT.go program

Moreover, `htmlT.go` will generate the following type of output in your Unix shell, which is mainly debugging information:

```
$ go run htmlT.go htmlT.db html.gohtml
Emptying database table.
Populating htmlT.db
Host: localhost:8080 Path: /
Host: localhost:8080 Path: /favicon.ico
Host: localhost:8080 Path: /123
```

If you want to see the HTML output of the program from the Unix shell, you can use the `wget(1)` utility as follows:

```
$ wget -qO- http://localhost:8080
<!doctype html>
<html lang="en">
<head>
    <meta charset="UTF-8">
            <title>Doing Maths in Go!</title>
            <style>
                html {
                        font-size: 14px;
                }
                table, th, td {
                    border: 2px solid blue;
                }
            </style>
    </head>
    <body>


<table>
    <thead>
            <tr>
```

Both `text/template` and `html/template` are powerful packages that can save you a lot of time, so I suggest that you use them when they fit the requirements of your applications.

Basic SQLite3 commands

This section will help you find your way to SQLite3 by presenting some simple commands.

You can create a new table as follows:

```
sqlite> CREATE TABLE data (
   ...> number INTEGER PRIMARY KEY,
   ...> double INTEGER,
   ...> square INTEGER );
```

You can check the contents of a table named `data` as follows:

```
sqlite> select * from data;
20|40|400
21|42|441
22|44|484
```

You can completely delete an existing table using the following command:

```
$ DROP TABLE ATABLE;
```

You can delete the data from an existing table as follows:

```
$ DELETE FROM ATABLE;
```

Finally, if you want to delete an entire SQLite3 database, you can simply delete its database file using the `rm(1)` Unix command.

Additional resources

You will find the following resources very useful:

- Visit the documentation page of the `syscall` standard Go package at https://golang.org/pkg/syscall/. This is one of the biggest Go documentation pages that I have ever seen!
- Visit the documentation page of the `text/template` package, which can be found at https://golang.org/pkg/text/template/.
- Similarly, go to https://golang.org/pkg/html/template/ for the documentation of the `html/template` package.
- Visit the home page of SQLite3 at https://www.sqlite.org/.

- Watch the Writing Beautiful Packages in Go video by **Mat Ryer** at `https://www.youtube.com/watch?v=cAWlv2SeQus`.
- If you want to know about **Plan 9**, look at `https://plan9.io/plan9/`.
- Take the time to look at the `find(1)` command line tool by visiting its man page (`man 1 find`).

Exercises

- Seek out more information about the actual implementation of the `fmt.Printf()` function.
- Can you write a function that sorts three `int` values? Try to write two versions of the function: one with named returned values and another without named return values. Which one do you think is better?
- Can you modify the Go code of `htmlT.go` in order to use `text/template` instead of `html/template`?
- Can you modify the Go code of `htmlT.go` in order to use either `https://github.com/feyeleanor/gosqlite3` or the `https://github.com/phf/go-sqlite3` package for communicating with the SQLite3 database?
- Write a Go program like `htmlT.go` that reads data from a MySQL database. Write down the code changes that you made.

Summary

In this chapter, we presented two primary topics: Go functions and Go packages. We offered you ample advice about developing good Go packages. We also talked about the `text/template` and `html/template` packages, which allow you to create plain text and HTML output based on templates. Last, we talked about the `syscall` standard Go package that offers advanced features.

In the next chapter, we will discuss two important Go features: interfaces and reflection. Additionally, we will talk about object-oriented programming in Go and Go type methods. All of these topics are pretty advanced, and you might find them difficult at first. However, learning more about them will unquestionably make you a better Go programmer!

7 Reflection and Interfaces for All Seasons

In the previous chapter, we talked about developing packages and functions in Go. We also discussed how to work with text and HTML templates with the help of the `text/template` and `html/template` packages. Finally, we explained the use of the `syscall` package.

In this chapter, you will learn three very interesting, handy, and somewhat advanced Go concepts: reflection, interfaces, and type methods. Although Go interfaces are used all of the time, reflection is not, mainly because it is usually unnecessary for your programs to make use of it. Furthermore, you will learn about type assertions and **object-oriented programming** (**OOP**) in Go.

In this chapter, you will learn the following topics:

- Type methods
- Go Interfaces
- Type assertions
- Developing and using your own interfaces
- Object-oriented programming the Go way
- Reflection and the `reflect` standard Go package

Type methods

A Go **type method** is a function with a special receiver argument. You declare methods as ordinary functions with an additional parameter that appears in front of the function name. This particular parameter connects the function to the type of that extra parameter. As a result, that parameter is called the **receiver of the method**.

The following Go code represents the implementation of the `Close()` function as found in `https://golang.org/src/os/file_plan9.go`:

```go
func (f *File) Close() error {
    if err := f.checkValid("close"); err != nil {
        return err
    }
    return f.file.close()
}
```

The `Close()` function is a type method because there is that `(f *File)` parameter in front of its name and after the `func` keyword. The `f` parameter is called the receiver of the method; in object-oriented programming terminology this process can be described as sending a message to an **object**. In Go, the receiver of a method is defined using a regular variable name—usually using a single letter—without the need to use a dedicated keyword, such as `this` or `self`.

Now let's review a complete example using the Go code of the `methods.go` file, which will be presented in four parts.

The first part of `methods.go` is shown in the following Go code:

```go
package main

import (
    "fmt"
)

type twoInts struct {
    X int64
    Y int64
}
```

In the preceding Go code, you can see the definition of a new structure with two fields named `twoInts`.

The second code segment of `methods.go` follows next:

```go
func regularFunction(a, b twoInts) twoInts {
    temp := twoInts{X: a.X + b.X, Y: a.Y + b.Y}
    return temp
}
```

In this part, you define a new function named regularFunction() that accepts two parameters of type twoInts and returns just one twoInts value.

The third part of the methods.go program contains the following Go code:

```
func (a twoInts) method(b twoInts) twoInts {
    temp := twoInts{X: a.X + b.X, Y: a.Y + b.Y}
    return temp
}
```

The method() function is equivalent to the regularFunction() function defined in the preceding part of methods.go. However, the method() function is a type method, and you will need to learn a different way for calling it in a moment.

The really interesting thing here is that the implementation of method() is exactly the same as the implementation of regularFunction().

The last code segment of methods.go is as follows:

```
func main() {
    i := twoInts{X: 1, Y: 2}
    j := twoInts{X: -5, Y: -2}
    fmt.Println(regularFunction(i, j))
    fmt.Println(i.method(j))
}
```

As you can see, the way that you are calling a type method (i.method(j)) is different from the way that you would call a conventional function (regularFunction(i, j)).

Executing methods.go will create the following output:

```
$ go run methods.go
{-4 0}
{-4 0}
```

Notice that type methods are also associated with interfaces, which will be the subject of the next section. As a result, you will see more type methods in a little while.

Go interfaces

Strictly speaking, a Go `interface` type defines the behavior of other types by specifying a set of methods that need to be implemented. For a type to satisfy an interface, it needs to implement all of the methods required by that interface. Put simply, interfaces are **abstract types** that define a set of functions that need to be implemented so that a type can be considered an instance of the interface. When this happens, we say that the type satisfies this interface. So, an interface is two things: a set of methods and a type, and it is used for defining the behavior of other types.

The biggest advantage that you receive from having and using an interface is that you can pass a variable of a type that implements that particular interface to any function that expects a parameter of that specific interface. Without that amazing capability, interfaces would only be a formality without any practical or real benefit.

 If you find yourselves defining an interface and its implementation in the same Go package, then you might have been using interfaces the wrong way!

Two very common Go interfaces are `io.Reader` and `io.Writer`, which are used in file input and output operations. More specifically, `io.Reader` is used for reading a file, whereas `io.Writer` is used for writing to a file.

The definition of `io.Reader`, as found in https://golang.org/src/io/io.go, is as follows:

```
type Reader interface {
    Read(p []byte) (n int, err error)
}
```

In order for a type to satisfy the `io.Reader` interface, you will need to implement the `Read()` method as described in the interface.

Similarly, the definition of `io.Writer`, as found in https://golang.org/src/io/io.go, is the following:

```
type Writer interface {
    Write(p []byte) (n int, err error)
}
```

To satisfy the `io.Writer` interface, you will just need to implement a single method named `Write()`.

Each one of the `io.Reader` and `io.Writer` interfaces requires the implementation of just one method, yet both interface are very powerful. Most likely, their power comes from their simplicity! Generally speaking, interfaces should be as simple as possible.

In the following subsections, you will learn how to define an interface on your own and how to use it in other Go packages. Notice that it is not necessary for an interface to be fancy or impressive, as long as it does what you want it to do.

Put simply, interfaces should be utilized when there is a need for making sure that certain conditions will be met and certain behaviors will be anticipated from a Go element.

About type assertion

A **type assertion** is the `x.(T)` notation where x is of interface type and T is a type. Additionally, the actual value stored in x is of type T, and T must satisfy the interface type of x! The following paragraphs, as well as the code example, will clarify this relatively eccentric definition of type assertion.

Type assertions help you do *two things*. The first thing is to check whether an interface value keeps a particular type. When used this way, a type assertion returns two values: the underlying value and a `bool` value. Although the underlying value is what you might want to use, the Boolean value tells you whether the type assertion was successful or not!

The second thing a type assertion does is to allow you to use the concrete value stored in an interface or assign it to a new variable. This means that if there is an `int` variable in an interface, you can get that value using type assertion.

However, if a type assertion is not successful and you do not handle that failure on your own, your program will panic. Now study the Go code of the `assertion.go` program, which will be presented in two parts. The first part contains the following Go code:

```
package main

import (
    "fmt"
)
```

```
func main() {
    var myInt interface{} = 123

    k, ok := myInt.(int)
    if ok {
        fmt.Println("Success:", k)
    }

    v, ok := myInt.(float64)
    if ok {
        fmt.Println(v)
    } else {
        fmt.Println("Failed without panicking!")
    }
```

First, you declare the `myInt` variable, which has dynamic type `int` and value `123`. Then, you use type assertion two times to test the interface of the `myInt` variable—once for `int` and once for `float64`.

As the `myInt` variable does not contain a `float64` value, the `myInt.(float64)` type assertion will fail unless handled properly. Fortunately, in this case the correct use of the `ok` variable will save your program from panicking.

The second part of the `assertion.go` program is shown in the following Go code:

```
    i := myInt.(int)
    fmt.Println("No cheking:", i)

    j := myInt.(bool)
    fmt.Println(j)
}
```

There are two type assertions taking place here. The first type assertion is successful, so there will be no problem with that. But let's review this particular type assertion a bit further. The type of variable `i` will be `int`, and its value will be `123`, which is the value stored in `myInt`. So, as `int` satisfies the `myInt` interface, which in this case occurs because the `myInt` interface requires no functions in order to be implemented, the value of `myInt.(int)` is an `int` value.

However, the second type assertion, which is `myInt.(bool)`, will trigger a panic because the underlying value of `myInt` is not Boolean (`bool`). Therefore, executing `assertion.go` will generate the following output:

```
$ go run assertion.go
Success: 123
Failed without panicking!
No cheking: 123
panic: interface conversion: interface {} is int, not bool
goroutine 1 [running]:
main.main()
        /Users/mtsouk/Desktop/masterGo/ch/ch7/code/assertion.go:25 +0x1d9
exit status 2
```

Go states pretty clearly the reason for panicking: `interface {} is int, not bool`.

Generally speaking, when working with interfaces, expect to use type assertions as well. You will see more type assertions in the `useInterface.go` program that will be presented in a short while.

Developing your own interfaces

In this section, you will learn how to develop your own interfaces, This is a relatively easy process as long as you know what you want to develop.

The technique will be illustrated using the Go code of `myInterface.go`, which will be presented below. The interface that will be created will help you work with geometric shapes of the plane.

The Go code of `myInterface.go` follows next:

```
package myInterface

type Shape interface {
    Area() float64
    Perimeter() float64
}
```

The definition of the `shape` interface is truly simple and straightforward, as it requires that you implement just two functions named `Area()` and `Perimeter()`, which both return a `float64` value. The first function will be used for calculating the area of a shape in the plane and the second one will be used for calculating the perimeter of a shape in the plane. After that, you will need to install the `myInterface.go` package and make it available to the current user. As you already know, the installation process involves the execution of the following Unix commands:

```
$ mkdir ~/go/src/myInterface
$ cp myInterface.go ~/go/src/myInterface
$ go install myInterface
```

Using a Go interface

This subsection will teach you how to use the interface defined in `myInterface.go` in a Go program named `useInterface.go`, which will be presented in five parts.

The first part of `useInterface.go` is shown in the following Go code:

```
package main

import (
    "fmt"
    "math"
    "myInterface"
)

type square struct {
    X float64
}

type circle struct {
    R float64
}
```

As the desired interface is defined in its own package, it should come as no surprise that you are importing the `myInterface` package.

The second code portion of `useInterface.go` contains the following code:

```
func (s square) Area() float64 {
    return s.X * s.X
}

func (s square) Perimeter() float64 {
    return 4 * s.X
}
```

In this part, you implement the `shape` interface for the `square` type.

The third part contains the following Go code:

```
func (s circle) Area() float64 {
    return s.R * s.R * math.Pi
}

func (s circle) Perimeter() float64 {
    return 2 * s.R * math.Pi
}
```

In this part, you implement the `shape` interface for the `circle` type.

The fourth part of `useInterface.go` is shown in the following Go code:

```
func Calculate(x myInterface.Shape) {
    _, ok := x.(circle)
    if ok {
        fmt.Println("Is a circle!")
    }

    v, ok := x.(square)
    if ok {
        fmt.Println("Is a square:", v)
    }

    fmt.Println(x.Area())
    fmt.Println(x.Perimeter())
}
```

Thus, in the preceding code, you implement one function that requires a single `shape` parameter (`myInterface.Shape`). The magic here should be obvious once you understand that it requires any `shape` parameter, which is any parameter whose type implements the `shape` interface!

The code at the beginning of the function shows how you can differentiate between the various data types that implement the desired interface. In the second block, you see how you can find the values stored in a `square` parameter. You can use this technique for any type that implements the `myInterface.Shape` interface.

The last code segment includes the following code:

```go
func main() {
    x := square{X: 10}
    fmt.Println("Perimeter:", x.Perimeter())
    Calculate(x)
    y := circle{R: 5}
    Calculate(y)
}
```

In this part, you see how you can use both `circle` and `square` variables as parameters to the `Calculate()` function that you implemented earlier.

If you execute `useInterface.go`, you will get the following as output:

```
$ go run useInterface.go
Perimeter: 40
Is a square: {10}
100
40
Is a circle!
78.53981633974483
31.41592653589793
```

Using switch with interface and data types

In this subsection, you will learn how to use the `switch` statement to differentiate between different data types using the Go code of `switch.go`, which will be presented in four parts. The Go code of `switch.go` is partially based on `useInterface.go`, but it will add another type named `rectangle` and it will not need to implement the methods of any interface.

The first part of the `switch.go` program follows next:

```go
package main

import (
    "fmt"
)
```

Chapter 7

As the code in `switch.go` will not work with the interface defined in `myInterface.go`, there is no need to import the `myInterface` package.

The second part of `switch.go` is where you define the three types that will be used in the program:

```
type square struct {
    X float64
}

type circle struct {
    R float64
}

type rectangle struct {
    X float64
    Y float64
}
```

All three types are pretty simple.

The third code segment of `switch.go` comes with the following Go code:

```
func tellInterface(x interface{}) {
    switch v := x.(type) {
    case square:
        fmt.Println("This is a square!")
    case circle:
        fmt.Printf("%v is a circle!\n", v)
    case rectangle:
        fmt.Println("This is a rectangle!")
    default:
        fmt.Printf("Unknown type %T!\n", v)
    }
}
```

Here you can see the implementation of a function named `tellInterface()` with a single parameter named x and type `interface{}`.

The trick presented here will help you differentiate between the different data types of the x parameter. All of the magic is performed by the use of the `x.(type)` statement that returns the type of the x element. The `%v` verb used in `fmt.Printf()` allows you to acquire the value of the type.

The last part of `switch.go` contains the implementation of the `main()` function:

```
func main() {
    x := circle{R: 10}
    tellInterface(x)
    y := rectangle{X: 4, Y: 1}
    tellInterface(y)
    z := square{X: 4}
    tellInterface(z)
    tellInterface(10)
}
```

Executing `switch.go` will generate the following type of output:

```
$ go run switch.go
{10} is a circle!
This is a rectangle!
This is a square!
Unknown type int!
```

Reflection

Reflection is an advanced Go feature that allows you to dynamically learn the type of an arbitrary object as well as information about its structure. Go offers the `reflect` package for working with reflection. What you should remember is that you will most likely not need to use reflection in each one of your Go programs.

The first two questions that might arise are why is reflection necessary and when should I use it?

Reflection is necessary for the implementation of packages such as `fmt`, `text/template`, and `html/template`. In the `fmt` package, reflection saves you from having to deal explicitly with every data type that exists. However, even if you had the patience to write code for working with every data type that you knew, you still would not be able to work with all possible types! In this case, reflection makes it possible for the methods of the `fmt` package to find the structure and to work with new types!

Consequently, you might need to use reflection when you want to be as generic as possible, or when you want to make sure that you will be able to deal with data types that do not exist at the time that you are writing your code. Additionally, reflection is handy when working with values of types that do not implement a common interface.

Reflection helps you work with unknown types and unknown values of types.

The stars of the `reflect` package are two types named `reflect.Value` and `reflect.Type`. The former type is used for storing values of any type, whereas the latter type is used for representing Go types.

A simple Reflection example

This section will present a relatively simple reflection example in order to start feeling comfortable with this advanced Go feature.

The name of the Go source file is `reflection.go`, and it will be presented in four parts. The purpose of `reflection.go` is to examine an "unknown" structure variable and to find out more about it at runtime. For added interest, the program will define two new `struct` types. Based on these two types, it will also define two new variables; however, it will examine only one of them. If the program has no command-line arguments, it will examine the first one; otherwise, it will explore the second one. For practical purposes, this means that the program will not know in advance the kind of `struct` variable it will have to process.

The first part of `reflection.go` contains the following Go code:

```
package main

import (
    "fmt"
    "os"
    "reflect"
)

type a struct {
    X int
    Y float64
    Z string
}

type b struct {
    F int
    G int
    H string
```

```
    I float64
}
```

In this part of the program, you can see the definition of the `struct` data types that will be used.

The second code segment of `reflection.go` is as follows:

```
func main() {
    x := 100
    xRefl := reflect.ValueOf(&x).Elem()
    xType := xRefl.Type()
    fmt.Printf("The type of x is %s.\n", xType)
```

The preceding Go code presents a small and simplistic reflection example. First you declare a variable named x, and then you call the `reflect.ValueOf(&x).Elem()` function. Next you call `xRefl.Type()` in order to get the type of the variable, which is stored in xType. These three lines of code illustrate how you can get the data type of a variable using reflection. However, if all you care about is the data type of a variable, you can just call `reflect.TypeOf(x)` instead.

The third code portion of `reflection.go` contains the following Go code:

```
    A := a{100, 200.12, "Struct a"}
    B := b{1, 2, "Struct b", -1.2}
    var r reflect.Value

    arguments := os.Args
    if len(arguments) == 1 {
        r = reflect.ValueOf(&A).Elem()
    } else {
        r = reflect.ValueOf(&B).Elem()
    }
```

In this part, you declare two variables named A and B. The type of the A variable is a, and the type of the B variable is b. The type of the r variable should be `reflect.Value` because this is what the `reflect.ValueOf()` function returns. The `Elem()` method returns the value contained in the reflection interface (`reflect.Value`).

The last part of the `reflection.go` code follows next:

```
    iType := r.Type()
    fmt.Printf("i Type: %s\n", iType)
    fmt.Printf("The %d fields of %s are:\n", r.NumField(), iType)

    for i := 0; i < r.NumField(); i++ {
```

```
            fmt.Printf("Field name: %s ", iType.Field(i).Name)
            fmt.Printf("with type: %s ", r.Field(i).Type())
            fmt.Printf("and value %v\n", r.Field(i).Interface())
      }
}
```

In this part of the program, you use the appropriate functions of the `reflect` package in order to obtain the desired information. The `NumField()` method returns the number of fields in a `reflect.Value` structure, whereas the `Field()` function returns the field of the structure that is specified by its parameter. The `Interface()` function returns the value of a field of the `reflect.Value` structure as an interface.

Executing `reflection.go` two times will generate the following output:

```
$ go run reflection.go 1
The type of x is int.
i Type: main.b
The 4 fields of main.b are:
Field name: F with type: int and value 1
Field name: G with type: int and value 2
Field name: H with type: string and value Struct b
Field name: I with type: float64 and value -1.2
$ go run reflection.go
The type of x is int.
i Type: main.a
The 3 fields of main.a are:
Field name: X with type: int and value 100
Field name: Y with type: float64 and value 200.12
Field name: Z with type: string and value Struct a
```

It is important to see that Go uses its internal representation to print the data types of variables A and B, which is `main.a` and `main.b`, respectively. However, this is not the case with variable x, which is an `int`.

A more advanced reflection example

In this section, we will explore more advanced uses of reflection as illustrated in relatively small code blocks using the Go code of `advRefl.go`.

The `advRefl.go` program will be presented in five parts. The first part of the code follows next:

```
package main

import (
    "fmt"
    "os"
    "reflect"
)

type t1 int
type t2 int
```

Note that although both `t1` and `t2` types are based on `int` and therefore are essentially the same type as `int`, Go treats them as totally different types. Their internal representation after Go parses the code of the program will be `main.t1` and `main.t2`, respectively.

The second code portion of `advRefl.go` is as follows:

```
type a struct {
    X    int
    Y    float64
    Text string
}

func (a1 a) compareStruct(a2 a) bool {
    r1 := reflect.ValueOf(&a1).Elem()
    r2 := reflect.ValueOf(&a2).Elem()

    for i := 0; i < r1.NumField(); i++ {
        if r1.Field(i).Interface() != r2.Field(i).Interface() {
            return false
        }
    }
    return true
}
```

In this code segment, we define a Go structure type named `a` and implement a Go function named `compareStruct()`. The purpose of this function is to find out whether two variables of the `a` type are exactly the same or not. As you can see, `compareStruct()` uses Go code from `reflection.go` to perform its task.

The third code segment of `advRefl.go` is shown in the following Go code:

```go
func printMethods(i interface{}) {
    r := reflect.ValueOf(i)
    t := r.Type()
    fmt.Printf("Type to examine: %s\n", t)

    for j := 0; j < r.NumMethod(); j++ {
        m := r.Method(j).Type()
        fmt.Println(t.Method(j).Name, "-->", m)
    }
}
```

The `printMethods()` function prints the methods of a variable. The variable type that will be used in `advRefl.go` for illustrating `printMethods()` will be `os.File`.

The fourth code segment of `advRefl.go` contains the following Go code:

```go
func main() {
    x1 := t1(100)
    x2 := t2(100)
    fmt.Printf("The type of x1 is %s\n", reflect.TypeOf(x1))
    fmt.Printf("The type of x2 is %s\n", reflect.TypeOf(x2))

    var p struct{}
    r := reflect.New(reflect.ValueOf(&p).Type()).Elem()
    fmt.Printf("The type of r is %s\n", reflect.TypeOf(r))
```

The last code portion of `advRefl.go` is as follows:

```go
    a1 := a{1, 2.1, "A1"}
    a2 := a{1, -2, "A2"}

    if a1.compareStruct(a1) {
        fmt.Println("Equal!")
    }

    if !a1.compareStruct(a2) {
        fmt.Println("Not Equal!")
    }

    var f *os.File
    printMethods(f)
}
```

As you will see in a short while, the `a1.compareStruct(a1)` call returns `true` because we are comparing `a1` with itself, whereas the `a1.compareStruct(a2)` call will return `false` because the `a1` and `a2` variables have different values.

Executing `advRefl.go` will create the following output:

```
$ go run advRefl.go
The type of x1 is main.t1
The type of x2 is main.t2
The type of r is reflect.Value
Equal!
Not Equal!
Type to examine: *os.File
Chdir --> func() error
Chmod --> func(os.FileMode) error
Chown --> func(int, int) error
Close --> func() error
Fd --> func() uintptr
Name --> func() string
Read --> func([]uint8) (int, error)
ReadAt --> func([]uint8, int64) (int, error)
Readdir --> func(int) ([]os.FileInfo, error)
Readdirnames --> func(int) ([]string, error)
Seek --> func(int64, int) (int64, error)
Stat --> func() (os.FileInfo, error)
Sync --> func() error
Truncate --> func(int64) error
Write --> func([]uint8) (int, error)
WriteAt --> func([]uint8, int64) (int, error)
WriteString --> func(string) (int, error)
```

You can see that the type of the `r` variable, which is returned by `reflect.New()`, will be `reflect.Value`. Additionally, the output of the `printMethods()` method tells us that the `*os.File` type supports a plethora of methods, such as `Chdir()`, `Chmod()`, and so on.

The three disadvantages of reflection

Without a doubt, reflection is a powerful Go feature. However, as it happens with all tools, reflection should be used sparingly for three main reasons.

The first reason is that extensive use of reflection will make your programs hard to read and maintain. A potential solution to this problem is good documentation; however, developers are famous for not having the time to write the required documentation.

The second reason is that the Go code which uses reflection will make your programs slower. Generally speaking, Go code that is made to work with a particular data type will always be faster than Go code that uses reflection to work dynamically with any Go data type. Additionally, such dynamic code will make it difficult for tools to refactor or analyze your code.

The last reason is that reflection errors cannot be caught at build time and are reported at runtime as a panic. This means that reflection errors can potentially crash your programs! This can happen months or even years after the development of a Go program! One solution to this problem is extensive testing before a dangerous function call. However, this will add even more Go code to your programs, which will make them even slower.

Object-oriented programming in Go!

You should know by now that Go lacks inheritance, but that it supports composition and that Go interfaces provide a kind of polymorphism. So, although Go is not an object-oriented programming language, it has some features that let you mimic object-oriented programming.

If you really want to develop applications using object-oriented methodology, then choosing Go might not be your best option. As I am not really into **Java**, I would suggest looking at **C++** or **Python** instead. On the other hand, Go prohibits you from building complicated and deep type hierarchies that are hard to work with and maintain!

First let me explain the two techniques that will be used in the Go program in this section. The first technique uses methods in order to associate a function with a type. This means that, in some way, you can consider that the function and type construct an object.

Reflection and Interfaces for All Seasons

With the second technique, you embed a type into a new structure type in order to create a kind of hierarchy.

There is also a third technique where you use a Go interface to make two or more elements—*objects of the same class*. This technique will not be illustrated in this section, as it has been demonstrated earlier in this chapter. The key point here is that a Go interface allows you to define a common behavior between different elements in a way such that all of these different elements share the characteristics of an object. This might allow you to say that these different elements are objects of the same class. However, objects and classes of an actual object-oriented programming language can do many more things.

The first two techniques will be illustrated in the program ooo.go, which will be presented in four parts. The first code segment of ooo.go contains the following Go code:

```go
package main

import (
    "fmt"
)

type a struct {
    XX int
    YY int
}

type b struct {
    AA string
    XX int
}
```

The second part of the ooo.go program is as follows:

```go
type c struct {
    A a
    B b
}
```

Composition allows you to create a structure in your Go elements using multiple struct types. In this case, data type c groups an a variable and a b variable.

The third portion of ooo.go is shown in the following Go code:

```go
func (A a) A() {
    fmt.Println("Function A() for A")
}

func (B b) A() {
    fmt.Println("Function A() for B")
}
```

The two methods defined here can have the same name (A()) because they have different function headers. The first method works with a variables, whereas the second method works with b variables. This technique allows you to share the same function name between multiple types.

The last part of ooo.go follows next:

```go
func main() {
    var i c
    i.A.A()
    i.B.A()
}
```

All the Go code in ooo.go is pretty simplistic when compared with the code of an object-oriented programming language, which would implement **abstract classes** and **inheritance**. However, it is more than adequate for generating types and elements with a structure in them, as well as having different data types with the same method names.

Executing ooo.go will generate the following output:

```
$ go run ooo.go
Function A() for A
Function A() for B
```

As the following code illustrates, however, composition is not inheritance, and the `first` type knows nothing about the changes made to the `shared()` function by the `second` type:

```go
package main

import (
    "fmt"
)

type first struct{}

func (a first) F() {
```

```
        a.shared()
}

func (a first) shared() {
    fmt.Println("This is shared() from first!")
}

type second struct {
    first
}

func (a second) shared() {
    fmt.Println("This is shared() from second!")
}

func main() {
    first{}.F()
    second{}.shared()
    i := second{}
    j := i.first
    j.F()
}
```

Note that type `second` embeds type `first`, and the two types share a function, named `shared()`.

Saving the former Go code as `goCoIn.go` and executing it will generate the following output:

```
$ go run goCoIn.go
This is shared() from first!
This is shared() from second!
This is shared() from first!
```

Although the calls to `first{}.F()` and `second{}.shared()` generated the expected results, the call to `j.F()` still calls `first.shared()` instead of the `second.shared()` function, despite the fact that type `second` changed the implementation of the `shared()` function. This is called **method overriding** in object-oriented terminology.

Note that the `j.F()` call can be written as `(i.first).F()` or as `(second{}.first).F()` without the need to define too many variables. Breaking it into three lines of code makes it a little easier to understand!

Additional resources

You will find the following two resources very handy:

- Visit the documentation page of the reflect Go standard package, which can be found at `https://golang.org/pkg/reflect/`. This package has many more capabilities than the ones presented in this chapter.
- If you are really into reflection or you want to learn more about it, you can refer to the *reflectwalk* library by *Mitchell Hashimoto* at `https://github.com/mitchellh/reflectwalk`. The `reflectwalk` library allows you to walk complex values in Go using reflection. If you have the time, explore the Go code of the library!

Exercises

- Write your own interface, and use it in another Go program. Then state why your interface is useful.
- Write an interface for calculating the volume of shapes with three dimensions, such as cubes and spheres.
- Write an interface for calculating the length of line segments and the distance between two points in the plane.
- Explore reflection using your own example.
- How does reflection work on Go maps?
- If you are good in mathematics, try to write an interface that implements the four basic mathematical operations for both real numbers and complex numbers. Do not use the `complex64` and `complex128` standard Go types—define your own structure for supporting complex numbers.

Summary

In this chapter, you learned about interfaces, which are like contracts, and also about type methods, type assertion, and reflection in Go. You also learned about creating Go code that follows the principles of object-oriented programming.

Although reflection is a very powerful Go feature, it might slow down your Go programs because it adds a layer of complexity at runtime. Additionally, your Go programs could crash if you use reflection carelessly.

If you will remember just one thing from this chapter, it should be that Go is not an object-oriented programming language, but that it can mimic some of the functionality offered by object-programming languages, such as Java and C++. This means that if you plan to develop software using the object-oriented paradigm all of the time, it would be best if you choose a programming language other than Go. Nevertheless, object-oriented programming is not a panacea, and you might create a better, cleaner, and more robust design if you choose a programming language such as Go!

Although there might have been more theory in this chapter than you expected, the next chapter will pay off your patience, as it will address systems programming in Go. File I/O, working with Unix system files, handling Unix signals, and supporting Unix pipes will be discussed. Additionally, the next chapter will talk about using the `flag` package to support multiple command-line arguments and options in your command-line tools, traversing directory structures, Unix file permissions, and some advanced uses of the functionality offered by the `syscall` standard Go package.

If you are really into systems programming with Go, it might be a good idea to read my book, *Go Systems Programming, Packt Publishing, 2017*.

8
Telling a Unix System What to Do

In the previous chapter, we talked about two advanced, but somewhat theoretical, Go topics: interfaces and reflection. The Go code that you will find in this chapter is anything but theoretical!

The subject of this chapter is **systems programming** because, after all, Go is a serious systems programming language that was born out of frustration; its spiritual fathers were unsatisfied with the programming language choices they had for creating systems software, so they decided to create a new programming language!

If you are really into **Systems programming** with Go, then *Go Systems Programming, Packt Publishing, 2017*, truly will definitely help you along in your journey. However, this chapter contains some interesting and somewhat advanced topics that cannot be found in my Go Systems Programming book.

In this chapter of *Mastering Go*, you will learn about the following topics:

- Unix Processes
- The `flag` package
- The use of `io.Reader` and `io.Writer` interfaces
- Handling Unix **signals** in Go with the help of the `os/signal` package
- Supporting Unix **pipes** in your Unix system utilities
- Reading text files
- Reading CSV files
- Writing to files
- The `bytes` package

- Advanced uses of the `syscall` package
- Traversing directory structures
- Unix file permissions

About Unix processes

Strictly speaking, a **process** is an execution environment that contains instructions, user data and system data parts, and other types of resources that are obtained during runtime. On the other hand, a **program** is a binary file that contains instructions and data that are used for initializing the instruction and user data parts of a process. Each running Unix process is uniquely identified by an unsigned integer, which is called the **process ID** of the process.

There are three categories of processes: user processes, daemon processes, and kernel processes. **User processes** run in user space and usually have no special access rights. **Kernel processes** are executed in kernel space only and can fully access all kernel data structures. **Daemon processes** are programs that can be found in the user space and run in the background without the need for a terminal.

 The C way for creating new processes involves calling the `fork()` system call. The return value of `fork()` allows the programmer to differentiate between the parent and child process. In contrast, Go does not support a similar functionality but offers goroutines.

The flag package

Flags are specially-formatted strings that are passed into a program to control its behavior. Dealing with flags on your own might become very difficult if you want to support multiple flags. Thus, if you are developing Unix system command-line utilities, you will find the `flag` package very interesting and useful-if not downright irreplaceable. Among its other characteristics, the `flag` package makes no assumptions about the order of the command-line arguments and options, and it prints helpful messages in case there is an error in the way the command-line utility was executed.

> The biggest advantage of the flag package is that it is part of the standard Go library, which means that it has been extensively tested and debugged.

I will present two Go programs that use the flag package: a simple one and a more advanced one.

The first one, named simpleFlag.go, will be offered in four parts. The simpleFlag.go program will recognize two command-line options: the first one will be a Boolean option and the second one will require an integer value.

The first part of simpleFlag.go contains the following Go code:

```
package main

import (
    "flag"
    "fmt"
)
```

The second code portion from simpleFlag.go is as follows:

```
func main() {
    minusK := flag.Bool("k", true, "k")
    minusO := flag.Int("O", 1, "O")
    flag.Parse()
```

The flag.Bool("k", true, "k") statement defines a Boolean command-line option named k with the default value of true. The last parameter of the statement is the usage string that will be displayed with the usage information of the program. Similarly, the flag.Int() function adds support for an integer command-line option.

> You always need to call flag.Parse() after defining the command-line options that you want to support.

The third part of the simpleFlag.go program contains the following Go code:

```
    valueK := *minusK
    valueO := *minusO
    valueO++
```

In the preceding Go code, you see how you can obtain the values of your options. The good thing here is that the `flag` package automatically converts the input associated with the `flag.Int()` flag to an integer value. This means that you do not have to do that on your own. Additionally, the `flag` package makes sure that it was given an acceptable integer value.

The remaining Go code from `simpleFlag.go` follows next:

```go
    fmt.Println("-k:", valueK)
    fmt.Println("-O:", valueO)
}
```

After getting the values of the desired parameters, you are now ready to use them.

Interacting with `simpleFlag.go` will create the following type of output:

```
$ go run simpleFlag.go -O 100
-k: true
-O: 101
$ go run simpleFlag.go -O=100
-k: true
-O: 101
$ go run simpleFlag.go -O=100 -k
-k: true
-O: 101
$ go run simpleFlag.go -O=100 -k false
-k: true
-O: 101
$ go run simpleFlag.go -O=100 -k=false
-k: false
-O: 101
```

If there is an error in the way `simpleFlag.go` was executed, you will get the following type of error message from the `flag` package:

```
$ go run simpleFlag.go -O=notAnInteger
invalid value "notAnInteger" for flag -O: strconv.ParseInt: parsing
"notAnInteger": invalid syntax
Usage of /var/folders/sk/ltk8cnw50lzdtr2hxcj5sv2m0000gn/T/go-
build020625525/command-line arguments/_obj/exe/simpleFlag:
  -O int
    O (default 1)
  -k  k (default true)
exit status 2
```

[286]

Notice the convenient usage message that is automatically printed when there is an error in the command-line options given to your program.

Now it is time to present a more realistic and advanced program that uses the `flag` package. Its name is `funWithFlag.go`, and it will be presented in five parts. The `funWithFlag.go` utility will recognize various kinds of options, including one that accepts multiple values which are separated by commas. Additionally, it will illustrate how you can access the command-line arguments that are located at the end of the executable and that do not belong to any option.

The `flag.Var()` function used in `funWithFlag.go` creates a flag of any type that satisfies the `flag.Value` interface, which is defined as follows:

```
type Value interface {
    String() string
    Set(string) error
}
```

The first part of `funWithFlag.go` contains the following Go code:

```
package main

import (
    "flag"
    "fmt"
    "strings"
)

type NamesFlag struct {
    Names []string
}
```

The `NamesFlag` structure will be used in a short while for the `flag.Value` interface.

The second part of `funWithFlag.go` is as follows:

```
func (s *NamesFlag) GetNames() []string {
    return s.Names
}

func (s *NamesFlag) String() string {
    return fmt.Sprint(s.Names)
}
```

Telling a Unix System What to Do

The third code portion from `funWithFlag.go` contains the following code:

```go
func (s *NamesFlag) Set(v string) error {
    if len(s.Names) > 0 {
        return fmt.Errorf("Cannot use names flag more than once!")
    }

    names := strings.Split(v, ",")
    for _, item := range names {
        s.Names = append(s.Names, item)
    }
    return nil
}
```

First, the `Set()` method makes sure that the related command-line option is not already set. After that, it gets the input and separates its arguments using the `strings.Split()` function. Then the arguments are saved in the `Names` field of the `NamesFlag` structure.

The fourth code segment from `funWithFlag.go` is shown in the following Go code:

```go
func main() {
    var manyNames NamesFlag
    minusK := flag.Int("k", 0, "An int")
    minusO := flag.String("o", "Mihalis", "The name")
    flag.Var(&manyNames, "names", "Comma-separated list")

    flag.Parse()
    fmt.Println("-k:", *minusK)
    fmt.Println("-o:", *minusO)
```

The last part of the `funWithFlag.go` utility follows next:

```go
    for i, item := range manyNames.GetNames() {
        fmt.Println(i, item)
    }

    fmt.Println("Remaining command-line arguments:")
    for index, val := range flag.Args() {
        fmt.Println(index, ":", val)
    }
}
```

The `flag.Args()` slice keeps the command-line arguments left, while the `manyNames` variable holds the values from the `flag.Var()` command-line option.

Executing `funWithFlag.go` will create the following type of output:

```
$ go run funWithFlag.go -names=Mihalis,Jim,Athina 1 two Three
-k: 0
-o: Mihalis
0 Mihalis
1 Jim
2 Athina
Remaining command-line arguments:
0 : 1
1 : two
2 : Three
$ go run funWithFlag.go -Invalid=Marietta 1 two Three
flag provided but not defined: -Invalid
Usage of funWithFlag:
  -k int
        An int
  -names value
        Comma-separated list
  -o string
        The name (default "Mihalis")
exit status 2
$ go run funWithFlag.go -names=Marietta -names=Mihalis
invalid value "Mihalis" for flag -names: Cannot use names flag more than once!
Usage of funWithFlag:
  -k int
        An int
  -names value
        Comma-separated list
  -o string
        The name (default "Mihalis")
exit status 2
```

> Unless you are developing a trivial command-line utility that requires no command-line options, you will most likely need to use the `flag` package for processing the command-line arguments of your program.

[289]

The io.Reader and io.Writer interfaces

As stated in the previous chapter, compliance with the `io.Reader` interface requires the implementation of the `Read()` method, whereas if you want to satisfy the `io.Writer` interface guidelines, you will need to implement the `Write()` method. Both these interfaces are very popular in Go, and we will put them to use in a little while.

Buffered and unbuffered file input and output

Buffered file input and output happens when there is a buffer for temporarily storing data before reading data or writing data. Thus, instead of reading a file byte by byte, you read many bytes at once. You put it in a buffer and wait for someone to read it in the desired way. Unbuffered file input and output happens when there is no buffer to temporarily store data before actually reading or writing it.

The next question that you might ask is how to decide when to use buffered and when to use unbuffered file input and output. When dealing with critical data, unbuffered file input and output is generally a better choice because buffered reads might result in out-of-date data and buffered writes might result in data loss when the power of your computer is interrupted. However, most of the time, there is no definitive answer to that question. This means that you can use whatever makes your tasks easier to implement.

The bufio package

As the name suggests, the `bufio` package is about buffered input and output. However, the `bufio` package still uses (internally) the `io.Reader` and `io.Writer` objects, which it wraps in order to create the `bufio.Reader` and `bufio.Writer` objects, respectively.

As you will see in the forthcoming sections, the `bufio` package is very popular for reading text files.

Reading text files

A text file is the most common kind of file that you can find on a Unix system. In this section, you will learn how to read text files in three ways: line by line, word by word, and character by character. As you will see, reading a text file line by line is the easiest method to access a text file, while reading a text file word by word is the most difficult method of all.

If you look closely at the `byLine.go`, `byWord.go`, and `byCharacter.go` programs, you will see many similarities in their Go code. First, all three utilities read the input file line by line. Second, all three utilities have the same `main()` function, with the exception of the function that is called in the `for` loop of the `main()` function. Last, the three functions that process the input text files are almost identical, except for the part that implements the actual functionality of the function.

Reading a text file line by line

Going line by line is the most common method of reading a text file. This is the main reason it is being shown first. The Go code of `byLine.go`, which will be presented in three parts, will help you understand the technique.

The first code segment from `byLine.go` is shown in the following Go code:

```
package main

import (
    "bufio"
    "flag"
    "fmt"
    "io"
    "os"
)
```

As you can see by the presence of the `bufio` package, we will use buffered input.

The second part of `byLine.go` contains the following Go code:

```
func lineByLine(file string) error {
    var err error

    f, err := os.Open(file)
    if err != nil {
        return err
```

```go
            }
            defer f.Close()

            r := bufio.NewReader(f)
            for {
                line, err := r.ReadString('\n')
                if err == io.EOF {
                    break
                } else if err != nil {
                    fmt.Printf("error reading file %s", err)
                    break
                }
                fmt.Print(line)
            }
            return nil
    }
```

All the magic happens in the `lineByLine()` function. After making sure that you can open the given filename for reading, you create a new reader using `bufio.NewReader()`. Then you use that reader with `bufio.ReadString()` in order to read the input file line by line. The trick is done by the parameter of `bufio.ReadString()`, which is a character that tells `bufio.ReadString()` to keep reading until that character is found. Constantly calling `bufio.ReadString()` when that parameter is the newline character results in reading the input file line by line! Note that the use of `fmt.Print()` for printing the read line instead of `fmt.Println()` shows that the newline character is included in each input line.

The third part of `byLine.go` follows next:

```go
    func main() {
        flag.Parse()
        if len(flag.Args()) == 0 {
            fmt.Printf("usage: byLine <file1> [<file2> ...]\n")
            return
        }

        for _, file := range flag.Args() {
            err := lineByLine(file)
            if err != nil {
                fmt.Println(err)
            }
        }
    }
```

Executing `byLine.go` and processing its output with `wc(1)` will generate the following type of output:

```
$ go run byLine.go /tmp/swtag.log /tmp/adobegc.log | wc
    4761   88521  568402
```

The following command will verify the accuracy of the preceding output:

```
$ wc /tmp/swtag.log /tmp/adobegc.log
    131     693    8440 /tmp/swtag.log
   4630   87828  559962 /tmp/adobegc.log
   4761   88521  568402 total
```

Reading a text file word by word

The technique presented in this subsection will be demonstrated by the `byWord.go` file, which is shown in four parts. As you will see in the Go code, separating the words of a line can be tricky. The first part of this utility is as follows:

```go
package main

import (
    "bufio"
    "flag"
    "fmt"
    "io"
    "os"
    "regexp"
)
```

The second code portion of `byWord.go` is shown in the following Go code:

```go
func wordByWord(file string) error {
    var err error
    f, err := os.Open(file)
    if err != nil {
        return err
    }
    defer f.Close()

    r := bufio.NewReader(f)
    for {
        line, err := r.ReadString('\n')
        if err == io.EOF {
            break
        } else if err != nil {
```

```
            fmt.Printf("error reading file %s", err)
            return err
    }
```

This part of the `wordByWord()` function is the same as the `lineByLine()` function of the `byLine.go` utility.

The third part of `byWord.go` is as follows:

```
        r := regexp.MustCompile("[^\\s]+")
        words := r.FindAllString(line, -1)
        for i := 0; i < len(words); i++ {
            fmt.Println(words[i])
        }
    }
    return nil
}
```

The remaining code of the `wordByWord()` function is totally new, and it uses regular expressions to separate the words found in each line of the input. The regular expression defined in the `regexp.MustCompile("[^\\s]+")` statement states that empty characters will separate one word from another.

The last code segment of `byWord.go` is as follows:

```
func main() {
    flag.Parse()
    if len(flag.Args()) == 0 {
        fmt.Printf("usage: byWord <file1> [<file2> ...]\n")
        return
    }

    for _, file := range flag.Args() {
        err := wordByWord(file)
        if err != nil {
            fmt.Println(err)
        }
    }
}
```

Executing `byWord.go` will produce the following type of output:

```
$ go run byWord.go /tmp/adobegc.log
01/08/18
20:25:09:669
|
[INFO]
```

You can verify the validity of byWord.go with the help of the wc(1) utility:

```
$ go run byWord.go /tmp/adobegc.log | wc
   91591   91591  559005
$ wc /tmp/adobegc.log
    4831   91591  583454 /tmp/adobegc.log
```

As you can see, the number of words calculated by wc(1) is the same as the number of lines and words that you took from the execution of byWord.go.

Reading a text file character by character

In this section, you will learn how to read a text file character by character, which is a pretty rare requirement unless you want to create a text editor. The relevant Go code will be saved as byCharacter.go, which will be presented in four parts.

The first part of byCharacter.go is shown in the following Go code:

```
package main

import (
    "bufio"
    "flag"
    "fmt"
    "io"
    "os"
)
```

As you can see, you will not need to use regular expressions for this task.

The second code segment from byCharacter.go is as follows:

```
func charByChar(file string) error {
    var err error
    f, err := os.Open(file)
    if err != nil {
        return err
    }
    defer f.Close()

    r := bufio.NewReader(f)
    for {
        line, err := r.ReadString('\n')
        if err == io.EOF {
            break
```

```
        } else if err != nil {
            fmt.Printf("error reading file %s", err)
            return err
        }
```

The third part of byCharacter.go is where the logic of the program is found:

```
        for _, x := range line {
            fmt.Println(string(x))
        }
    }
    return nil
}
```

Here you take each line that you read and split it using range. The range returns two values: You discard the first, which is the location of the current character in the line variable, and you use the second. However, that value is a not a character-that is the reason why you have to convert it into a character using the string() function.

Note that due to the fmt.Println(string(x)) statement, each character is printed in a distinct line, which means that the output of the program will be large. If you want a more compressed output, you should use the fmt.Print() function instead.

The last part of byCharacter.go contains the following Go code:

```
func main() {
    flag.Parse()
    if len(flag.Args()) == 0 {
        fmt.Printf("usage: byChar <file1> [<file2> ...]\n")
        return
    }

    for _, file := range flag.Args() {
        err := charByChar(file)
        if err != nil {
            fmt.Println(err)
        }
    }
}
```

The execution of byCharacter.go will generate the following type of output:

```
$ go run byCharacter.go /tmp/adobegc.log
0
1
/
0
```

```
8
/
1
8
```

Note that the Go code presented here can be used for counting the number of characters found in the input file, which can help you implement a Go version of the handy wc(1) command-line utility.

Reading from /dev/random

In this section, you will learn how to read from the /dev/random system device. The purpose of the /dev/random system device is to generate random data, which you might use for testing your programs or, in this case, you will plant the seed for a random number generator. Getting data from /dev/random can be a little bit tricky, and this is the main reason for specifically discussing it here.

On a macOS High Sierra machine, the /dev/random file has the following permissions:

```
$ ls -l /dev/random
crw-rw-rw-  1 root  wheel   14,   0 Jan  8 20:24 /dev/random
```

Similarly, on a Debian Linux machine, the /dev/random system device has the following Unix file permissions:

```
$ ls -l /dev/random
crw-rw-rw- 1 root root 1, 8 Jan 13 12:19 /dev/random
```

This means that the /dev/random file has analogous file permissions on both Unix variants. The only difference between these two Unix variants is the group that owns the file, which is wheel on macOS and root on Debian Linux, respectively.

The name of the program for this topic is devRandom.go, and it will be presented in three parts. The first part of the program is as follows:

```
package main

import (
    "encoding/binary"
    "fmt"
    "os"
)
```

Telling a Unix System What to Do

In order to read from /dev/random, you will need to import the encoding/binary standard Go package, because /dev/random returns binary data that needs to be decoded.

The second code portion of devRandom.go follows next:

```go
func main() {
    f, err := os.Open("/dev/random")
    defer f.Close()

    if err != nil {
        fmt.Println(err)
        return
    }
```

You open /dev/random as usual because everything in Unix is a file.

The last code segment of devRandom.go is shown in the following Go code:

```go
    var seed int64
    binary.Read(f, binary.LittleEndian, &seed)
    fmt.Println("Seed:", seed)
}
```

You need the binary.Read() function, which requires three parameters, in order to read from the /dev/random system device. The value of the second parameter (binary.LittleEndian) specifies that you want to use the **little endian** byte order. The other option is binary.BigEndian, which is used when your computer is using the **big endian** byte order.

Executing devRandom.go will generate the following type of output:

```
$ go run devRandom.go
Seed: -2044736418491485077
$ go run devRandom.go
Seed: -5174854372517490328
$ go run devRandom.go
Seed: 7702177874251412774
```

Reading the amount of data you want from a file

In this section, you will learn how to read exactly the amount of data you want. This technique is particularly useful when reading binary files, where you have to decode the data you read in a particular way. Nevertheless, this technique still works with text files.

The logic behind this technique is not that difficult: You create a **byte slice** with the size you need and use that byte slice for reading. To make this more interesting, this functionality is going to be implemented as a function with two parameters. One parameter will be used for specifying the amount of data that you want you read, and the other parameter, which will have the `*os.File` type, will be used for accessing the desired file. The return value of that function will be the data you have read.

The name of the Go program for this topic will be `readSize.go`, and it will be presented in four parts. The utility will accept a single parameter, which will be the size of the byte slice.

This particular program, when used with the present technique, can help you copy any file using the buffer size you want.

The first part of `readSize.go` has the expected preamble:

```
package main

import (
    "fmt"
    "io"
    "os"
    "strconv"
)
```

The second part of `readSize.go` contains the following Go code:

```
func readSize(f *os.File, size int) []byte {
    buffer := make([]byte, size)

    n, err := f.Read(buffer)
    if err == io.EOF {
        return nil
    }

    if err != nil {
```

[299]

```
            fmt.Println(err)
            return nil
    }

    return buffer[0:n]
}
```

This is the function discussed earlier. Although the code is straightforward, there is one point that needs an explanation. The `io.Reader.Read()` method returns two parameters: the number of bytes read as well as an `error` variable. The `readSize()` function uses the former return value of `io.Read()` in order to return a byte slice of that size. Although this is a tiny detail, and it is only significant when you reach the end of the file, it ensures that the output of the utility will be same as the input and that it will not contain any extra characters. Finally, there is code that checks for `io.EOF`, which is an error that signifies that you have reached the end of a file. When that kind of error occurs, the function returns.

The third code portion of this utility is as follows:

```
func main() {
    arguments := os.Args
    if len(arguments) != 3 {
        fmt.Println("<buffer size> <filename>")
        return
    }

    bufferSize, err := strconv.Atoi(os.Args[1])
    if err != nil {
        fmt.Println(err)
        return
    }

    file := os.Args[2]
    f, err := os.Open(file)
    if err != nil {
        fmt.Println(err)
        return
    }
    defer f.Close()
```

The last code segment of readSize.go is as follows:

```
for {
    readData := readSize(f, bufferSize)
    if readData != nil {
        fmt.Print(string(readData))
    } else {
        break
    }
}
}
```

So, here you keep reading your input file until readSize() returns an error or nil.

Executing readSize.go by telling it to process a binary file and handling its output with wc(1) will validate the correctness of the program:

```
$ go run readSize.go 1000 /bin/ls | wc
      80    1032   38688
$ wc /bin/ls
      80    1032   38688 /bin/ls
```

Why are we using binary format?

In the previous section, the readSize.go utility illustrated how you can read a file byte by byte, which is a technique that best applies to binary files. So, you might ask why read data in binary format when text formats are so much easier to understand? The main reason is space reduction. Imagine that you want to store the number 20 as a string to a file. It is easy to understand that you will need two bytes for storing 20 using ASCII characters, one for storing 2 and another for storing 0. Storing 20 in binary format requires just one byte, since 20 can be represented as 00010100 in binary or as 0x14 in hexadecimal.

This difference might look insignificant when you are dealing with small amounts of data, but it could be pretty substantial when dealing with data found in applications such as database servers.

Reading CSV files

CSV files are plain text files with a format. In this section, you will learn how to read a text file that contains points of a plane, which means that each line will contain a pair of coordinates. Additionally, you are also going to use an external Go library named Glot, which will help you create a plot of the points that you read from the CSV file. Note that **Glot** uses **Gnuplot**, which means that you will need to install Gnuplot on your Unix machine in order to use Glot.

The name of the source file for this topic is `CSVplot.go`, and it is going to be presented in five parts. The first code segment is as follows:

```
package main

import (
    "encoding/csv"
    "fmt"
    "github.com/Arafatk/glot"
    "os"
    "strconv"
)
```

The second part of `CSVplot.go` is shown in the following Go code:

```
func main() {
    if len(os.Args) != 2 {
        fmt.Println("Need a data file!")
        return
    }

    file := os.Args[1]
    _, err := os.Stat(file)
    if err != nil {
        fmt.Println("Cannot stat", file)
        return
    }
```

In this part, you see a technique for checking whether a file already exists or not using the powerful `os.Stat()` function.

The third part of CSVplot.go is as follows:

```go
f, err := os.Open(file)
if err != nil {
    fmt.Println("Cannot open", file)
    fmt.Println(err)
    return
}
defer f.Close()

reader := csv.NewReader(f)
reader.FieldsPerRecord = -1
allRecords, err := reader.ReadAll()
if err != nil {
    fmt.Println(err)
    return
}
```

The fourth code segment of CSVplot.go is shown in the following Go code:

```go
xP := []float64{}
yP := []float64{}
for _, rec := range allRecords {
    x, _ := strconv.ParseFloat(rec[0], 64)
    y, _ := strconv.ParseFloat(rec[1], 64)
    xP = append(xP, x)
    yP = append(yP, y)
}

points := [][]float64{}
points = append(points, xP)
points = append(points, yP)
fmt.Println(points)
```

Here you convert the string values you read into numbers and put them into a slice with two dimensions named points.

The last part of CSVplot.go contains the following Go code:

```go
dimensions := 2
persist := true
debug := false
plot, _ := glot.NewPlot(dimensions, persist, debug)

plot.SetTitle("Using Glot with CSV data")
plot.SetXLabel("X-Axis")
plot.SetYLabel("Y-Axis")
```

```
        style := "circle"
        plot.AddPointGroup("Circle:", style, points)
        plot.SavePlot("output.png")
}
```

In the preceding Go code, you saw how you can create a PNG file with the help of the Glot library and its `glot.SavePlot()` function.

As you can guess, you will need to download the Glot Go library before being able to compile and execute the `CSVplot.go` source code, which requires the execution of the following command from your favorite Unix shell:

```
$ go get github.com/Arafatk/glot
```

You can see the results of the preceding command by looking inside ~/go directory:

```
$ ls -l ~/go/pkg/darwin_amd64/github.com/Arafatk/
total 240
-rw-r--r--  1 mtsouk  staff   119750 Jan 22 22:12 glot.a
$ ls -l ~/go/src/github.com/Arafatk/glot/
total 120
-rw-r--r--  1 mtsouk  staff     1818 Jan 22 22:12 README.md
-rw-r--r--  1 mtsouk  staff     6092 Jan 22 22:12 common.go
-rw-r--r--  1 mtsouk  staff      552 Jan 22 22:12 common_test.go
-rw-r--r--  1 mtsouk  staff     3162 Jan 22 22:12 core.go
-rw-r--r--  1 mtsouk  staff      138 Jan 22 22:12 core_test.go
-rw-r--r--  1 mtsouk  staff     3049 Jan 22 22:12 function.go
-rw-r--r--  1 mtsouk  staff      511 Jan 22 22:12 function_test.go
-rw-r--r--  1 mtsouk  staff     4955 Jan 22 22:12 glot.go
-rw-r--r--  1 mtsouk  staff      220 Jan 22 22:12 glot_test.go
-rw-r--r--  1 mtsouk  staff    10536 Jan 22 22:12 pointgroup.go
-rw-r--r--  1 mtsouk  staff      378 Jan 22 22:12 pointgroup_test.go
```

The CSV data file containing the points that will be plotted has the following format:

```
$ cat /tmp/dataFile
1,2
2,3
3,3
4,4
5,8
6,5
-1,12
-2,10
-3,10
-4,10
```

Executing `CSVplot.go` will generate the following kind of output:

```
$ go run CSVplot.go /tmp/doesNoExist
Cannot stat /tmp/doesNoExist
$ go run CSVplot.go /tmp/dataFile
[[1 2 3 4 5 6 -1 -2 -3 -4] [2 3 3 4 8 5 12 10 10 10]]
```

You can see the results of `CSVplot.go` in a much better format in the following figure:

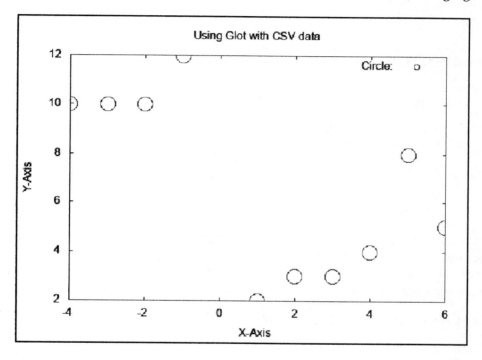

The type of graphical output you can get from Glot.

Writing to a file

Generally speaking, you can use the functionality of the `io.Writer` interface for writing data to files on a disk. Nevertheless, the Go code of `save.go` will show you five ways to write data to a file. The `save.go` program will be presented in six parts.

The first part of `save.go` is as follows:

```
package main

import (
    "bufio"
    "fmt"
    "io"
    "io/ioutil"
    "os"
)
```

The second code portion of `save.go` is shown in the following Go code:

```
func main() {
    s := []byte("Data to write\n")

    f1, err := os.Create("f1.txt")
    if err != nil {
        fmt.Println("Cannot create file", err)
        return
    }
    defer f1.Close()
    fmt.Fprintf(f1, string(s))
```

Notice that the `s` byte slice will be used in every line that involves writing presented in this Go program. Additionally, the `fmt.Fprintf()` function used here can help you write data to your own log files using the format you want. In this case, `fmt.Fprintf()` writes your data to the file identified by `f1`.

The third part of `save.go` contains the following Go code:

```
    f2, err := os.Create("f2.txt")
    if err != nil {
        fmt.Println("Cannot create file", err)
        return
    }
    defer f2.Close()
    n, err := f2.WriteString(string(s))
    fmt.Printf("wrote %d bytes\n", n)
```

In this case, `f2.WriteString()` is used for writing your data to a file.

The fourth code segment of `save.go` follows next:

```
f3, err := os.Create("f3.txt")
if err != nil {
    fmt.Println(err)
    return
}
w := bufio.NewWriter(f3)
n, err = w.WriteString(string(s))
fmt.Printf("wrote %d bytesn", \n)
w.Flush()
```

In this case, `bufio.NewWriter()` opens a file for writing and `bufio.WriteString()` writes the data.

The fifth part of `save.go` will teach you another method for writing to a file:

```
f4 := "f4.txt"
err = ioutil.WriteFile(f4, s, 0644)
if err != nil {
    fmt.Println(err)
    return
}
```

This method needs just a single function call named `ioutil.WriteFile()` for writing your data, and it does not require the use of `os.Create()`.

The last code segment of `save.go` is as follows:

```
f5, err := os.Create("f5.txt")
if err != nil {
    fmt.Println(err)
    return
}
n, err = io.WriteString(f5, string(s))
if err != nil {
    fmt.Println(err)
    return
}
fmt.Printf("wrote %d bytes\n", n)
}
```

The last technique uses `io.WriteString()` to write the desired data to a file.

Executing `save.go` will create the following type of output:

```
$ go run save.go
wrote 14 bytes
wrote 14 bytes
wrote 14 bytes
$ ls -l f?.txt
-rw-r--r--  1 mtsouk  staff  14 Jan 23 20:30 f1.txt
-rw-r--r--  1 mtsouk  staff  14 Jan 23 20:30 f2.txt
-rw-r--r--  1 mtsouk  staff  14 Jan 23 20:30 f3.txt
-rw-r--r--  1 mtsouk  staff  14 Jan 23 20:30 f4.txt
-rw-r--r--  1 mtsouk  staff  14 Jan 23 20:30 f5.txt
$ cat f?.txt
Data to write
Data to write
Data to write
Data to write
Data to write
```

The next section will show you how to save data to a file with the help of a specialized function of a package that is in the standard Go library.

Loading and saving data on disk

Do you remember the `keyValue.go` application from Chapter 4, *The Uses of Composite Types*? Well, it was far from complete, so in this section you will learn how to save the data of a **key-value store** on disk and how to load it back into memory when you next start your application.

We are going to create two new functions, one named `save()` for saving data to disk and another named `load()` for retrieving the data from disk. Thus, we will only present the code differences between `keyValue.go` and `kvSaveLoad.go` using the `diff(1)` Unix command-line utility.

> The `diff(1)` Unix command-line utility can be very convenient when you want to spot the differences between two text files. You can learn more about it by executing `man 1 diff` at the command-line of your favorite Unix shell.

If you think about the task to be implemented here, you will recognize that what you need is an easy way to save the contents of a Go map to disk, as well as a way to load the data from a file and put it into a Go map.

The process of converting your data into a stream of bytes is called **serialization**. The process of reading a data file and converting it into an object is called **deserialization**. The encoding/gob standard Go package will be used for the program kvSaveLoad.go. It will help the serialization and deserialization process. The encoding/gob package uses the **gob format** for storing its data. The official name for such formats is **stream encoding**. The good thing about the gob format is that Go does all of the dirty work, so that you don't have to worry about the encoding and decoding stages.

Other Go packages that can help you serialize and deserialize data are encoding/xml, which uses the **XML format**, and encoding/json, which uses the **JSON format**.

The following output will reveal the code changes between kvSaveLoad.go and keyValue.go without including the implementations of save() and load() functions that will be fully presented here:

```
$ diff keyValue.go kvSaveLoad.go
4a5
>       "encoding/gob"
16a18,55
> var DATAFILE = "/tmp/dataFile.gob"
> func save() error {
>
>       return nil
> }
>
> func load() error {
>
> }
59a99,104
>
>       err := load()
>       if err != nil {
>               fmt.Println(err)
>       }
>
88a134,137
>                       err = save()
>                       if err != nil {
>                               fmt.Println(err)
>                       }
```

An important part of the `diff(1)` output is the definition of the `DATAFILE` global variable that holds the path to the file that is used by the key-value store. Apart from this, you see where the `load()` function is called as well as the point where the `save()` function is called. The `load()` function is used first in the `main()` function, while the `save()` function is executed when the user issues the `STOP` command.

Each time you execute `kvSaveLoad.go`, the program checks whether there is data to be read by trying to read the default data file. If there is no data file to read, you will start with an empty key-value store. When the program is about to terminate, it writes all of its data on the disk using the `save()` function.

The `save()` function has the following implementation:

```
func save() error {
    fmt.Println("Saving", DATAFILE)
    err := os.Remove(DATAFILE)
    if err != nil {
        fmt.Println(err)
    }

    saveTo, err := os.Create(DATAFILE)
    if err != nil {
        fmt.Println("Cannot create", DATAFILE)
        return err
    }
    defer saveTo.Close()

    encoder := gob.NewEncoder(saveTo)
    err = encoder.Encode(DATA)
    if err != nil {
        fmt.Println("Cannot save to", DATAFILE)
        return err
    }
    return nil
}
```

Note that the first thing the `save()` function does is to delete the existing data file using the `os.Remove()` function in order to create it later on.

One of the most critical tasks the `save()` function does is to make sure that you can actually create and write to the desired file. Although there are many ways to do this, the `save()` function uses the simplest way, which is by checking the `error` value returned by the `os.Create()` function. If that value is not `nil`, then there is a problem and the `save()` function finishes without saving any data.

The `load()` function is implemented as follows:

```go
func load() error {
    fmt.Println("Loading", DATAFILE)
    loadFrom, err := os.Open(DATAFILE)
    defer loadFrom.Close()
    if err != nil {
        fmt.Println("Empty key/value store!")
        return err
    }

    decoder := gob.NewDecoder(loadFrom)
    decoder.Decode(&DATA)
    return nil
}
```

One of the tasks of the `load()` function is to make sure that the file which you are trying to read is actually there and that you can read it without any problems. Once again, the `load()` function uses the simplest approach, which is to look at the return value of the `os.Open()` function. If the `error` value returned is equal to `nil`, then everything is fine.

It is also important to close the file after reading the data from it, as it will be overwritten by the `save()` function later on. The release of the file is accomplished by the `defer loadFrom.Close()` statement.

Executing `kvSaveLoad.go` will generate the following type of output:

```
$ go run kvSaveLoad.go
Loading /tmp/dataFile.gob
Empty key/value store!
open /tmp/dataFile.gob: no such file or directory
ADD 1 2 3
ADD 4 5 6
STOP
Saving /tmp/dataFile.gob
remove /tmp/dataFile.gob: no such file or directory
$ go run kvSaveLoad.go
Loading /tmp/dataFile.gob
PRINT
key: 1 value: {2 3 }
```

```
key: 4 value: {5 6 }
DELETE 1
PRINT
key: 4 value: {5 6 }
STOP
Saving /tmp/dataFile.gob
rMacBook:code mtsouk$ go run kvSaveLoad.go
Loading /tmp/dataFile.gob
PRINT
key: 4 value: {5 6 }
STOP
Saving /tmp/dataFile.gob
$ ls -l /tmp/dataFile.gob
-rw-r--r--  1 mtsouk  wheel  80 Jan 22 11:22 /tmp/dataFile.gob
$ file /tmp/dataFile.gob
/tmp/dataFile.gob: data
```

In Chapter 13, *Network Programming - Building Servers and Clients,* you are going to see the final version of the key-value store, which will be able to operate over a TCP/IP connection and will serve multiple network clients using **goroutines**.

The strings package revisited

We first talked about the handy strings package back in Chapter 4, *The Uses of Composite Types*. This section will address the functions of the strings package that are related to file input and output.

The first part of str.go is shown in the following Go code:

```
package main

import (
    "fmt"
    "io"
    "os"
    "strings"
)
```

The second code segment of str.go is as follows:

```go
func main() {
    r := strings.NewReader("test")
    fmt.Println("r length:", r.Len())
```

The strings.NewReader() function creates a read-only Reader from a string. The strings.Reader object implements the io.Reader, io.ReaderAt, io.Seeker, io.WriterTo, io.ByteScanner, and io.RuneScanner interfaces.

The third part of str.go follows next:

```go
    b := make([]byte, 1)
    for {
        n, err := r.Read(b)
        if err == io.EOF {
            break
        }

        if err != nil {
            fmt.Println(err)
            continue
        }
        fmt.Printf("Read %s Bytes: %d\n", b, n)
    }
```

Here you can see how you to use strings.Reader as an io.Reader interface in order to read a string byte by byte using the Read() function.

The last code portion of str.go contains the following Go code:

```go
    s := strings.NewReader("This is an error!\n")
    fmt.Println("r length:", s.Len())
    n, err := s.WriteTo(os.Stderr)

    if err != nil {
        fmt.Println(err)
        return
    }
    fmt.Printf("Wrote %d bytes to os.Stderr\n", n)
}
```

In this code segment, you can see how you can write to standard error with the help of the `strings` package.

Executing `str.go` will generate the following output:

```
$ go run str.go
r length: 4
Read t Bytes: 1
Read e Bytes: 1
Read s Bytes: 1
Read t Bytes: 1
r length: 18
This is an error!
Wrote 18 bytes to os.Stderr
$ go run str.go 2>/dev/null
r length: 4
Read t Bytes: 1
Read e Bytes: 1
Read s Bytes: 1
Read t Bytes: 1
r length: 18
Wrote 18 bytes to os.Stderr
```

About the bytes package

The `bytes` standard Go package contains functions for working with **byte slices** in the same way that the `strings` standard Go package helps you work with **strings**. The name of the Go source code file is `bytes.go`, and it will be presented in three code portions.

 This section is not included in my *Go Systems Programming, Packt Publishing, 2017* book.

The first part of `bytes.go` follows next:

```
package main

import (
    "bytes"
    "fmt"
    "io"
    "os"
)
```

The second code portion of bytes.go contains the following Go code:

```go
func main() {
    var buffer bytes.Buffer
    buffer.Write([]byte("This is"))
    fmt.Fprintf(&buffer, " a string!\n")
    buffer.WriteTo(os.Stdout)
    buffer.WriteTo(os.Stdout)
```

First, you create a new bytes.Buffer variable, and you put data into it using buffer.Write() and fmt.Fprintf(). Then you call buffer.WriteTo() two times.

The first buffer.WriteTo() call will print the contents of the buffer variable. However, the second call to buffer.WriteTo() has nothing to print because the buffer variable will be empty after the first buffer.WriteTo() call.

The last part of bytes.go is as follows:

```go
    buffer.Reset()
    buffer.Write([]byte("Mastering Go!"))
    r := bytes.NewReader([]byte(buffer.String()))
    fmt.Println(buffer.String())
    for {
        b := make([]byte, 3)
        n, err := r.Read(b)
        if err == io.EOF {
            break
        }

        if err != nil {
            fmt.Println(err)
            continue
        }
        fmt.Printf("Read %s Bytes: %d\n", b, n)
    }
}
```

The Reset() method resets the buffer variable, and the Write() method puts some data into it again. Then you create a new reader using bytes.NewReader(), and after that you use the Read() method of the io.Reader interface to read the data found in the buffer variable.

Executing `bytes.go` will create the following type of output:

```
$ go run bytes.go
This is a string!
Mastering Go!
Read Mas Bytes: 3
Read ter Bytes: 3
Read ing Bytes: 3
Read  Go Bytes: 3
Read ! Bytes: 1
```

File permissions

A popular topic in Unix systems programming is Unix file permissions. In this section, you will learn how to print the permissions of any file, provided that you have adequate Unix permission to do so! The name of the program is `permissions.go`, and it will be presented in three parts.

The first part of `permissions.go` contains the following Go code:

```go
package main

import (
    "fmt"
    "os"
)
```

The second code segment of `permissions.go` is shown in the following Go code:

```go
func main() {
    arguments := os.Args
    if len(arguments) == 1 {
        fmt.Printf("usage: permissions filename\n")
        return
    }
```

The last part of this utility follows next:

```go
    filename := arguments[1]
    info, _ := os.Stat(filename)
    mode := info.Mode()
    fmt.Println(filename, "mode is", mode.String()[1:10])
}
```

The call to `os.Stat(filename)` returns a big structure with lots of data. As we are only interested in the permissions of the file, we will call the `Mode()` method and print its output. Actually, we are printing a part of the output denoted by `mode.String()[1:10]` because this is where the data that interests us is found.

Executing `permissions.go` will create the following type of output:

```
$ go run permissions.go /tmp/adobegc.log
/tmp/adobegc.log mode is rw-rw-rw-
$ go run permissions.go /dev/random
/dev/random mode is crw-rw-rw
```

The output of the `ls(1)` utility verifies the correctness of `permissions.go`:

```
$ ls -l /dev/random /tmp/adobegc.log
crw-rw-rw-  1 root  wheel    14,    0 Jan  8 20:24 /dev/random
-rw-rw-rw-  1 root  wheel       583454 Jan 16 19:12 /tmp/adobegc.log
```

Handling Unix signals

Go provides the `os/signal` package to programmers to help them work with signals. This section will show you how to use it for Unix signal handling.

First let's present some useful information about Unix signals. Have you ever pressed *Ctrl* + *C* in order to stop a running program? If your answer is *Yes*, then you are already familiar with signals because *Ctrl* + *C* sends the **SIGINT signal** to a program. Strictly speaking, Unix **Signals** are software interrupts that can be accessed either by name or by number, and they offer a way to handle asynchronous events on a Unix system. There are two ways to send a signal: by name or by number. Generally speaking, it is safer to send a signal by name because you are is less likely to send the wrong signal accidentally.

A program cannot handle all of the available signals: some signals cannot be caught, nor can they be ignored. The `SIGKILL` and `SIGSTOP` signals cannot be caught, blocked, or ignored. The reason for this is that they provide the kernel and the root user a way of stopping any process they want. The `SIGKILL` signal, which is also known by the number 9, is usually called in extreme conditions where you need to act fast. Thus, it is the only signal that is usually called by number, simply because it is quicker to do so.

signal.SIGINFO is not available on Linux machines, which means that if you find it in a Go program that you want to run on a Linux machine, you need to replace it with another signal, or your Go program will not be able to compile and execute.

The most common way to send a signal to a process is by using the kill(1) utility. By default, kill(1) sends the SIGTERM signal. If you want to find all of the supported signals on your Unix machine, you should execute the kill -l command.

If you try to send a signal to a process without having the required permissions, kill(1) will not do the job and you will get an error message similar to the following:

```
$ kill 1210
-bash: kill: (1210) - Operation not permitted
```

Handling two signals

In this subsection, you will learn how to handle two signals in a Go program using the code found in handleTwo.go, which will be presented in four parts. The signals that will be handled by handleTwo.go are SIGINFO and SIGINT, which in Golang are named syscall.SIGINFO and os.Interrupt, respectively.

If you look at the documentation of the os package, you will find that the only two signals that are guaranteed to be present on all systems are syscall.SIGKILL and syscall.SIGINT, which in Go are also defined as os.Kill and os.Interrupt, respectively.

The first part of handleTwo.go contains the following Go code:

```
package main

import (
    "fmt"
    "os"
    "os/signal"
    "syscall"
    "time"
)
```

The second part of the `handleTwo.go` program follows next:

```
func handleSignal(signal os.Signal) {
    fmt.Println("handleSignal() Caught:", signal)
}
```

The `handleSignal()` function will be used for handling the `syscall.SIGINFO` signal, while the `os.Interrupt` signal will be handled inline.

The third code segment of `handleTwo.go` is shown in the following Go code:

```
func main() {
    sigs := make(chan os.Signal, 1)
    signal.Notify(sigs, os.Interrupt, syscall.SIGINFO)
    go func() {
        for {
            sig := <-sigs
            switch sig {
            case os.Interrupt:
                fmt.Println("Caught:", sig)
            case syscall.SIGINFO:
                handleSignal(sig)
                return
            }
        }
    }()
```

This technique works as follows: first you define a **channel** named `sigs` that helps you pass data around. Then you call `signal.Notify()` in order to state the signals that interest you. Next, you implement an **anonymous function** that runs as a goroutine in order to act when you receive any one of the signals you care about. You will have to wait for Chapter 9, *Go Concurrency - Goroutines, Channels, and Pipelines,* to learn more about goroutines and channels.

The last portion of the `handleTwo.go` program is as follows:

```
    for {
        fmt.Printf(".")
        time.Sleep(20 * time.Second)
    }
}
```

The `time.Sleep()` call is used to prohibit the program from terminating, as it has no real work to do. In an actual application, there would be no need to use similar code.

As we need the process ID of a program in order to send signals to it using the `kill(1)` utility, we will first compile `handleTwo.go` and run the executable file instead of using `go run handleTwo.go`. Working with `handleTwo` will generate the following type of output:

```
$ go build handleTwo.go
$ ls -l handleTwo
-rwxr-xr-x  1 mtsouk  staff  2005200 Jan 18 07:49 handleTwo
$ ./handleTwo
.^CCaught: interrupt
.Caught: interrupt
handleSignal() Caught: information request
.Killed: 9
```

Note that you will need an additional terminal in order to interact with `handleTwo.go` and obtain the preceding output. You will execute the following commands in that Terminal:

```
$ ps ax | grep ./handleTwo | grep -v grep
47988 s003  S+     0:00.00 ./handleTwo
$ kill -s INT 47988
$ kill -s INFO 47988
$ kill -s USR1 47988
$ kill -9 47988
```

The first command is used for finding the process ID of the `handleTwo` executable, while the remaining commands are used for sending the desired signals to that process. The `SIGUSR1` signal is ignored, and it does not appear in the output.

The problem with `handleTwo.go` is that if it gets a signal which is not programmed to handle, it will ignore it. Thus, in the next section you will see a technique that uses a relatively different approach in order to handle signals in a more efficient way.

Handling all signals

In this subsection, you will learn how to handle all signals but respond only to the ones that really interest you. This is a much better and safer technique than the one presented in the previous subsection. The technique will be illustrated using the Go code of `handleAll.go`, which will be presented in four parts.

The first part of `handleAll.go` contains the following Go code:

```
package main

import (
    "fmt"
    "os"
    "os/signal"
    "syscall"
    "time"
)

func handle(signal os.Signal) {
    fmt.Println("Received:", signal)
}
```

The second code segment from `handleAll.go` is shown in the following Go code:

```
func main() {
    sigs := make(chan os.Signal, 1)
    signal.Notify(sigs)
```

So, all the magic happens due to the `signal.Notify(sigs)` statement. As no signals are specified, all incoming signals will be handled!

You are allowed to call `signal.Notify()` multiple times in the same program using different channels and the same signals. In that case, each relevant channel will receive a copy of the signals which it was programmed to handle!

The third code portion of the `handleAll.go` utility follows next:

```
    go func() {
        for {
            sig := <-sigs
            switch sig {
            case os.Interrupt:
                handle(sig)
            case syscall.SIGTERM:
                handle(sig)
                os.Exit(0)
            case syscall.SIGUSR2:
                fmt.Println("Handling syscall.SIGUSR2!")
            default:
```

Telling a Unix System What to Do

```
            fmt.Println("Ignoring:", sig)
        }
    }
}()
```

It is very convenient to use one of the signals in order to exit your program. This gives you the opportunity to do some housekeeping in your program when needed. In this case, the `syscall.SIGTERM` signal is used for that purpose. This does not prevent you from using `SIGKILL` to kill a program, though!

The remaining Go code for `handleAll.go` follows:

```
for {
    fmt.Printf(".")
    time.Sleep(30 * time.Second)
}
}
```

You still need to call `time.Sleep()` to prevent your program from terminating immediately.

Once again, it would be better to build an executable file for `handleAll.go` using the `go build` tool. Executing `handleAll` and interacting with it from another terminal will generate the following type of output:

```
$ go build handleAll.go
$ ls -l handleAll
-rwxr-xr-x  1 mtsouk  staff  2005216 Jan 18 08:25 handleAll
$ ./handleAll
.Ignoring: hangup
Handling syscall.SIGUSR2!
Ignoring: user defined signal 1
Received: interrupt
^CReceived: interrupt
Received: terminated
```

The commands issued from the second terminal are as follows:

```
$ ps ax | grep ./handleAll | grep -v grep
49902 s003   S+      0:00.00 ./handleAll
$ kill -s HUP 49902
$ kill -s USR2 49902
$ kill -s USR1 49902
$ kill -s INT 49902
$ kill -s TERM 49902
```

Programming Unix pipes in Go

According to the Unix philosophy, Unix command-line utilities should do one job and perform that job well! In practice, this means that instead of developing huge utilities that do lots of jobs, you should develop multiple smaller programs, which when combined, should perform the desired job. The most common way for two or more Unix command-line utilities to communicate is by using pipes. In a **Unix pipe**, the output of a command-line utility becomes the input of another command-line utility. This process may involve more than two programs! The symbol that is used for Unix pipes is the | character.

Pipes have two serious limitations: first, they usually communicate in one direction, and second, they can only be used between processes that have a common ancestor. The general idea behind the implementation of Unix pipes is that if you do not have a file to process, you should wait to get your input from standard input. Similarly, if you are not told to save your output to a file, you should write your output to standard output, either for the user to see it or for another program to process it.

In Chapter 1, *Go and the Operating System*, you learned how to read from standard input and how to write to standard output. If you have doubts about these two operations, it would be a good time to review the Go code of stdOUT.go and stdIN.go.

Implementing the cat(1) utility in Go

In this section, you will see a Go version of the cat(1) utility. You will most likely be surprised by the length of the program!

The source code of cat.go will be presented in three parts. The first part of cat.go follows next:

```
package main

import (
    "bufio"
    "fmt"
    "io"
    "os"
)
```

The second code segment of cat.go contains the following Go code:

```go
func printFile(filename string) error {
    f, err := os.Open(filename)
    if err != nil {
        return err
    }
    defer f.Close()
    scanner := bufio.NewScanner(f)
    for scanner.Scan() {
        io.WriteString(os.Stdout, scanner.Text())
        io.WriteString(os.Stdout, "\n")
    }
    return nil
}
```

In this part, you can see the implementation of a function whose purpose is to print the contents of a file in the standard output.

The last part of cat.go is as follows:

```go
func main() {
    filename := ""
    arguments := os.Args
    if len(arguments) == 1 {
        io.Copy(os.Stdout, os.Stdin)
        return
    }

    for i := 1; i < len(arguments); i++ {
        filename = arguments[i]
        err := printFile(filename)
        if err != nil {
            fmt.Println(err)
        }
    }
}
```

The preceding code contains all of the magic of cat.go, because this is where you define how the program will behave. First of all, if you execute cat.go without any command-line arguments, then the program will just copy standard input to standard output as defined by the io.Copy(os.Stdout, os.Stdin) statement. However, if there are command-line arguments, then the program will process them all in the same order that they were given.

Executing `cat.go` will create the following type of output:

```
$ go run cat.go
Mastering Go!
Mastering Go!
1 2 3 4
1 2 3 4
```

However, things get really interesting if you execute `cat.go` using Unix pipes:

```
$ go run cat.go /tmp/1.log /tmp/2.log | wc
    2367   44464  279292
$ go run cat.go /tmp/1.log /tmp/2.log | go run cat.go | wc
    2367   44464  279292
```

`cat.go` is also able to print multiple files on your screen:

```
$ go run cat.go 1.txt 1 1.txt
    2367   44464  279292
    2367   44464  279292
open 1: no such file or directory
    2367   44464  279292
    2367   44464  279292
```

Please note that if you try to execute `cat.go` as `go run cat.go cat.go` and expect that you will get the contents of `cat.go` on your screen, the process will fail and you will get the following error message instead:

```
package main: case-insensitive file name collision: "cat.go" and "cat.go"
```

The reason for this is that Go does not understand that the second `cat.go` should be used as a command-line argument to the `go run cat.go` command. Instead `go run` tries to compile `cat.go` twice, which causes the error message. The solution to this problem is to execute `go build cat.go` first, and then use `cat.go` or any other Go source file as the argument to the generated executable file.

Traversing directory trees

This section will present a relatively simple version of the `find(1)` command-line utility written in Go.

Telling a Unix System What to Do

The name of the program is goFind.go, and it will be presented in four parts. The code of goFind.go will support two command-line options. The -d option prints a star character in front of paths that are directories, while the -f parameter prints a plus character in front of paths that are regular files.

The first code portion of goFind.go is shown in the following Go code:

```go
package main

import (
    "flag"
    "fmt"
    "os"
    "path/filepath"
)
```

As you can see, the goFind.go program uses the flag package for efficiently dealing with its command-line parameters.

The second part of goFind.go contains the following code:

```go
var minusD bool = false
var minusF bool = false

func walk(path string, info os.FileInfo, err error) error {
    fileInfo, err := os.Stat(path)
    if err != nil {
        return err
    }

    mode := fileInfo.Mode()
    if mode.IsRegular() && minusF {
        fmt.Println("+", path)
        return nil
    }

    if mode.IsDir() && minusD {
        fmt.Println("*", path)
        return nil
    }

    fmt.Println(path)
    return nil
}
```

As both `minusD` and `minusF` should be accessible from everywhere in the program, including the `walk()` function, I have decided to make both of them global variables. The `IsDir()` function helps you identify directories and `IsRegular()` helps you identify regular files.

The third code segment of `goFind.go` follows next:

```go
func main() {
    starD := flag.Bool("d", false, "Signify directories")
    plusF := flag.Bool("f", false, "Signify regular files")
    flag.Parse()
    flags := flag.Args()

    Path := "."
    if len(flags) == 1 {
        Path = flags[0]
    }
}
```

The last part of `goFind.go` is shown in the following Go code:

```go
    minusD = *starD
    minusF = *plusF

    err := filepath.Walk(Path, walk)
    if err != nil {
        fmt.Println(err)
        os.Exit(1)
    }
}
```

Executing `goFind.go` will create the following type of output:

```
$ go run goFind.go -d -f /tmp/
* /tmp/
+ /tmp/.keystone_install_lock
* /tmp/5580C65A-E7E2-4B27-AD91-506F85545E1D
* /tmp/569A57CB-8FD3-4879-A6A3-B86116CB0116
+ /tmp/ExmanProcessMutex
+ /tmp/adobegc.log
/tmp/com.adobe.AdobeIPCBroker.ctrl-mtsouk
* /tmp/com.apple.launchd.h3Izgq45dz
/tmp/com.apple.launchd.h3Izgq45dz/Listeners
* /tmp/lilo.46843
+ /tmp/swtag.log
/tmp/textmate-501.sock
$ go run goFind.go -f /tmp/
/tmp/
```

Telling a Unix System What to Do

```
+ /tmp/.keystone_install_lock
/tmp/5580C65A-E7E2-4B27-AD91-506F85545E1D
/tmp/569A57CB-8FD3-4879-A6A3-B86116CB0116
+ /tmp/ExmanProcessMutex
+ /tmp/adobegc.log
/tmp/com.adobe.AdobeIPCBroker.ctrl-mtsouk
/tmp/com.apple.launchd.h3Izgq45dz
/tmp/com.apple.launchd.h3Izgq45dz/Listeners
/tmp/lilo.46843
+ /tmp/swtag.log
/tmp/textmate-501.sock
$ go run goFind.go /tmp/
/tmp/
/tmp/.keystone_install_lock
/tmp/5580C65A-E7E2-4B27-AD91-506F85545E1D
/tmp/569A57CB-8FD3-4879-A6A3-B86116CB0116
/tmp/ExmanProcessMutex
/tmp/adobegc.log
/tmp/com.adobe.AdobeIPCBroker.ctrl-mtsouk
/tmp/com.apple.launchd.h3Izgq45dz
/tmp/com.apple.launchd.h3Izgq45dz/Listeners
/tmp/lilo.46843
/tmp/swtag.log
/tmp/textmate-501.sock
```

The output of `goFind.go` contains every kind of file that can be found under the specified root directory, including sockets and symbolic links. If you make the necessary changes to the `walk()` function, you can tell `goFind.go` to print only the information that really interests you. This is left as an exercise for you to do it on your own.

Now imagine how difficult it would be to implement the same program using the standard library functions of the C programming language!

Using eBPF from Go

eBPF stands for enhanced Berkeley Packet Filter, and it is an in-kernel virtual machine that is integrated into the Linux kernel and can be used for Linux tracing. In order to be able to use eBPF, you will need to have a kernel compiled with the `CONFIG_BPF_SYSCALL` option, which is automatically activated on Ubuntu Linux.

eBPF works on Linux machines with relatively new kernel versions but does not work on macOS or Mac OS X machines.

You can learn more about eBPF at `https://github.com/iovisor/bcc` and about eBPF and Go at `https://kinvolk.io/blog/2016/11/introducing-gobpf---using-ebpf-from-go/`. The `gobpf` package can be found at `https://github.com/iovisor/gobpf/`.

Unfortunately, further discussion of eBPF and Go is beyond the scope of this book.

About syscall.PtraceRegs

You might have assumed that you are done dealing with the `syscall` standard Go package, however you are mistaken! In this section, we will work with `syscall.PtraceRegs`, which is a structure that holds information about the state of the registers.

 This section is not included in my *Go Systems Programming book, Packt Publishing*, 2017.

You will learn now how to print the values of all of the following registers on your screen using the Go code of `ptraceRegs.go`, which will be presented in four parts. The star of the `ptraceRegs.go` utility is the `syscall.PtraceGetRegs()` function-there are also the `syscall.PtraceSetRegs()`, `syscall.PtraceAttach()`, `syscall.PtracePeekData()`, and `syscall.PtracePokeData()` functions that can help you work with registers, but these functions will not be used in `ptraceRegs.go`.

The first part of the `ptraceRegs.go` utility follows next:

```
package main

import (
    "fmt"
    "os"
    "os/exec"
    "syscall"
    "time"
)
```

Telling a Unix System What to Do

The second code portion of `ptraceRegs.go` is shown in the following Go code:

```go
func main() {
    var r syscall.PtraceRegs
    cmd := exec.Command(os.Args[1], os.Args[2:]...)

    cmd.Stdout = os.Stdout
    cmd.Stderr = os.Stderr
```

The third part of `ptraceRegs.go` contains the following Go code:

```go
    cmd.SysProcAttr = &syscall.SysProcAttr{Ptrace: true}
    err := cmd.Start()
    if err != nil {
        fmt.Println("Start:", err)
        return
    }

    err = cmd.Wait()
    fmt.Printf("State: %v\n", err)
    wpid := cmd.Process.Pid
```

In the preceding Go code, you call an external command, which is specified in the command-line arguments of the program, and you find its process ID, which will be used in the `syscall.PtraceGetRegs()` call. The `&syscall.SysProcAttr{Ptrace: true}` statement specifies that you want to use ptrace on the child process.

The last code segment of `ptraceRegs.go` follows next:

```go
    err = syscall.PtraceGetRegs(wpid, &r)
    if err != nil {
        fmt.Println("PtraceGetRegs:", err)
        return
    }
    fmt.Printf("Registers: %#v\n", r)
    fmt.Printf("R15=%d, Gs=%d\n", r.R15, r.Gs)

    time.Sleep(2 * time.Second)
}
```

Here you call `syscall.PtraceGetRegs()`, and you print the results that are stored in the `r` variable, which should be passed as a pointer.

Executing `ptraceRegs.go` on a macOS High Sierra machine will generate the following output:

```
$ go run ptraceRegs.go
# command-line-arguments
./ptraceRegs.go:11:8: undefined: syscall.PtraceRegs
./ptraceRegs.go:14:9: undefined: syscall.PtraceGetRegs
```

This means that this program will not work on machines running macOS and OS X.

Executing `ptraceRegs.go` on a Debian Linux machine will create the following output:

```
$ go version
go version go1.3.3 linux/amd64
$ go run ptraceRegs.go echo "Mastering Go!"
State: stop signal: trace/breakpoint trap
Registers: syscall.PtraceRegs{R15:0x0, R14:0x0, R13:0x0, R12:0x0, Rbp:0x0,
Rbx:0x0, R11:0x0, R10:0x0, R9:0x0, R8:0x0, Rax:0x0, Rcx:0x0, Rdx:0x0,
Rsi:0x0, Rdi:0x0, Orig_rax:0x3b, Rip:0x7f8b1200e130, Cs:0x33, Eflags:0x200,
Rsp:0x7ffd9d53f320, Ss:0x2b, Fs_base:0x0, Gs_base:0x0, Ds:0x0, Es:0x0,
Fs:0x0, Gs:0x0}
R15=0, Gs=0
Mastering Go!
```

You can also find the list of registers in the documentation page of the `syscall` package.

Tracing system calls

This section will present a pretty advanced technique that uses the `syscall` package and allows you to monitor the system calls executed in a Go program.

This section is not included in my *Go Systems Programming* book, Packt Publishing, 2017.

Telling a Unix System What to Do

The name of the Go utility is `traceSyscall.go`, and it is going to be presented in five code segments. The first part of `traceSyscall.go` follows next:

```go
package main

import (
    "bufio"
    "fmt"
    "os"
    "os/exec"
    "strings"
    "syscall"
)

var maxSyscalls = 0

const SYSCALLFILE = "SYSCALLS"
```

You will learn more about the purpose of the `SYSCALLFILE` variable in a short while.

The second code segment from `traceSyscall.go` is the following:

```go
func main() {
    var SYSTEMCALLS []string
    f, err := os.Open(SYSCALLFILE)
    defer f.Close()
    if err != nil {
        fmt.Println(err)
        return
    }

    scanner := bufio.NewScanner(f)
    for scanner.Scan() {
        line := scanner.Text()
        line = strings.Replace(line, " ", "", -1)
        line = strings.Replace(line, "SYS_", "", -1)
        temp := strings.ToLower(strings.Split(line, "=")[0])
        SYSTEMCALLS = append(SYSTEMCALLS, temp)
        maxSyscalls++
    }
```

Note that the information of the `SYSCALLS` text file is taken from the documentation of the `syscall` package, and it associates each system call with a number, which is the internal Go representation of the system call. This file is mainly used for printing the names of the system calls used by the program that is being traced.

The format of the SYSCALLS text file is as follows:

```
SYS_READ = 0
SYS_WRITE = 1
SYS_OPEN = 2
SYS_CLOSE = 3
SYS_STAT = 4
```

After reading the text file, the program creates a slice named SYSTEMCALLS for storing that information.

The third part of traceSyscall.go is as follows:

```go
COUNTER := make([]int, maxSyscalls)
var regs syscall.PtraceRegs
cmd := exec.Command(os.Args[1], os.Args[2:]...)

cmd.Stdin = os.Stdin
cmd.Stdout = os.Stdout
cmd.Stderr = os.Stderr
cmd.SysProcAttr = &syscall.SysProcAttr{Ptrace: true}

err = cmd.Start()
err = cmd.Wait()
if err != nil {
    fmt.Println("Wait:", err)
}

pid := cmd.Process.Pid
fmt.Println("Process ID:", pid)
```

The COUNTER slice stores the number of times each system call is found in the program that is being traced.

The fourth code segment of traceSyscall.go contains the following Go code:

```go
before := true
forCount := 0
for {
    if before {
        err := syscall.PtraceGetRegs(pid, &regs)
        if err != nil {
            break
        }
        if regs.Orig_rax > uint64(maxSyscalls) {
            fmt.Println("Unknown:", regs.Orig_rax)
            return
```

```
            }
            COUNTER[regs.Orig_rax]++
            forCount++
        }
        err = syscall.PtraceSyscall(pid, 0)
        if err != nil {
            fmt.Println("PtraceSyscall:", err)
            return
        }
        _, err = syscall.Wait4(pid, nil, 0, nil)
        if err != nil {
            fmt.Println("Wait4:", err)
            return
        }
        before = !before
    }
```

The `syscall.PtraceSyscall()` function tells Go to continue the execution of the program that is being traced, but to stop when that program enters or exits a system call, which is exactly what we want! As each system call is traced before being called and right after it has finished its job, we use the `before` variable in order to count each system call only once.

The last part of `traceSyscall.go` follows next:

```
    for i, x := range COUNTER {
        if x != 0 {
            fmt.Println(SYSTEMCALLS[i], "->", x)
        }
    }
    fmt.Println("Total System Calls:", forCount)
}
```

In this part, we print the contents of the COUNTER slice. The SYSTEMCALLS slice is used here for finding out the name of a system call when we know its numerical Go representation.

Executing `traceSyscall.go` on a macOS High Sierra machine will create the following output:

```
$ go run traceSyscall.go
# command-line-arguments
./traceSyscall.go:36:11: undefined: syscall.PtraceRegs
./traceSyscall.go:57:11: undefined: syscall.PtraceGetRegs
./traceSyscall.go:70:9: undefined: syscall.PtraceSyscall
```

Once again, the `traceSyscall.go` utility will not run on macOS and Mac OS X.

Executing the same program on a Debian Linux machine will create the following output:

```
$ go run traceSyscall.go ls /tmp/
Wait: stop signal: trace/breakpoint trap
Process ID: 5657
go-build084836422  test.go  upload_progress_cache
read -> 11
write -> 1
open -> 37
close -> 27
stat -> 1
fstat -> 25
mmap -> 39
mprotect -> 16
munmap -> 4
brk -> 3
rt_sigaction -> 2
rt_sigprocmask -> 1
ioctl -> 2
access -> 9
execve -> 1
getdents -> 2
getrlimit -> 1
statfs -> 2
arch_prctl -> 1
futex -> 1
set_tid_address -> 1
openat -> 1
set_robust_list -> 1
Total System Calls: 189
```

At the end of the program, `traceSyscall.go` prints the number of times each system call was called in the program! The correctness of `traceSyscall.go` is verified by the output of the `strace -c` utility.

```
$ strace -c ls /tmp
test.go  upload_progress_cache
% time     seconds  usecs/call     calls    errors syscall
------ ----------- ----------- --------- --------- ----------------
  0.00    0.000000           0        11           read
  0.00    0.000000           0         1           write
  0.00    0.000000           0        37        13 open
  0.00    0.000000           0        27           close
  0.00    0.000000           0         1           stat
  0.00    0.000000           0        25           fstat
```

```
  0.00    0.000000           0        39             mmap
  0.00    0.000000           0        16             mprotect
  0.00    0.000000           0         4             munmap
  0.00    0.000000           0         3             brk
  0.00    0.000000           0         2             rt_sigaction
  0.00    0.000000           0         1             rt_sigprocmask
  0.00    0.000000           0         2             ioctl
  0.00    0.000000           0         9         9   access
  0.00    0.000000           0         1             execve
  0.00    0.000000           0         2             getdents
  0.00    0.000000           0         1             getrlimit
  0.00    0.000000           0         2         2   statfs
  0.00    0.000000           0         1             arch_prctl
  0.00    0.000000           0         1             futex
  0.00    0.000000           0         1             set_tid_address
  0.00    0.000000           0         1             openat
  0.00    0.000000           0         1             set_robust_list
------ ----------- ----------- --------- ---------   ---------------
100.00    0.000000                   189        24   total
```

User ID and group ID

In the last section of this chapter, you will learn how to find the user ID of the current user as well as the group IDs to which the current user belongs. Both user ID and group IDs are positive integers kept in Unix system files.

The name of the utility is ids.go, and it will be presented in two parts. The first part of the utility follows next:

```
package main

import (
    "fmt"
    "os"
    "os/user"
)

func main() {
    fmt.Println("User id:", os.Getuid())
```

Finding the user ID of the current user is as simple as calling the os.Getuid() function.

The second part of ids.go is as follows:

```go
var u *user.User
u, _ = user.Current()
fmt.Print("Group ids: ")
groupIDs, _ := u.GroupIds()
for _, i := range groupIDs {
    fmt.Print(i, " ")
}
fmt.Println()
}
```

On the other hand, finding the group IDs to which a user belongs is a much trickier task.

Executing ids.go will generate the following type of output:

```
$ go run ids.go
User id: 501
Group ids: 20 701 12 61 79 80 81 98 33 100 204 250 395 398 399
```

Additional resources

You will find the following web links very useful:

- Read the documentation page of the io package, which can be found at https://golang.org/pkg/io/.
- You can learn more about the Glot plotting library by visiting its official web page at https://github.com/Arafatk/glot.
- You can learn more about the encoding/binary standard package by visiting https://golang.org/pkg/encoding/binary/.
- Check the documentation page of the encoding/gob package, which can be found at https://golang.org/pkg/encoding/gob/.
- You can learn about eBPF at http://www.brendangregg.com/ebpf.html. You can also watch https://www.youtube.com/watch?v=JRFNIKUROPE and https://www.youtube.com/watch?v=w8nFRoFJ6EQ.
- You can learn about **Endianness** in many places including https://en.wikipedia.org/wiki/Endianness.
- Visit the documentation page of the flag package, which can be found at https://golang.org/pkg/flag/.

Exercises

- Write a Go program that takes three arguments, the name of a text file, and two strings. This utility should then replace every occurrence of the first string in the file with the second string. For reasons of security, the final output will be printed on screen, which means that the original text file will remain intact.
- Use the `encoding/gob` package to serialize and deserialize a Go map as well as a slice of structures.
- Create a Go program that handles any three signals you choose.
- Create a utility in Go that replaces all tab characters found in a text file with a given number of spaces that is specified as a command-line parameter to the program. Once again, the output will be printed on the screen.
- Develop a utility that reads a text file line by line and removes the space characters from each line using the `strings.TrimSpace()` function.
- Modify `kvSaveLoad.go` in order to support a single command-line argument, which is the filename that you will use both to load and save your data.
- Can you create a Go version of the `wc(1)` utility? Look at the manual page of `wc(1)` to find out about the command-line options that it supports.
- Modify the code of `goFind.go` in order to print regular files only. This means that it should not print directories, sockets, and symbolic links.
- Can you write a program that uses **Glot** to plot a function?
- Modify `traceSyscall.go` in order to display each system call at the time it is being traced.
- Modify `cat.go` just to do `io.Copy(os.Stdout, f)` in order to copy the contents of a file straight out instead of scanning it all.
- You can also use `bufio.NewScanner()` and `bufio.ScanWords` to read a line word by word. Find out how, and create a new version of the `byWord.go` utility.

Summary

This amazing chapter talked about many interesting topics including reading files, writing to files, and the use of the `flag` package, Nevertheless, there are many more topics related to systems programming not mentioned in this chapter, such as working with directories, copying, deleting, and renaming files, dealing with Unix users, groups and Unix processes, working with environment variables such as PATH, changing Unix file permissions, generating **sparse files**, reading and saving JSON data, locking and creating files, using and rotating your own log files, as well as the information found in the structure returned by the `os.Stat()` call.

At the end of this chapter, we presented two advanced utilities written in Go. The first one allowed you to inspect the state of the registers, while the second one showed you a technique that allows you to trace the system calls of any program.

I am pretty sure that the next chapter will also astonish you, as it talks about goroutines, channels, and pipelines, which are unique and powerful Go features.

9
Go Concurrency – Goroutines, Channels, and Pipelines

The previous chapter discussed systems programming in Go, which encompasses the Go functions and techniques that allow you to communicate with the operating system. Two of the areas of systems programming that were not covered in the previous chapter are concurrent programming and creating and manage multiple threads. Both these topics will be addressed in this chapter and the next one.

Go offers its own unique and innovative way of achieving concurrency, which comes in the form of **goroutines** and **channels**. Goroutines are the smallest Go entities that can be executed on their own in a Go program, while channels can get data from goroutines in a concurrent and efficient way and thus allow goroutines to have a point of reference and communicate with each other. Everything in Go is executed using goroutines; this makes perfect sense since Go is a concurrent programming language by design. Therefore, when a Go program starts its execution, its single goroutine calls the `main()` function, which starts the actual program execution.

The contents and the code of this chapter will be pretty simple, and you should have no problem following and understanding them, as I left the more advanced parts of goroutines and channels for `Chapter 10`, *Go Concurrency – Advanced Topics*.

In this chapter, you will learn the following topics:

- The differences between processes, threads, and goroutines
- The Go scheduler
- Concurrency and parallelism
- Creating goroutines

- Creating channels
- Reading or receiving data from a channel
- Writing or sending data to a channel
- Creating pipelines
- Waiting for your goroutines to finish

About processes, threads, and goroutines

A **process** is an execution environment that contains instructions and user-data and system-data parts, as well as other types of resources that are obtained during runtime, whereas a **program** is a file that contains the instructions and data that are used for initializing the instruction and user-data parts of a process.

A **thread** is a smaller and lighter entity than a process or a program. Threads are created by processes and have their own flow of control and stack. A quick and simplistic way to differentiate a thread from a process is to consider a process as the running binary file and a thread as a subset of a process.

A **goroutine** is the minimum Go entity that can be executed concurrently. The use of the word *minimum* is very important here, as goroutines are not autonomous entities like Unix processes—goroutines live in threads that live in Unix processes. The main advantage of goroutines is that they are extremely lightweight and running thousands or hundreds of thousands of them is no problem.

The good thing is that goroutines are lighter than threads, which, in turn are lighter than processes. In practice, this means that a process can have multiple threads as well as lots of goroutines, whereas a goroutine needs the environment of a process in order to exist. So, in order to create a goroutine, you will need to have a process with at least one thread-Unix takes care of the process and thread management, while Go and the developer need to take care of the goroutines.

Now that you know the basics of processes, programs, threads, and goroutines, let's talk a little bit about the **Go scheduler**.

The Go scheduler

The Unix kernel scheduler is responsible for the execution of the threads of a program. On the other hand, the Go runtime has its own scheduler, which is responsible for the execution of the goroutines using a technique known as **m:n scheduling,** where *m* goroutines are executed using *n* operating system threads using multiplexing. The **Go scheduler** is the Go component responsible for the way and the order in which the goroutines of a Go program get executed. This makes the Go scheduler a really important part of the Go programming language, as everything in a Go program is executed as a goroutine.

Be aware that as the Go scheduler only deals with the goroutines of a single program, its operation is much simpler, cheaper, and faster than the operation of the kernel scheduler.

Chapter 10, *Go Concurrency–Advanced Topics*, will talk about the way the Go scheduler operates in much more detail.

Concurrency and parallelism

It is a very common misconception that **concurrency** is the same thing as **parallelism**—this is just not true! Parallelism is the simultaneous execution of multiple entities of some kind, whereas concurrency is a way of structuring your components so that they can be executed independently when possible.

It is only when you build software components concurrently that you can safely execute them in parallel, when and if your operating system and your hardware permit it. The **Erlang** programming language did this a long time ago—long before CPUs had multiple cores and computers had lots of RAM.

In a valid concurrent design, adding concurrent entities makes the whole system run faster because more things can run in parallel. So, the desired parallelism comes from a better concurrent expression and implementation of the problem. The developer is responsible for taking concurrency into account during the design phase of a system and benefit from a potential parallel execution of the components of the system. So, the developer should not think about parallelism, but about breaking things into independent components that solve the initial problem when combined.

Even if you cannot run your functions in parallel on a Unix machine, a valid concurrent design will still improve the design and the maintainability of your programs. In other words, concurrency is better than parallelism!

Goroutines

You can define a new goroutine using the `go` keyword followed by a function name or the full definition of an **anonymous function**. The `go` keyword makes the function call to return immediately, while the function starts running in the background as a goroutine and the rest of the program continues its execution.

However, as you will see in a moment, you cannot control or make any assumptions about the order in which your goroutines are going to be executed because this depends on the scheduler of the operating system, the Go scheduler, and the load of the operating system.

Creating a goroutine

In this subsection, you will learn two ways for creating goroutines. The first one is using regular functions, while the second method is using anonymous functions—these two ways are equivalent.

The name of the program covered in this section is `simple.go`, and it is presented in three parts.

The first part of `simple.go` is the following Go code:

```go
package main

import (
    "fmt"
    "time"
)

func function() {
    for i := 0; i < 10; i++ {
        fmt.Print(i)
    }
    fmt.Println()
}
```

Apart from the `import` block, the preceding code defines a function named `function()` that will be used in a short while.

The second part of `simple.go` follows next:

```
func main() {
    go function()
```

The preceding code starts with `function()` as a goroutine. After this, the program continues its execution, while `function()` begins to run in the background.

The last code portion of `simple.go` is shown in the following Go code:

```
    go func() {
        for i := 10; i < 20; i++ {
            fmt.Print(i, " ")
        }
    }()

    time.Sleep(1 * time.Second)
}
```

With this code, you create a goroutine using an anonymous function. This method works best for relatively small functions. However, if you have lots of code, it is considered better practice to create a regular function and execute it using the `go` keyword.

As you will see in the next section, you can create multiple goroutines any way you wish, including using a `for` loop.

Executing `simple.go` two times will generate the following type of output:

```
$ go run simple.go
10 11 12 13 14 0123456789
15 16 17 18 19
$ go run simple.go
10 11 12 13 14 15 16 17 18 19 0123456789
```

Although what you really want from your programs is to generate the same output for the same input, the output you get from `simple.go` is not always the same. The preceding output supports the fact that you cannot control the order in which your goroutines will be executed without taking extra care. This means writing extra code specifically for this to occur. In Chapter 10, *Go Concurrency – Advanced Topics*, you will learn how to control the order in which your goroutines are executed as well as how to print the results of one goroutine before printing the results of the following one.

Creating multiple goroutines

In this subsection, you will learn how to create a variable number of goroutines. The program reviewed in this section is called `create.go`. It will be presented in four parts, and it will allow you to create a dynamic number of goroutines. The number of goroutines will be given as a command-line argument to the program, which uses the `flag` package to process its command-line argument.

The first code part of `create.go` follows next:

```
package main

import (
    "flag"
    "fmt"
    "time"
)
```

The second code segment from `create.go` contains the following Go code:

```
func main() {
    n := flag.Int("n", 10, "Number of goroutines")
    flag.Parse()

    count := *n
    fmt.Printf("Going to create %d goroutines.\n", count)
```

The preceding code reads the value of the `n` command-line option, which determines the number of goroutines that will be created.

The third code portion of `create.go` follows next:

```
    for i := 0; i < count; i++ {
        go func(x int) {
            fmt.Printf("%d ", x)
        }(i)
    }
```

A `for` loop is used for creating the desired number of goroutines. Once again, you should remember that you cannot make any assumptions about the order in which they are going to be created and executed.

The last part of the Go code from `create.go` is the following:

```
    time.Sleep(time.Second)
    fmt.Println("\nExiting...")
}
```

The purpose of the `time.Sleep()` statement is to give the goroutines enough time to finish their jobs so that their output can be seen on the screen. In a real program, you will not need a `time.Sleep()` statement, as you both want to finish as soon as possible and, moreover, you will learn a better technique to make your program wait for the various goroutines to finish before the `main()` function returns.

Executing `create.go` multiple times will generate the following type of output:

```
$ go run create.go -n 100
Going to create 100 goroutines.
5 3 2 4 19 9 0 1 7 11 10 12 13 14 15 31 16 20 17 22 8 18 28 29 21 52 30 45
25 24 49 38 41 46 6 56 57 54 23 26 53 27 59 47 69 66 51 44 71 48 74 33 35
73 39 37 58 40 50 78 85 86 90 67 72 91 32 64 65 95 75 97 99 93 36 60 34 77
94 61 88 89 83 84 43 80 82 87 81 68 92 62 55 98 96 63 76 79 42 70
Exiting...
$ go run create.go -n 100
Going to create 100 goroutines.
2 5 3 16 6 7 8 9 1 22 10 12 13 17 11 18 15 14 19 20 31 23 26 21 29 24 30 25
37 32 36 38 35 33 45 41 43 42 40 39 34 44 48 46 47 56 53 50 0 49 55 59 58
28 54 27 60 4 57 51 52 64 61 65 72 62 63 67 69 66 74 73 71 75 89 70 76 84
85 68 79 80 93 97 83 82 99 78 88 91 92 77 81 95 94 98 87 90 96 86
Exiting...
```

Once again, you can see that the output is nondeterministic and messy in the sense that you will have to search the output to find what you are looking for. Additionally, if you do not use a suitable delay in the `time.Sleep()` call, you will not be able to see the output of the goroutines. The `time.Second` might be OK for now, but this kind of code can cause nasty and unpredictable bugs down the road.

In the next section, you will learn how to give your goroutines enough time to finish what they are doing before your program ends without the need to call `time.Sleep()`.

Waiting for your goroutines to finish

This section will present a way that prevents the `main()` function from ending while it is waiting for its goroutines to finish, using the `sync` package. The logic of the `syncGo.go` program will be based on `create.go`, which was presented in the previous section.

The first part of `syncGo.go` is as follows:

```
package main

import (
    "flag"
    "fmt"
    "sync"
)
```

As you can see in the preceding code block, there is no need for importing and using the `time` package, as we will use the functionality of the `sync` package and wait for as long as necessary for all the goroutines to end.

In Chapter 10, *Go Concurrency – Advanced Topics,* you will study two techniques for timing out goroutines when they are taking longer than expected.

The second code segment of `syncGo.go` is shown in the following Go code:

```
func main() {
    n := flag.Int("n", 20, "Number of goroutines")
    flag.Parse()
    count := *n
    fmt.Printf("Going to create %d goroutines.\n", count)

    var waitGroup sync.WaitGroup
```

In the preceding Go code, you define a `sync.WaitGroup` variable. If you look at the source code of the `sync` Go package, and more specifically at the `waitgroup.go` file that is located inside the `sync` directory, you will see that the `sync.WaitGroup` type is nothing more than a structure with three fields:

```
type WaitGroup struct {
    noCopy noCopy
    state1 [12]byte
    sema   uint32
}
```

The output of `syncGo.go` will reveal more information about the way the `sync.WaitGroup` variables work. The number of goroutines that belong to a `sync.WaitGroup` group is defined by one or multiple calls to the `sync.Add()` function.

The third part of `syncGo.go` contains the following Go code:

```go
fmt.Printf("%#v\n", waitGroup)
for i := 0; i < count; i++ {
    waitGroup.Add(1)
    go func(x int) {
        defer waitGroup.Done()
        fmt.Printf("%d ", x)
    }(i)
}
```

Here you create the desired number of goroutines using a `for` loop. (You could have used multiple sequential Go statements instead.)

Each call to `sync.Add()` increases a counter in a `sync.WaitGroup` variable. Notice that it is really important to call `sync.Add(1)` before the `go` statement in order to prevent any race conditions. When each goroutine finishes its job, the `sync.Done()` function will be executed, which will decrease the same counter.

The last code portion of `syncGo.go` follows next:

```go
    fmt.Printf("%#v\n", waitGroup)
    waitGroup.Wait()
    fmt.Println("\nExiting...")
}
```

The `sync.Wait()` call blocks until the counter in the relevant `sync.WaitGroup` variable is zero, giving your goroutines time to finish.

Executing `syncGo.go` will create the following output:

```
$ go run syncGo.go
Going to create 20 goroutines.
sync.WaitGroup{noCopy:sync.noCopy{}, state1:[12]uint8{0x0, 0x0, 0x0, 0x0,
0x0, 0x0, 0x0, 0x0, 0x0, 0x0, 0x0, 0x0}, sema:0x0}
sync.WaitGroup{noCopy:sync.noCopy{}, state1:[12]uint8{0x0, 0x0, 0x0, 0x0,
0x14, 0x0, 0x0, 0x0, 0x0, 0x0, 0x0, 0x0}, sema:0x0}
19 7 8 9 10 11 12 13 14 15 16 17 0 1 2 5 18 4 6 3
Exiting...
$ go run syncGo.go -n 30
Going to create 30 goroutines.
sync.WaitGroup{noCopy:sync.noCopy{}, state1:[12]uint8{0x0, 0x0, 0x0, 0x0,
```

```
              0x0, 0x0, 0x0, 0x0, 0x0, 0x0, 0x0, 0x0}, sema:0x0}
              1 0 4 5 17 7 8 9 10 11 12 13 2 sync.WaitGroup{noCopy:sync.noCopy{},
              state1:[12]uint8{0x0, 0x0, 0x0, 0x0, 0x17, 0x0, 0x0, 0x0, 0x0, 0x0, 0x0,
              0x0}, sema:0x0}
              29 15 6 27 24 25 16 22 14 23 18 26 3 19 20 28 21
              Exiting...
              $ go run syncGo.go -n 30
              Going to create 30 goroutines.
              sync.WaitGroup{noCopy:sync.noCopy{}, state1:[12]uint8{0x0, 0x0, 0x0, 0x0,
              0x0, 0x0, 0x0, 0x0, 0x0, 0x0, 0x0, 0x0}, sema:0x0}
              sync.WaitGroup{noCopy:sync.noCopy{}, state1:[12]uint8{0x0, 0x0, 0x0, 0x0,
              0x1e, 0x0, 0x0, 0x0, 0x0, 0x0, 0x0, 0x0}, sema:0x0}
              29 1 7 8 2 9 10 11 12 4 13 15 0 6 5 22 25 23 16 28 26 20 19 24 21 14 3 17
              18 27
              Exiting...
```

The output of `syncGo.go` still varies from execution to execution, especially if you are dealing with a large number of goroutines. Most of the time this is acceptable; however, there are times when this is not the desired behavior. Additionally, when the number of goroutines is 30, some of the goroutines have finished their job before the second `fmt.Printf("%#v\n", waitGroup)` statement. Finally, note that one of the elements of the `state1` field in `sync.WaitGroup` is the one that holds the counter which increases and decreases according to the `sync.Add()` and `sync.Done()` calls.

What if the number of Add() and Done() calls do not agree?

When the number of the `sync.Add()` and `sync.Done()` calls are equal, everything will be fine in your programs. However, this section will tell you what will happen when these two numbers do not agree with each other.

If you have executed more `sync.Add()` calls than `sync.Done()` calls, in this case, by adding a `waitGroup.Add(1)` statement before the first `fmt.Printf("%#v\n", waitGroup)` statement of the `syncGo.go` program, then the output of the `go run` command will be similar to the following:

```
              $ go run syncGo.go
              Going to create 20 goroutines.
              sync.WaitGroup{noCopy:sync.noCopy{}, state1:[12]uint8{0x0, 0x0, 0x0, 0x0,
              0x1, 0x0, 0x0, 0x0, 0x0, 0x0, 0x0, 0x0}, sema:0x0}
              sync.WaitGroup{noCopy:sync.noCopy{}, state1:[12]uint8{0x0, 0x0, 0x0, 0x0,
              0x15, 0x0, 0x0, 0x0, 0x0, 0x0, 0x0, 0x0}, sema:0x0}
              19 10 11 12 13 17 18 8 5 4 6 14 1 0 7 3 2 15 9 16 fatal error: all
```

```
goroutines are asleep - deadlock!

goroutine 1 [semacquire]:
sync.runtime_Semacquire(0xc4200120bc)
        /usr/local/Cellar/go/1.9.3/libexec/src/runtime/sema.go:56 +0x39
sync.(*WaitGroup).Wait(0xc4200120b0)
        /usr/local/Cellar/go/1.9.3/libexec/src/sync/waitgroup.go:131 +0x72
main.main()
        /Users/mtsouk/Desktop/masterGo/ch/ch9/code/syncGo.go:28 +0x2d7
exit status 2
```

The error message is pretty clear: `fatal error: all goroutines are asleep - deadlock!` So, this happened because you told your program to wait for n+1 goroutines by calling the `sync.Add(1)` function n+1 times while only n `sync.Done()` statements were executed by your n goroutines! As a result, the `sync.Wait()` call will wait indefinitely for one or more calls to `sync.Done()` without any luck, which is obviously a deadlock situation.

If you have made fewer `sync.Add()` calls than `sync.Done()` calls, which can be emulated by adding a `waitGroup.Done()` statement after the `for` loop of the `syncGo.go` program, then the `go run` output will be similar to the following:

```
$ go run syncGo.go
Going to create 20 goroutines.
sync.WaitGroup{noCopy:sync.noCopy{}, state1:[12]uint8{0x0, 0x0, 0x0, 0x0,
0x0, 0x0, 0x0, 0x0, 0x0, 0x0, 0x0, 0x0}, sema:0x0}
sync.WaitGroup{noCopy:sync.noCopy{}, state1:[12]uint8{0x0, 0x0, 0x0, 0x0,
0x12, 0x0, 0x0, 0x0, 0x0, 0x0, 0x0, 0x0}, sema:0x0}
19 6 1 2 9 7 8 15 13 0 14 16 17 3 11 4 5 12 18 10 panic: sync: negative
WaitGroup counter
goroutine 22 [running]:
sync.(*WaitGroup).Add(0xc4200120b0, 0xffffffffffffffff)
        /usr/local/Cellar/go/1.9.3/libexec/src/sync/waitgroup.go:75 +0x134
sync.(*WaitGroup).Done(0xc4200120b0)
        /usr/local/Cellar/go/1.9.3/libexec/src/sync/waitgroup.go:100 +0x34
main.main.func1(0xc4200120b0, 0x11)
        /Users/mtsouk/Desktop/masterGo/ch/ch9/code/syncGo.go:25 +0xd8
created by main.main
        /Users/mtsouk/Desktop/masterGo/ch/ch9/code/syncGo.go:21 +0x206
exit status 2
```

Once again, the root of the problem is stated pretty clearly: `panic: sync: negative WaitGroup counter`.

Although the error messages are very descriptive in both cases and will help you solve the real problem, you should be very careful with the number of `sync.Add()` and `sync.Done()` calls that you put into your programs. Additionally, notice that in the second error case (`panic: sync: negative WaitGroup counter`), the problem might not appear all of the time.

Channels

A **channel** is a communication mechanism that allows goroutines to exchange data, among other things. However, there are some definite rules here. First, each channel allows the exchange of a particular data type, which is also called the **element type** of the channel, and second, for a channel to operate properly, you will need someone to receive what is sent via the channel. You should declare a new channel using the `chan` keyword, and you can close a channel using the `close()` function.

Finally, a very important detail: when you are using a channel as a function parameter, you can specify its direction; that is, whether it is going to be used for sending or receiving. In my opinion, if you know the purpose of a channel in advance, you should use this capability because it will make your programs more robust as well as safer because you will not be able to send data accidentally to a channel from which you should only receive data or receive data from a channel to which you should only be sending data. As a result, if you declare that a channel function parameter will be used for reading only and you try to write to it, you will get an error message that will most likely save you from nasty bugs. We will talk about this later on in this chapter.

 Although you will learn many things about channels in this chapter, you will have to wait for `Chapter 10`, *Go Concurrency –Advanced Topics*, to fully understand the power and flexibility that channels offer to the Go developer.

Writing to a channel

The code in this subsection will teach you how to write to a channel. Writing the x value to the c channel is as easy as writing `c <- x`. The arrow shows the direction of the value, and you will have no problem with this statement as long as both x and c have the same type. The example code in this section is saved in `writeCh.go`, and it will be presented in three parts.

The first code segment from writeCh.go is as follows:

```
package main

import (
    "fmt"
    "time"
)

func writeToChannel(c chan int, x int) {
    fmt.Println(x)
    c <- x
    close(c)
    fmt.Println(x)
}
```

The chan keyword is used for declaring that the c function parameter will be a channel, and it should be followed by the type of the channel (int). The c <- x statement allows you to write the x value to the c channel, and the close() function closes the channel; that is, it makes communicating with it impossible.

The second part of writeCh.go contains the following Go code:

```
func main() {
    c := make(chan int)
```

In the preceding code, you will find the definition of a channel variable which is named c, and for the first time in this chapter you're using the make() function as well as the chan keyword. All channels have a type associated with them.

The remaining code from writeCh.go is as follows:

```
    go writeToChannel(c, 10)
    time.Sleep(1 * time.Second)
}
```

Here you execute the writeToChannel() function as a goroutine and call time.Sleep() in order to give enough time to the writeToChannel() function to execute.

Executing writeCh.go will create the following output:

```
$ go run writeCh.go
10
```

The strange thing here is that the `writeToChannel()` function printed the given value only once. The cause of this unexpected output is that the second `fmt.Println(x)` statement is never executed. The reason for this is pretty simple once you understand how channels work: the `c <- x` statement is blocking the execution of the rest of the `writeChannel()` function because nobody is reading what was written to the `c` channel. Therefore, when the `time.Sleep(1 * time.Second)` statement finishes, the program terminates without waiting for `writeChannel()`.

The next section will illustrate how to read data from a channel.

Reading from a channel

In this subsection, you will learn how to read from a channel. You can read a single value from a channel named c by executing `<-c`. In this case, the direction is from the channel to the outer world.

The name of the program that I will use to help you understand how to read from a channel is `readCh.go`, and it will be presented in three parts.

The first code segment from `readCh.go` is shown in the following Go code:

```go
package main

import (
    "fmt"
    "time"
)

func writeToChannel(c chan int, x int) {
    fmt.Println("1", x)
    c <- x
    close(c)
    fmt.Println("2", x)
}
```

The implementation of the `writeToChannel()` function is the same as before.

The second part of readCh.go follows next:

```
func main() {
    c := make(chan int)
    go writeToChannel(c, 10)
    time.Sleep(1 * time.Second)
    fmt.Println("Read:", <-c)
    time.Sleep(1 * time.Second)
```

In the preceding code, you read from the c channel using the <-c notation. If you want to store that value to a variable named k instead of just printing it, you can use a k := <-c statement. The second time.Sleep(1 * time.Second) statement gives you the time to read from the channel.

The last code portion of readCh.go contains the following Go code:

```
    _, ok := <-c
    if ok {
        fmt.Println("Channel is open!")
    } else {
        fmt.Println("Channel is closed!")
    }
}
```

In the preceding code, you can see a technique for determining whether a given channel is open or not. The current Go code works fine when the channel is closed; however, if the channel were open, the Go code presented here would have discarded the read value of the channel because of the use of the _ character in the _, ok := <-c statement. Use a proper variable name instead of _ if you also want to read the value of the channel in case it is open.

Executing readCh.go will generate the following output:

```
$ go run readCh.go
1 10
Read: 10
2 10
Channel is closed!
$ go run readCh.go
1 10
2 10
Read: 10
Channel is closed!
```

Although the output is still not deterministic, both the `fmt.Println(x)` statements of the `writeToChannel()` function are executed because the channel is unblocked when you read from it.

Channels as function parameters

Although neither `readCh.go` nor `writeCh.go` used this feature, Go allows you to specify the direction of a channel when used as a function parameter; that is, whether it will be used for reading or writing. These two types of channels are called **unidirectional channels**, whereas, by default, channels are bidirectional.

Examine the Go code of the following two functions:

```
func f1(c chan int, x int) {
    fmt.Println(x)
    c <- x
}
func f2(c chan<- int, x int) {
    fmt.Println(x)
    c <- x
}
```

Although both functions implement the same functionality, their definitions are slightly different. The difference is created by the <- symbol found on the right of the `chan` keyword in the definition of the `f2()` function. This denotes that the `c` channel can be used for writing only. If the code of a Go function attempts to read from a write-only channel **(send-only channel)** parameter, the Go compiler will generate the following error message:

```
# command-line-arguments
a.go:19:11: invalid operation: range in (receive from send-only type chan<- int)
```

Similarly, you can have the following function definitions:

```
func f1(out chan<- int64, in <-chan int64) {
    fmt.Println(x)
    c <- x
}

func f2(out chan int64, in chan int64) {
    fmt.Println(x)
    c <- x
}
```

The definition of f2() combines a read-only channel named in with a write-only channel named out. If you accidentally try to write and close a read-only channel (**receive-only channel**) parameter of a function, you will get the following error message:

```
# command-line-arguments
a.go:13:7: invalid operation: out <- i (send to receive-only type <-chan
int)
a.go:15:7: invalid operation: close(out) (cannot close receive-only
channel)
```

Pipelines

A **pipeline** is a virtual method for connecting goroutines and channels so that the output of one goroutine becomes the input of another goroutine, using channels to transfer your data.

One of the benefits that you get from using pipelines is that there is a constant data flow in your program, as no goroutine and channel have to wait for everything to be completed in order to start their execution. Additionally, you are using fewer variables and therefore less memory space because you do not have to save everything as a variable. Finally, the use of pipelines simplifies the design of the program and improves its maintainability.

Pipelines will be illustrated using the code of pipeline.go. This program will be presented in six parts. The task performed by the pipeline.go program is to generate random numbers in a given range and stop when any number in the random sequence appears a second time. However, before terminating, the program will print the sum of all random numbers that appeared up to the point where the first random number appeared a second time. You will need three functions for connecting the channels of the program. The logic of the program is found in these three functions, but the data flows in the channels of the pipeline.

This program will have two channels. The first channel (channel A) will be used for getting the random numbers from the first function and sending them to the second function. The second channel (channel B) will be used by the second function for sending the acceptable random numbers to the third function. The third function will be responsible for getting the data from channel B, calculating it, and presenting the results.

The first code segment of `pipeline.go` contains the following Go code:

```
package main

import (
    "fmt"
    "math/rand"
    "os"
    "strconv"
    "time"
)

var CLOSEA = false

var DATA = make(map[int]bool)
```

As the `second()` function will need a way to tell the `first()` function to close the first channel, I will use a global variable named `CLOSEA` for that. The `CLOSEA` variable is only checked by the `first()` function, and it can only be altered by the `second()` function.

The second part of `pipeline.go` as shown in the following Go code:

```
func random(min, max int) int {
    return rand.Intn(max-min) + min
}

func first(min, max int, out chan<- int) {
    for {
        if CLOSEA {
            close(out)
            return
        }
        out <- random(min, max)
    }
}
```

The preceding code presents the implementation of two functions named `random()` and `first()`. You are already familiar with the `random()` function that generates random numbers in a given range. However, the `first()` function is really interesting as it keeps running using a `for` loop until a Boolean variable (`CLOSEA`) becomes `true`. In that case, it will close its `out` channel.

The third code segment of `pipeline.go` is as follows:

```go
func second(out chan<- int, in <-chan int) {
    for x := range in {
        fmt.Print(x, " ")
        _, ok := DATA[x]
        if ok {
            CLOSEA = true
        } else {
            DATA[x] = true
            out <- x
        }
    }
    fmt.Println()
    close(out)
}
```

The `second()` function receives data from the `in` channel and keeps sending it to the `out` channel. However, as soon as the `second()` function finds a random number that already exists in the `DATA` map, it sets the `CLOSEA` global variable to `true` and stops sending any more numbers to the `out` channel. After that, it closes the `out` channel.

The fourth code portion of `pipeline.go` is shown in the following Go code:

```go
func third(in <-chan int) {
    var sum int
    sum = 0
    for x2 := range in {
        sum = sum + x2
    }
    fmt.Printf("The sum of the random numbers is %d\n", sum)
}
```

The `third()` function keeps reading from the `in` function parameter channel. When that channel is closed by the `second()` function, the `for` loop will stop getting any more data and the function will display its output. At this point, it should become clear that the `second()` function controls many things.

The fifth code segment of `pipeline.go` is as follows:

```
func main() {
    if len(os.Args) != 3 {
        fmt.Println("Need two integer parameters!")
        os.Exit(1)
    }

    n1, _ := strconv.Atoi(os.Args[1])
    n2, _ := strconv.Atoi(os.Args[2])

    if n1 > n2 {
        fmt.Printf("%d should be smaller than %d\n", n1, n2)
        return
    }
```

The last part of the of the `pipeline.go` program follows next:

```
    rand.Seed(time.Now().UnixNano())
    A := make(chan int)
    B := make(chan int)

    go first(n1, n2, A)
    go second(B, A)
    third(B)
}
```

Here you define the required channels, and you execute two goroutines and one function. The `third()` function is what prevents `main()` from returning, because it is not executed as a goroutine.

Executing `pipeline.go` will produce the following type of output:

```
$ go run pipeline.go 1 10
2 2
The sum of the random numbers is 2
$ go run pipeline.go 1 10
9 7 8 4 3 3
The sum of the random numbers is 31
$ go run pipeline.go 1 10
1 6 9 7 1
The sum of the random numbers is 23
$ go run pipeline.go 10 20
16 19 16
The sum of the random numbers is 35
$ go run pipeline.go 10 20
10 16 17 11 15 10
```

```
The sum of the random numbers is 69
$ go run pipeline.go 10 20
12 11 14 15 10 15
The sum of the random numbers is 62
```

The important point here is that although the `first()` function keeps generating random numbers at its own pace and the `second()` function will print all of them on your screen, the unwanted random numbers, that is, the random numbers that have already appeared, will not be sent to the `third()` function, and therefore, will not be included in the final sum!

Additional resources

Visit the following useful resources:

- The documentation page of the `sync` package, which can be found at https://golang.org/pkg/sync/.

- The documentation page of the `sync` package once more. Pay close attention to the `sync.Mutex` and `sync.RWMutex` types that will appear in the next chapter.

Exercises

- Create a pipeline that reads text files, finds the number of occurrences of a given phrase in each text file, and calculates the total number of occurrences of the phrase in all files.

- Create a pipeline for calculating the sum of the squares of all of the natural numbers in a given range.

- Remove the `time.Sleep(1 * time.Second)` statement from the `simple.go` program and see what happens. Why does this happen?

- Modify the Go code of `pipeline.go` in order to create a pipeline with five functions and the appropriate number of channels.

- Modify the Go code of `pipeline.go` in order to find out what will happen when you forget to close the `out` channel of the `first()` function.

Summary

In this chapter, you learned many unique Go features, including goroutines, channels, and pipelines. Additionally, you found out how to give your goroutines enough time to finish their jobs using the functionality offered by the `sync` package. Finally, you learned that channels can be used as parameters to Go functions. This allows developers to create pipelines where your data can flow.

The next chapter will continue talking about Go concurrency by introducing the formidable `select` keyword. This keyword helps Go channels perform many interesting jobs, and I think that you will be truly amazed by its power. After this, you will see two techniques that allow you to time out one or more goroutines that are stalled for some reason. Afterwards, you will learn about nil channels, signal channels, channel of channels, and buffered channels, as well as the `context` package.

You will also learn about **shared memory** in the next chapter, which is the traditional way of sharing information among the threads of the same Unix process which also applies to goroutines. Nevertheless, shared memory is not that popular among Go programmers because Go offers better, safer, and faster ways for goroutines to exchange data.

10
Go Concurrency – Advanced Topics

The previous chapter introduced goroutines, which is the most important feature of Go, channels, and pipelines. This chapter will continue from the point where the previous one left off in order to learn more about goroutines, channels, and the `select` keyword before discussing shared variables and the `sync.Mutex` and `sync.RWMutex` types. This chapter also includes code examples that demonstrate the use of signal channels, buffered channels, nil channels, and channels of channels. Additionally, early on in this chapter, you will learn two techniques for timing out a goroutine after a given amount of time, because nobody can guarantee that all goroutines will finish before a desired time. The chapter will end by examining race conditions, the `context` standard Go package, and worker pools.

In this chapter of *Mastering Go*, you will learn the following topics:

- The `select` keyword
- How the Go scheduler works
- Two techniques that allow you to time-out a goroutine which takes longer than expected to finish
- Signals channels
- Buffered channels
- Nil channels
- Monitor goroutines
- Channels of channels
- Shared memory and mutexes
- The `sync.Mutex` and the `sync.RWMutex` types
- The `context` package and its functionality
- Worker pools
- Detecting race conditions

The Go scheduler revisited

A **scheduler** is responsible for distributing the amount of work to be done over the available resources in an efficient way. In this section, we will examine the way that the Go scheduler operates in much greater depth than in the previous chapter. As you already know, Go works using the **m:n scheduler** (or **M:N scheduler**) that schedules goroutines—which are lighter than OS threads—using OS threads. First, though, let's review the necessary theory and the definition of some useful terms.

Go uses the **fork-join concurrency** model. The fork part of the model states that a child branch can be created at any point of a program. Analogously, the join part of the Go concurrency model is where the child branch will end and join with its parent. Among other things, both sync.Wait() statements and channels that collect the results of goroutines are join points, whereas any new goroutine creates a child branch.

The fork phase of the fork-join model and the fork(2) C system call are two totally different things.

The **fair scheduling strategy**, which is pretty straightforward and has a simple implementation, shares evenly all the load between the available processors. At first, this might look like the perfect strategy because it does not have to take many things into consideration while keeping all processors equally occupied. It turns out that this is not exactly the case because most of the distributed tasks usually depend on other tasks. Therefore, at the end of the day, some processors are underutilized, or equivalently, some processors are utilized more than others.

A goroutine in Go is a **task**, whereas everything after the calling statement of a goroutine is a **continuation**. In the **work stealing strategy** used by Go scheduler, a (logical) processor that is underutilized looks for additional work from other processors. When it finds such jobs, it steals them from the other processor(s), hence the name, *work stealing strategy*. Additionally, the work-stealing algorithm of Go queues and steals continuations. A **stalling join**, as is suggested by its name, is a point where a thread of execution stalls at a join and starts looking for other work to do. Although both task stealing and continuation stealing have stalling joins, continuations happen more often than tasks; therefore, the Go algorithm works with continuations rather than tasks.

The main disadvantage of continuation stealing is that it requires extra work from the compiler of the programming language. Fortunately, Go provides that extra help and therefore uses **continuation stealing** in its work-stealing algorithm.

One of the benefits of continuation stealing is that you get the same results when using just functions or a single thread with multiple goroutines. This makes perfect sense as only one thing is executed at any given time in both cases.

Now, let's return back to the **m:n scheduling** algorithm used in Go. Strictly speaking, at any time, you have *m* goroutines that are executed, and therefore scheduled to run, on *n* OS threads, using the most GOMAXPROCS number of logical processors. You will learn about GOMAXPROCS shortly.

The Go scheduler works using three main kinds of entities: OS threads (**M**) that are related to the operating system in use, goroutines (**G**), and **logical processors** (**P**). The number of processors that can be used by a Go program is specified by the **GOMAXPROCS environment variable**—at any given time there are most GOMAXPROCS processors.

The following figure illustrates this point:

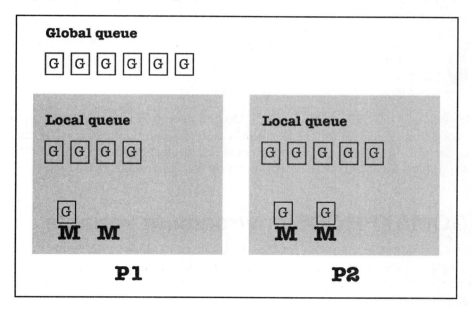

How the Go scheduler works

What the illustration tells us is that there are two different kinds of queues: a global queue and a local queue attached to each logical processor. Goroutines from the global queue are assigned to the queue of a logical processor in order to be executed. As a result, the Go scheduler needs to check the global queue in order to avoid executing goroutines that are only located at the local queue of each logical processor. However, the global queue is not checked all of the time, which means that it does not have an advantage over the local queues. Additionally, each logical processor can have multiple threads, and the stealing occurs between the local queues of the available logical processors. Finally, keep in mind that the Go scheduler is allowed to create more OS threads when needed. OS threads are pretty expensive, however, which means that dealing with OS threads too much might slow down your Go applications.

Keep in mind that using more goroutines in a program is not a panacea for performance, as more goroutines in addition to the various calls to `sync.Add()`, `sync.Wait()`, and `sync.Done()` might slow down your program due to the extra work that needs to be done by the Go scheduler.

The Go scheduler, as most Go components, is always evolving, which means that the people who work on the Go scheduler continually try to improve its performance by making small changes in the way it works. The core principles, however, remain the same.

Are all these details useful? I think they are! Do you need to know all of these in order to write Go code that uses goroutines? Absolutely not! However, knowing what occurs behind the scenes can definitely help you when strange things start happening, or if you are curious about how the Go scheduler works. It will certainly make you a better developer!

The GOMAXPROCS environment variable

The GOMAXPROCS environment variable (and Go function) allows you to limit the number of operating system threads that can execute user-level Go code simultaneously. Starting with Go version 1.5, the default value of GOMAXPROCS should be the number of cores available in your Unix machine.

If you decide to assign a value to GOMAXPROCS that is less than the number of the cores in your Unix machine, you might affect the performance of your program. However, using a GOMAXPROCS value that is larger than the number of the available cores will not necessarily make your program run faster.

You can programmatically discover the value of the GOMAXPROCS environment variable, that is, the relevant code. It is be found in the following `maxprocs.go` program:

```
package main

import (
    "fmt"
    "runtime"
)

func getGOMAXPROCS() int {
    return runtime.GOMAXPROCS(0)
}

func main() {
    fmt.Printf("GOMAXPROCS: %d\n", getGOMAXPROCS())
}
```

Executing `maxprocs.go` on a machine with an Intel i7 processor will produce the following output:

```
$ go run maxprocs.go
GOMAXPROCS: 8
```

You can modify the previous output by changing the value of the GOMAXPROCS environment variable prior to the execution of the program, however. The following commands are executed in the `bash(1)` Unix shell:

```
$ go version
go version go1.9.4 darwin/amd64
$ export GOMAXPROCS=800; go run maxprocs.go
GOMAXPROCS: 800
$ export GOMAXPROCS=4; go run maxprocs.go
GOMAXPROCS: 4
```

The select keyword

As you will learn in a short while, the `select` keyword is pretty powerful and it can do many things in a variety of situations. The `select` statement in Go looks like a `switch` statement for channels. In practice, this means that `select` allows a goroutine to wait on multiple communications operations. Therefore, the main benefit that you receive from `select` is that it gives you the power to work with multiple channels using a single `select` block. As a consequence, you can have nonblocking operations on channels.

The biggest problem when using multiple channels and the `select` keyword is **deadlocks**. This means that you should be extra careful during the design and the development process in order to avoid such deadlocks.

The Go code of `select.go` will clarify the use of the `select` keyword. This program will be presented in five parts.

The first part of `select.go` is shown in the following Go code:

```
package main

import (
    "fmt"
    "math/rand"
    "os"
    "strconv"
    "time"
)
```

The second code portion from `select.go` follows next:

```
func gen(min, max int, createNumber chan int, end chan bool) {
    for {
        select {
        case createNumber <- rand.Intn(max-min) + min:
        case <-end:
            close(end)
            return
        case <-time.After(4 * time.Second):
            fmt.Println("n\time.After()!")
        }
    }
}
```

So, what is really happening in the code of this `select` block? This particular `select` statement has three cases. Note that the `select` statements do not require a `default` branch. You can consider the third branch of the `select` statement as a clever `default` branch. The `time.After()` function will return after the specified time has passed, and therefore it will unblock the `select` statement in case all of the other channels are blocked.

A `select` statement is not evaluated sequentially, as all of its channels are examined simultaneously. If none of the channels in a `select` statement is ready, the `select` statement will block until one of the channels is ready. If multiple channels of a `select` statement is ready, then the Go runtime will make a random selection over the set of these ready channels. The Go runtime tries to make this random selection between these ready channels as uniformly and as fairly as possible.

The third part of `select.go` is as follows:

```
func main() {
    rand.Seed(time.Now().Unix())
    createNumber := make(chan int)
    end := make(chan bool)

    if len(os.Args) != 2 {
        fmt.Println("Please give me an integer!")
        return
    }
}
```

The fourth code portion of the `select.go` program contains the following Go code:

```
n, _ := strconv.Atoi(os.Args[1])
fmt.Printf("Going to create %d random numbers.\n", n)
go gen(0, 2*n, createNumber, end)

for i := 0; i < n; i++ {
    fmt.Printf("%d ", <-createNumber)
}
```

The reason for not examining the `error` value returned by `strconv.Atoi()` is to save some space. You should never do this in real applications.

The remaining Go code of the `select.go` program is as follows:

```
    time.Sleep(5 * time.Second)
    fmt.Println("Exiting...")
    end <- true
}
```

The main purpose of the `time.Sleep(5 * time.Second)` statement is to give the `time.After()` function of `gen()`, enough time to return, and therefore activate the relevant branch of the `select` statement.

The last statement of the `main()` function is what terminates the program by activating the `case <-end` branch of the `select` statement in `gen()` and executing the related Go code.

Executing `select.go` will generate the following output:

```
$ go run select.go 10
Going to create 10 random numbers.
13 17 8 14 19 9 2 0 19 5
time.After()!
Exiting...
```

The biggest advantage of `select` is that it can connect, orchestrate, and manage multiple channels. As channels connect goroutines, `select` connects channels that connect goroutines. Therefore, the `select` statement is one of the most important, if not the single most important, part of the Go concurrency model.

Timing out a goroutine

This section presents two very important techniques that will help you time out goroutines. Put simply, these two techniques will save you from having to wait forever for a goroutine to finish its job, and they will give you full control over the amount of time that you want to wait for a goroutine to end. Both techniques use the capabilities of the handy `select` keyword combined with the `time.After()` function that you experienced in the previous section.

Timing out a goroutine – take 1

The source code of the first technique will be saved in `timeOut1.go`, and it will be presented in four parts.

The first part of `timeOut1.go` is shown in the following Go code:

```go
package main

import (
    "fmt"
    "time"
)
```

The second code segment from `timeOut1.go` is as follows:

```go
func main() {
    c1 := make(chan string)
    go func() {
        time.Sleep(time.Second * 3)
        c1 <- "c1 OK"
    }()
```

The `time.Sleep()` call is used for emulating the time it will normally take for the function to finish its job. In this case, the anonymous function that is executed as a goroutine will take about 3 seconds (`time.Second * 3`) before writing a message to the `c1` channel.

The third code segment from `timeOut1.go` contains the following Go code:

```go
select {
    case res := <-c1:
        fmt.Println(res)
    case <-time.After(time.Second * 1):
        fmt.Println("timeout c1")
}
```

The purpose of the `time.After()` function call is to wait for the chosen time. In this case, you are not interested in the actual value returned by `time.After()`, but in the fact that the `time.After()` function call has ended, which means that the time has passed. In this case, as the value passed to the `time.After()` function is smaller than the value used in the `time.Sleep()` call that was executed as the goroutine in the previous code segment, you will most likely get a time-out message.

The remaining code from `timeOut1.go` follows next:

```go
c2 := make(chan string)
go func() {
    time.Sleep(3 * time.Second)
    c2 <- "c2 OK"
}()
```

```
            select {
            case res := <-c2:
                    fmt.Println(res)
            case <-time.After(4 * time.Second):
                    fmt.Println("timeout c2")
            }
    }
```

The preceding code both executes a goroutine that will take about 3 seconds to execute because of the `time.Sleep()` call, and it defines a timeout period of 4 seconds using `time.After(4 * time.Second)`. If the `time.After(4 * time.Second)` call returns after you get a value from the `c2` channel found in the first case of the `select` block, then there will not be any time out; otherwise, you will get a time out! However, in this case, the value of the `time.After()` call provides enough time to the `time.Sleep()` call to return, so you will most likely not get a time-out message here.

Executing `timeOut1.go` will generate the following output:

```
$ go run timeOut1.go
timeout c1
c2 OK
```

As expected, the first goroutine did not finish its job, whereas the second goroutine had enough time to finish.

Timing out a goroutine – take 2

The source code of the second technique will be saved in `timeOut2.go`, and it will be presented in five parts. This time, the time-out period is provided as a command-line argument to the program.

The first part of `timeOut2.go` is as follows:

```
package main

import (
    "fmt"
    "os"
    "strconv"
    "sync"
    "time"
)
```

The second code segment of timeOut2.go is shown in the following Go code:

```go
func timeout(w *sync.WaitGroup, t time.Duration) bool {
    temp := make(chan int)
    go func() {
        time.Sleep(5 * time.Second)
        defer close(temp)
        w.Wait()
    }()

    select {
    case <-temp:
        return false
    case <-time.After(t):
        return true
    }
}
```

In the preceding code, the time duration that will be used in the time.After() call is a parameter of the timeout() function, which means that it can vary. Once again, the select block implements the logic of the timeout. Additionally, the w.Wait() call will make the timeout() function wait for a matching sync.Done() function indefinitely in order to end. When the w.Wait() call returns, the first branch of the select statement will be executed.

The third code portion from timeOut2.go follows next:

```go
func main() {
    arguments := os.Args
    if len(arguments) != 2 {
        fmt.Println("Need a time duration!")
        return
    }

    var w sync.WaitGroup
    w.Add(1)

    t, err := strconv.Atoi(arguments[1])
    if err != nil {
        fmt.Println(err)
        return
    }
```

Go Concurrency – Advanced Topics

The fourth part of the `timeOut2.go` program follows next:

```
    duration := time.Duration(int32(t)) * time.Millisecond
    fmt.Printf("Timeout period is %s\n", duration)

    if timeout(&w, duration) {
        fmt.Println("Timed out!")
    } else {
        fmt.Println("OK!")
    }
```

The `time.Duration()` function converts an integer value into a `time.Duration` variable that you can use afterwards.

The remaining Go code from `timeOut2.go` is as follows:

```
    w.Done()
        if timeout(&w, duration) {
            fmt.Println("Timed out!")
        } else {
            fmt.Println("OK!")
        }
}
```

Once the `w.Done()` call is executed, the previous `timeout()` function will return. However, the second call to `timeout()` has no `sync.Done()` statement to wait for.

Executing `timeOut2.go` will generate the following type of output:

```
$ go run timeOut2.go 10000
Timeout period is 10s
Timed out!
OK!
```

In this execution of `timeOut2.go`, the time-out period is longer than the `time.Sleep(5 * time.Second)` call of the anonymous goroutine. However, without the required call to `w.Done()`, the anonymous goroutine cannot return and therefore the `time.After(t)` call will end first, so the `timeout()` function of the first `if` statement will return `true`. In the second `if` statement, the anonymous function does not have to wait for anything, so the `timeout()` function will return `false` because `time.Sleep(5 * time.Second)` will finish before `time.After(t)`:

```
$ go run timeOut2.go 100
Timeout period is 100ms
Timed out!
Timed out!
```

In the second program execution, however, the time-out period is too short, so both executions of `timeout()` do not have enough time to finish, therefore both will be timed out.

So, when defining a time-out period, make sure that you choose an appropriate value or your results might not be what you might expect.

Go channels revisited

Once the `select` keyword comes into play, Go channels can be used in several unique ways to do many more things than what you experienced in Chapter 9, *Go Concurrency – Goroutines, Channels, and Pipelines*. This section will reveal the many uses of Go channels.

It helps to remember that the zero value of the channel type is `nil`, and that if you send a message to a closed channel, the program will panic. However, if you try to read from a closed channel, you will get the zero value of the type of that channel. So, after closing a channel, you can no longer write to it, but you can still read from it.

In order to be able to close a channel, the channel must not be receive-only. Additionally, a `nil` channel always blocks, which means that trying to read or write from a `nil` channel will block. This property of channels can be very useful when you want to disable a branch of a `select` statement by assigning the `nil` value to a channel variable.

Go Concurrency – Advanced Topics

Finally, if you try to close a `nil` channel, your program will panic. This is best illustrated in the `closeNilChannel.go` program, which is presented next:

```
package main

func main() {
    var c chan string
    close(c)
}
```

Executing `closeNilChannel.go` will generate the following output:

```
$ go run closeNilChannel.go
panic: close of nil channel
goroutine 1 [running]:
main.main()
    /Users/mtsouk/closeNilChannel.go:5 +0x2a
exit status 2
```

Signal channels

A **signal channel** is one that is used just for signaling. Put simply, you can use a signal channel when you want to inform somebody else about something. Signal channels should not be used for transferring data.

> You should not confuse signal channels with **Unix signal handling**, which was discussed in `Chapter 8`, *Telling a Unix system What to Do*, because these two are totally different things.

You will see a code example that uses signal channels in the *Specifying the order of execution for your goroutines* section later in this chapter.

Buffered channels

The topic of this subsection is **buffered channels**. These are channels that allow the Go scheduler to put jobs in the queue quickly in order to be able to deal with more requests. Moreover, you can use buffered channels as **semaphores** in order to limit the throughput of your application.

The technique presented here works as follows:
All incoming requests are forwarded to a channel, which processes them one by one. When the channel is done processing a request, it sends a message to the original caller saying that it is ready to process a new one. So, the capacity of the buffer of the channel restricts the number of simultaneous requests that it can keep.

The technique will be presented with the help of the code found in `bufChannel.go`, which is broken into four parts.

The first part of the code of `bufChannel.go` is as follows:

```
package main

import (
    "fmt"
)
```

The second code segment of `bufChannel.go` contains the following Go code:

```
func main() {
    numbers := make(chan int, 5)
    counter := 10
```

The definition presented of the `numbers` channel gives it a place to store up to five integers.

The third part of the code of `bufChannel.go` is shown in the following Go code:

```
    for i := 0; i < counter; i++ {
        select {
            case numbers <- i:
            default:
                fmt.Println("Not enough space for", i)
        }
    }
}
```

In the preceding code, we tried to put 10 integers in the `numbers` channel. However, as the `numbers` channel has room for only 5 integers, you will not be able to store all 10 integers in it.

The remaining Go code of `bufChannel.go` follows next:

```
for i := 0; i < counter+5; i++ {
    select {
        case num := <-numbers:
            fmt.Println(num)
        default:
            fmt.Println("Nothing more to be done!")
            break
    }
}
```

In the preceding Go code, we tried to read the contents of the `numbers` channel using a `for` loop and a `select` statement. As long as there is something to read from the `numbers` channel, the first branch of the `select` statement will get executed. As long as the `numbers` channel is empty, the `default` branch will be executed.

Executing `bufChannel.go` will create the following type of output:

```
$ go run bufChannel.go
Not enough space for 5
Not enough space for 6
Not enough space for 7
Not enough space for 8
Not enough space for 9
0
1
2
3
4
Nothing more to be done!
Nothing more to be done!
Nothing more to be done!
Nothing more to be done!
Nothing more to be done!
Nothing more to be done!
Nothing more to be done!
Nothing more to be done!
Nothing more to be done!
Nothing more to be done!
```

Nil channels

In this subsection, you will learn about **nil channels**. These are a special kind of channel because they will always block. The use of the nil channels is illustrated in `nilChannel.go`, which will be presented in four code segments.

The first part of `nilChannel.go` is as follows:

```
package main

import (
    "fmt"
    "math/rand"
    "time"
)
```

The second code portion of `nilChannel.go` is shown in the following Go code:

```
func add(c chan int) {
    sum := 0
    t := time.NewTimer(time.Second)

    for {
        select {
        case input := <-c:
            sum = sum + input
        case <-t.c:
            c = nil
            fmt.Println(sum)
        }
    }
}
```

The `add()` function demonstrates how a nil channel is used. The `<-t.C` statement blocks the `C` channel of the `t` timer for the time that is specified in the `time.NewTimer()` call. Do not confuse the `c` channel, which is the parameter of the function, with the `t.C` channel, which belongs to the `t` timer. When the time expires, the timer sends a value to the `t.C` channel. This will trigger the execution of the relevant branch of the `select` statement, which will assign the `nil` value to the `c` channel and print the `sum` variable.

The third code segment of `nilChannel.go` follows next:

```go
func send(c chan int) {
    for {
        c <- rand.Intn(10)
    }
}
```

The purpose of the `send()` function is to generate random numbers and continue sending them to a channel for as long as the channel is open.

The remaining Go code of `nilChannel.go` is as follows:

```go
func main() {
    c := make(chan int)
    go add(c)
    go send(c)

    time.Sleep(3 * time.Second)
}
```

The `time.Sleep()` function is used for giving enough time to the two goroutines to operate.

Executing `nilChannel.go` will generate the following output:

```
$ go run nilChannel.go
13167523
$ go run nilChannel.go
12988362
```

Since the number of times that the first branch of the `select` statement in the `add()` function will be executed is not fixed, you will get different results from executing `nilChannel.go`.

Channel of channels

A **channel of channels** is a special kind of channel variable that works with channels instead of other types of variables. Nevertheless, you still have to declare a data type for a channel of channels. You can define a channel of channels using the `chan` keyword two times in a row, as shown in the following statement:

```go
c1 := make(chan chan int)
```

The other types of channels presented in this chapter are far more popular and handy than a channel of channels.

The use of channels of channels is illustrated using the code found in chSquare.go, which will be presented in four parts.

The first part of chSquare.go is as follows:

```
package main

import (
    "fmt"
    "os"
    "strconv"
    "time"
)

var times int
```

The second code portion from chSquare.go is shown in the following Go code:

```
func f1(cc chan chan int, f chan bool) {
    c := make(chan int)
    cc <- c
    defer close(c)

    sum := 0
    select {
    case x := <-c:
        for i := 0; i <= x; i++ {
            sum = sum + i
        }
        c <- sum
    case <-f:
        return
    }
}
```

After declaring a regular int channel, you will send that to the channel of channels variable. Then you use a select statement in order to be able to read data from the regular int channel or exit your function using the f signal channel.

Once you read a single value from the `c` channel, you start a `for` loop that calculates the sum of all integers from `0` up to the integer value that you just read. Next, you will send the calculated value to the `c int` channel and you are done.

The third part of `chSquare.go` contains the following Go code:

```go
func main() {
    arguments := os.Args
    if len(arguments) != 2 {
        fmt.Println("Need just one integer argument!")
        return
    }

    times, err := strconv.Atoi(arguments[1])
    if err != nil {
        fmt.Println(err)
        return
    }

    cc := make(chan chan int)
```

The last statement in the preceding code is where you declare a channel of channels variable named `cc`, which is the star of the program because everything depends on that variable: the `cc` variable is passed to the `f1()` function, and it will be used in the `for` loop which is coming next.

The remaining Go code of `chSquare.go` is as follows:

```go
    for i := 1; i < times+1; i++ {
        f := make(chan bool)
        go f1(cc, f)
        ch := <-cc
        ch <- i
        for sum := range ch {
            fmt.Print("Sum(", i, ")=", sum)
        }
        fmt.Println()
        time.Sleep(time.Second)
        close(f)
    }
}
```

The `f` channel is a **signal channel** used for ending the goroutine when the real work is finished. The `ch := <-cc` statement allows you to get a regular channel from the channel of channels variable in order to be able to send an `int` value to it using `ch <- i`. After this, you start reading from it using a `for` loop. Although the `f1()` function is programmed to send a single value back, you can also read multiple values. Note that each value of `i` is served by a different goroutine.

The type of a signal channel can be anything you want, including `bool`, which is used in the preceding code, and `struct{}`, which will be used in the signal channel in the next section. The main advantage of a `struct{}` signal channel is that no data can be sent to it, which can save you from bugs and misconceptions.

Executing `chSquare.go` will generate the following type of output:

```
$ go run chSquare.go 4
Sum(1)=1
Sum(2)=3
Sum(3)=6
Sum(4)=10
$ go run chSquare.go 6
Sum(1)=1
Sum(2)=3
Sum(3)=6
Sum(4)=10
Sum(5)=15
Sum(6)=21
```

Specifying the order of execution for your goroutines

Although you should not make any assumptions about the order in which your goroutines will be executed, there are times when you need to be able to control this order. This subsection illustrates such a technique using **signal channels**.

You might ask why choose to execute goroutines in a given order when simple functions could do the same job much more easily. The answer is simple: goroutines are able to operate concurrently and wait for other goroutines to end, whereas functions executed in sequence cannot do that!

Go Concurrency – Advanced Topics

The name of the Go program for this topic is `defineOrder.go`, and it will be presented in five parts.

The first part of `defineOrder.go` follows next:

```
package main

import (
    "fmt"
    "time"
)

func A(a, b chan struct{}) {
    <-a
    fmt.Println("A()!")
    time.Sleep(time.Second)
    close(b)
}
```

The `A()` function is blocked by the channel stored in the `a` parameter. Once that channel is unblocked in the `main()` function, the `A()` function will start working. Finally, it will close the `b` channel, which will unblock another function, in this case function `B()`.

The second code portion of `defineOrder.go` is shown in the following Go code:

```
func B(a, b chan struct{}) {
    <-a
    fmt.Println("B()!")
    close(b)
}
```

The logic in `B()` is the same as in the `A()` function. The function is blocked until the a channel is closed. Then it does its job and closes the b channel. Note that the a and b channels refer to the names of the parameters of the function.

The third code segment of `defineOrder.go` follows next:

```
func C(a chan struct{}) {
    <-a
    fmt.Println("C()!")
}
```

The `C()` function is blocked and waits for the `a` channel to close in order to start working.

The fourth part of defineOrder.go contains the following code:

```
func main() {
    x := make(chan struct{})
    y := make(chan struct{})
    z := make(chan struct{})
```

These three channels will be the parameters for the three functions.

The last code segment of defineOrder.go contains the following Go code:

```
    go C(z)
    go A(x, y)
    go C(z)
    go B(y, z)
    go C(z)

    close(x)
    time.Sleep(3 * time.Second)
}
```

Executing defineOrder.go will generate the desired output even though the C() function was called multiple times:

```
$ go run defineOrder.go
A()!
B()!
C()!
C()!
C()!
```

Calling the C() function multiple times as goroutines will work just fine because C() does not close any channels. However, if you call A() or B() more than once, you will most likely get an error message such as the following:

```
$ go run defineOrder.go
A()!
A()!
B()!
C()!
C()!
C()!
panic: close of closed channel
```

```
goroutine 7 [running]:
main.A(0xc420072060, 0xc4200720c0)
        /Users/mtsouk/Desktop/defineOrder.go:12 +0x9d
created by main.main
        /Users/mtsouk/Desktop/defineOrder.go:33 +0xfa
exit status
```

As you can see from the output, the A() function was called two times. However, as the A() function closes a channel, one of its goroutines will find that channel already closed and generate a panic situation. You will get a similar panic situation if you call B() more than once.

Shared memory and shared variables

Shared memory and **shared variables** are the most common ways for Unix threads to communicate with each other. A **Mutex** variable, which is an abbreviation for **mutual exclusion** variable, is mainly used for thread synchronization and for protecting shared data when multiple writes can occur at the same time. A mutex works like a **buffered channel** of capacity one, which allows at most one goroutine to access a shared variable at any given time. This means that there is no way for two or more goroutines to try to update that variable simultaneously.

A **critical section** of a concurrent program is the code that cannot be executed simultaneously by all processes, threads, or, in this case, by all goroutines. It is the code that needs to be protected by mutexes. Therefore, identifying the critical sections of your code will make the whole programming process so much simpler that you should pay attention to this task.

A critical section cannot be embedded in another critical section when both critical sections use the same sync.Mutex or sync.RWMutex variable. Put simply avoid, at almost any cost, spreading mutexes across functions because that makes it really hard to see whether you are embedding or not!

The next two subsections will illustrate the use of the sync.Mutex and sync.RWMutex types.

The sync.Mutex type

The `sync.Mutex` type is the Go implementation of a mutex. Its definition, which can be found in the `mutex.go` file of the `sync` directory, is as follows:

```
// A Mutex is a mutual exclusion lock.
// The zero value for a Mutex is an unlocked mutex.
//
// A Mutex must not be copied after first use.
type Mutex struct {
    state int32
    sema  uint32
}
```

The definition of the `sync.Mutex` type is nothing extraordinary. All of the interesting work is being done by the `sync.Lock()` and `sync.Unlock()` functions that can lock and unlock a `sync.Mutex` mutex, respectively. Locking a mutex means that nobody else can lock it until it has been released using the `sync.Unlock()` function.

The `mutex.go` program, which is going to be presented in five parts, illustrates the use of the `sync.Mutex` type.

The first code segment of `mutex.go` follows next:

```
package main

import (
    "fmt"
    "os"
    "strconv"
    "sync"
    "time"
)

var (
    m  sync.Mutex
    v1 int
)
```

The second part of `mutex.go` is shown in the following Go code:

```
func change(i int) {
    m.Lock()
    time.Sleep(time.Second)
    v1 = v1 + 1
    if v1%10 == 0 {
```

```
        v1 = v1 - 10*i
    }
    m.Unlock()
}
```

The critical section of this function is the Go code between the m.Lock() and m.Unlock() statements.

The third part of mutex.go contains the following Go code:

```
func read() int {
    m.Lock()
    a := v1
    m.Unlock()
    return a
}
```

Similarly, the critical section of this function is defined by the m.Lock() and m.Unlock() statements.

The fourth code segment of mutex.go follows next:

```
func main() {
    if len(os.Args) != 2 {
        fmt.Println("Please give me an integer!")
        return
    }

    numGR, err := strconv.Atoi(os.Args[1])
    if err != nil {
        fmt.Println(err)
        return
    }
    var waitGroup sync.WaitGroup
```

The last part of mutex.go is shown in the following Go code:

```
fmt.Printf("%d ", read())
    for i := 0; i < numGR; i++ {
        waitGroup.Add(1)
        go func(i int) {
            defer waitGroup.Done()
            change(i)
            fmt.Printf("-> %d", read())
        }(i)
    }
```

```
        waitGroup.Wait()
        fmt.Printf("-> %d\n", read())
}
```

Executing `mutex.go` will generate the following output:

```
$ go run mutex.go 21
0 -> 1-> 2-> 3-> 4-> 5-> 6-> 7-> 8-> 9-> -30-> -29-> -28-> -27-> -26->
-25-> -24-> -23-> -22-> -21-> -210-> -209-> -209
$ go run mutex.go 21
0 -> 1-> 2-> 3-> 4-> 5-> 6-> 7-> 8-> 9-> -130-> -129-> -128-> -127-> -126->
-125-> -124-> -123-> -122-> -121-> -220-> -219-> -219
$ go run mutex.go 21
0 -> 1-> 2-> 3-> 4-> 5-> 6-> 7-> 8-> 9-> -100-> -99-> -98-> -97-> -96->
-95-> -94-> -93-> -92-> -91-> -260-> -259-> -259
```

If you remove the `m.Lock()` and `m.Unlock()` statements from the `change()` function, the program will generate output similar to the following:

```
$ go run mutex.go 21
0 -> 1-> 6-> 7-> 5-> -60-> -59-> 9-> 2-> -58-> 3-> -52-> 4-> -57-> 8->
-55-> -90-> -54-> -89-> -53-> -56-> -51-> -89
$ go run mutex.go 21
0 -> 1-> 7-> 8-> 9-> 5-> -99-> 4-> 2-> -97-> -96-> 3-> -98-> -95-> -100->
-93-> -94-> -92-> -91-> -230-> 6-> -229-> -229
$ go run mutex.go 21
0 -> 3-> 7-> 8-> 9-> -120-> -119-> -118-> -117-> 1-> -115-> -114-> -116->
4-> 6-> -112-> 2-> -111-> 5-> -260-> -113-> -259-> -259
```

The reason for such a change in the output is that all goroutines are simultaneously changing the shared variable, which is the main reason that the output appears randomly generated.

What happens if you forget to unlock a mutex?

In this section, you will see what happens if you forget to unlock `sync.Mutex`. You will do this using the Go code of `forgetMutex.go`, which will be presented in two parts.

The first part of `forgetMutex.go` is shown in the following Go code:

```
package main

import (
    "fmt"
    "sync"
)
```

Go Concurrency – Advanced Topics

```go
var m sync.Mutex

func function() {
    m.Lock()
    fmt.Println("Locked!")
}
```

All of the problems in this program are caused because the developer forgot to release the lock on the `m sync.Mutex` mutex. However, if your program will call `function()` only once, then everything will look just fine!

The second part of `forgetMutex.go` is as follows:

```go
func main() {
    var w sync.WaitGroup

    go func() {
        defer w.Done()
        function()
    }()
    w.Add(1)

    go func() {
        defer w.Done()
        function()
    }()
    w.Add(1)

    w.Wait()
}
```

There is nothing wrong with the `main()` function that generates just two goroutines and waits for them to finish.

Executing `forgetMutex.go` will produce the following output:

```
$ go run forgetMutex.go
Locked!
fatal error: all goroutines are asleep - deadlock!
goroutine 1 [semacquire]:
sync.runtime_Semacquire(0xc42001209c)
        /usr/local/Cellar/go/1.9.4/libexec/src/runtime/sema.go:56 +0x39
sync.(*WaitGroup).Wait(0xc420012090)
        /usr/local/Cellar/go/1.9.4/libexec/src/sync/waitgroup.go:131 +0x72
main.main()
        /Users/mtsouk/forgetMutex.go:30 +0xb6
```

```
goroutine 5 [semacquire]:
sync.runtime_SemacquireMutex(0x115c6fc, 0x0)
        /usr/local/Cellar/go/1.9.4/libexec/src/runtime/sema.go:71 +0x3d
sync.(*Mutex).Lock(0x115c6f8)
        /usr/local/Cellar/go/1.9.4/libexec/src/sync/mutex.go:134 +0xee
main.function()
        /Users/mtsouk/forgetMutex.go:11 +0x2d
main.main.func1(0xc420012090)
        /Users/mtsouk/forgetMutex.go:20 +0x48
created by main.main
        /Users/mtsouk/forgetMutex.go:18 +0x58
exit status 2
```

So, forgetting to unlock a `sync.Mutex` mutex will create a panic situation even in the simplest kind of program. The same applies to the `sync.RWMutex` type of mutex that you will to work with in the next section.

The sync.RWMutex type

The `sync.RWMutex` type is another kind of mutex-actually, it is an improved version of `sync.Mutex`, which is defined in the `rwmutex.go` file of the `sync` directory as follows:

```
type RWMutex struct {
    w           Mutex  // held if there are pending writers
    writerSem   uint32 // semaphore for writers to wait for completing readers
    readerSem   uint32 // semaphore for readers to wait for completing writers
    readerCount int32  // number of pending readers
    readerWait  int32  // number of departing readers
}
```

In other words, `sync.RWMutex` is based on `sync.Mutex` with the necessary additions and improvements.

Now let's talk about how `sync.RWMutex` improves `sync.Mutex`. Although only one function is allowed to perform write operations using a `sync.RWMutex` mutex, you can have multiple readers owning a `sync.RWMutex` mutex. However, there is one thing of which you should be aware: until all of the readers of a `sync.RWMutex` mutex unlock that mutex, you cannot lock it for writing, which is the small price you have to pay for allowing multiple readers.

The functions that can help you work with a `sync.RWMutex` mutex are `RLock()` and `RUnlock()`, which are used for locking and unlocking the mutex, respectively, for reading purposes. The `Lock()` and `Unlock()` functions used in a `sync.Mutex` mutex should still be used when you want to lock and unlock a `sync.RWMutex` mutex for writing purposes. Thus, an `RLock()` function call that locks for reading purposes should be paired with an `RUnlock()` function call. Finally, it should be apparent that you should not make changes to any shared variables inside the `RLock()` and `RUnlock()` blocks of code.

The Go code found in `rwMutex.go` illustrates the use and usefulness of the `sync.RWMutex` type. The program is presented in six parts, and it contains two slightly different versions of the same function. The first one uses a `sync.RWMutex` mutex for reading, and the second one uses a `sync.Mutex` mutex for reading. The performance difference between the two functions will help you understand the benefits of the `sync.RWMutex` mutex better, when used for reading purposes.

The first part of `rwMutex.go` contains the following Go code:

```go
package main

import (
    "fmt"
    "os"
    "sync"
    "time"
)

var Password = secret{password: "myPassword"}

type secret struct {
    RWM      sync.RWMutex
    M        sync.Mutex
    password string
}
```

The `secret` structure holds a shared variable, a `sync.RWMutex` mutex, and a `sync.Mutex` mutex.

The second code portion of `rwMutex.go` is shown in the following code:

```go
func Change(c *secret, pass string) {
    c.RWM.Lock()
    fmt.Println("LChange")
    time.Sleep(10 * time.Second)
```

```
        c.password = pass
        c.RWM.Unlock()
}
```

The `Change()` function modifies a shared variable, which means that you need to use an exclusive lock, which is the reason for using the `Lock()` and `Unlock()` functions. You cannot get away with using exclusive locks when changing things!

The third part of `rwMutex.go` is as follows:

```
func show(c *secret) string {
        c.RWM.RLock()
        fmt.Print("show")
        time.Sleep(3 * time.Second)
        defer c.RWM.RUnlock()
        return c.password
}
```

The `show()` function uses the `RLock()` and `RUnlock()` functions because its critical section is used for reading a shared variable. So, although many goroutines can read the shared variable, no one can change it without using the `Lock()` and `Unlock()` functions. However, the `Lock()` function will be blocked for as long as there is someone reading that shared variable using the mutex.

The fourth code segment of `rwMutex.go` contains the following Go code:

```
func showWithLock(c *secret) string {
        c.M.Lock()
        fmt.Println("showWithLock")
        time.Sleep(3 * time.Second)
        defer c.M.Unlock()
        return c.password
}
```

The only difference between the code of the `showWithLock()` function and the code of the `show()` function is that the `showWithLock()` function uses an exclusive lock for reading, which means that only one `showWithLock()` function can read the `password` field of the `secret` structure.

The fifth part of `rwMutex.go` contains the following Go code:

```
func main() {
    var showFunction = func(c *secret) string { return "" }
    if len(os.Args) != 2 {
            fmt.Println("Using sync.RWMutex!")
            showFunction = show
```

```go
    } else {
        fmt.Println("Using sync.Mutex!")
        showFunction = showWithLock
    }

    var waitGroup sync.WaitGroup

    fmt.Println("Pass:", showFunction(&Password))
```

The remaining code of `rwMutex.go` follows next:

```go
    for i := 0; i < 15; i++ {
        waitGroup.Add(1)
        go func() {
            defer waitGroup.Done()
            fmt.Println("Go Pass:", showFunction(&Password))
        }()
    }

    go func() {
        waitGroup.Add(1)
        defer waitGroup.Done()
        Change(&Password, "123456")
    }()

    waitGroup.Wait()
    fmt.Println("Pass:", showFunction(&Password))
}
```

Executing `rwMutex.go` two times and using the `time(1)` command-line utility to benchmark the two versions of the program will generate the following type of output:

```
$ time go run rwMutex.go 10 >/dev/null
real    0m51.206s
user    0m0.130s
sys     0m0.074s
$ time go run rwMutex.go >/dev/null
real    0m22.191s
user    0m0.135s
sys     0m0.071s
```

Note that the `>/dev/null` at the end of the preceding commands is for omitting the output of the two commands. Thus, the version that uses the `sync.RWMutex` mutex is much faster than the version that uses `sync.Mutex`.

Sharing memory using goroutines

The last subsection of this topic illustrates how you can share data using a dedicated goroutine. Although shared memory is the traditional way that threads use to communicate with each other, Go comes with built-in synchronization features that allow a single goroutine to own a shared piece of data. This means that other goroutines must send messages to this single goroutine that owns the shared data, which prevents the corruption of the data. Such a goroutine is called a **monitor goroutine**. In Go terminology, this is *sharing by communicating instead of communicating by sharing.*

The technique will be illustrated using the `monitor.go` source file, which will be presented in five parts. The `monitor.go` program generates random numbers using a monitor goroutine.

The first part of `monitor.go` is as follows:

```
package main

import (
    "fmt"
    "math/rand"
    "os"
    "strconv"
    "sync"
    "time"
)

var readValue = make(chan int)
var writeValue = make(chan int)
```

The `readValue` channel is used for reading the random numbers, whereas the `writeValue` channel is used for getting new random numbers.

The second code portion of `monitor.go` is shown in the following code:

```
func set(newValue int) {
    writeValue <- newValue
}

func read() int {
    return <-readValue
}
```

Go Concurrency – Advanced Topics

The purpose of the `set()` function is to set the value of the shared variable, whereas the purpose of the `read()` function is to read the value of the saved variable.

The third code segment of the `monitor.go` program is as follows:

```
func monitor() {
    var value int
    for {
        select {
        case newValue := <-writeValue:
            value = newValue
            fmt.Printf("%d ", value)
        case readValue <- value:
        }
    }
}
```

All of the logic of the program can be found in the implementation of the `monitor()` function. Most specifically, the `select` statement orchestrates the operation of the entire program. When you have a read request, the `read()` function attempts to read from the `readValue` channel, which is controlled by the `monitor()` function. This returns the current value that is kept in the `value` variable. On the other hand, when you want to change the stored value, you call `set()`. This writes to the `writeValue` channel that is also handled by the `select` statement. As a result, no one can deal with the `value` shared variable without using the `monitor()` function.

The fourth code segment of `monitor.go` follows next:

```
func main() {
    if len(os.Args) != 2 {
        fmt.Println("Please give an integer!")
        return
    }
    n, err := strconv.Atoi(os.Args[1])
    if err != nil {
        fmt.Println(err)
        return
    }

    fmt.Printf("Going to create %d random numbers.\n", n)
    rand.Seed(time.Now().Unix())
    go monitor()
```

The last part of `monitor.go` contains the following Go code:

```go
var w sync.WaitGroup

    for r := 0; r < n; r++ {
        w.Add(1)
        go func() {
            defer w.Done()
            set(rand.Intn(10 * n))
        }()
    }
    w.Wait()
    fmt.Printf("\nLast value: %d\n", read())
}
```

Executing `monitor.go` generates the following output:

```
$ go run monitor.go 20
Going to create 20 random numbers.
89 88 166 42 149 89 20 84 44 178 184 28 52 121 62 91 31 117 140 106
Last value: 106
$ go run monitor.go 10
Going to create 10 random numbers.
30 16 66 70 65 45 31 57 62 26
Last value: 26
```

Personally, I prefer to use a monitor goroutine instead of the traditional shared memory techniques because the implementation that uses the monitor goroutine is safer and closer to the Go philosophy.

Catching race conditions

A **data race condition** is a situation where two or more running elements such as threads and goroutines try to take control or modify a shared resource or a variable of a program. Strictly speaking, a data race occurs when two or more instructions access the same memory address, where at least one of them performs a write operation.

Go Concurrency – Advanced Topics

Using the `-race` flag when running or building a Go source file will turn on the Go **race detector**, which will make the compiler create a modified version of a typical executable file. This modified version can record all accesses to shared variables as well as all synchronization events that take place, including calls to `sync.Mutex` and `sync.WaitGroup`. After analyzing the relevant events, the race detector prints a report that can help you identify potential problems so that you can correct them.

Look at the following Go code, which is saved as `raceC.go`. This program is presented in three parts.

The first part of `raceC.go` is as follows:

```go
package main

import (
    "fmt"
    "os"
    "strconv"
    "sync"
)

func main() {
    arguments := os.Args
    if len(arguments) != 2 {
        fmt.Println("Give me a natural number!")
        os.Exit(1)
    }
    numGR, err := strconv.Atoi(os.Args[1])
    if err != nil {
        fmt.Println(err)
        return
    }
```

The second part of `raceC.go` contains the following Go code:

```go
    var waitGroup sync.WaitGroup
    var i int

    k := make(map[int]int)
    k[1] = 12

    for i = 0; i < numGR; i++ {
        waitGroup.Add(1)
        go func() {
            defer waitGroup.Done()
            k[i] = i
```

```
        } ()
    }
```

The remaining Go code of `raceC.go` is as follows:

```
    k[2] = 10
    waitGroup.Wait()
    fmt.Printf("k = %#v\n", k)
}
```

As if it were not enough that many goroutines are accessing the `k` map at the same time, we added another statement that accesses the `k` map before calling the `sync.Wait()` function.

If you execute `raceC.go`, you will get the following output without any warning or error messages:

```
$ go run raceC.go 10
k = map[int]int{7:10, 2:10, 10:10, 1:12}
$ go run raceC.go 10
k = map[int]int{2:10, 10:10, 1:12, 8:8, 9:9}
$ go run raceC.go 10
k = map[int]int{10:10, 1:12, 6:7, 7:7, 2:10}
```

If you execute `raceC.go` only once, then everything will look normal despite the fact that you do not get what you would expect when printing the contents of the `k` map. However, executing `raceC.go` multiple times tells us that there is something wrong here, mainly because each execution generates a different output.

There are many more things that we can get from `raceC.go` and its unexpected output, if we decide to use the Go race detector to analyze it:

```
$ go run -race raceC.go 10
==================
WARNING: DATA RACE
Read at 0x00c42007c008 by goroutine 6:
  main.main.func1()
      /Users/mtsouk/ch/ch10/code/raceC.go:32 +0x69
Previous write at 0x00c42007c008 by main goroutine:
  main.main()
      /Users/mtsouk/ch/ch10/code/raceC.go:28 +0x27b
Goroutine 6 (running) created at:
  main.main()
      /Users/mtsouk/ch/ch10/code/raceC.go:30 +0x24b
==================
==================
WARNING: DATA RACE
Write at 0x00c420074180 by goroutine 7:
```

```
            runtime.mapassign_fast64()
                /usr/local/Cellar/go/1.9.3/libexec/src/runtime/hashmap_fast.go:598
    +0x0
      main.main.func1()
            /Users/mtsouk/ch/ch10/code/raceC.go:32 +0x90
    Previous write at 0x00c420074180 by goroutine 6:
      runtime.mapassign_fast64()
            /usr/local/Cellar/go/1.9.3/libexec/src/runtime/hashmap_fast.go:598
    +0x0
      main.main.func1()
            /Users/mtsouk/ch/ch10/code/raceC.go:32 +0x90
    Goroutine 7 (running) created at:
      main.main()
            /Users/mtsouk/ch/ch10/code/raceC.go:30 +0x24b
    Goroutine 6 (finished) created at:
      main.main()
            /Users/mtsouk/ch/ch10/code/raceC.go:30 +0x24b
    ==================
    k = map[int]int{3:4, 5:5, 9:9, 10:10, 1:1, 2:10, 4:4, 6:6, 7:7, 8:8}
    Found 2 data race(s)
    exit status 66
```

So, the race detector found two data races. Each one begins with the WARNING: DATA RACE message in its output.

The first **data race** happens inside main.main.func1(), which is called by the for loop that is called by a goroutine. The problem here is signified by the Previous write message. After examining the related code, it is easy to see that the actual problem is that the anonymous function takes no parameters, which means that the value of i that is used in the for loop cannot be deterministically discerned, as it keeps changing due to the for loop, which is a write operation.

The message of the second data race is Write at 0x00c420074180 by goroutine 7. If you read the relevant output, you will see that the data race is related to a write operation, and it is happening on a Go map by at least two goroutines. As the two goroutines have the same name (main.main.func1()), it indicates that we are talking about the same goroutine. Two goroutines trying to write the same variable is a data race condition!

The main.main.func1() notation is used by Go in order to name an anonymous function internally. If you had different anonymous functions, their names would have been different as well.

You might ask yourself, *What can I do now in order to correct the problems coming from the two data races?*

Well, you can rewrite the `main()` function of `raceC.go` as follows:

```
func main() {
    arguments := os.Args
    if len(arguments) != 2 {
        fmt.Println("Give me a natural number!")
        os.Exit(1)
    }
    numGR, err := strconv.Atoi(os.Args[1])
    if err != nil {
        fmt.Println(err)
        return
    }
    var waitGroup sync.WaitGroup
    var i int

    k := make(map[int]int)
    k[1] = 12

    for i = 0; i < numGR; i++ {
        waitGroup.Add(1)
        go func(j int) {
            defer waitGroup.Done()
            aMutex.Lock()
            k[j] = j
            aMutex.Unlock()
        }(i)
    }

    waitGroup.Wait()
    k[2] = 10
    fmt.Printf("k = %#v\n", k)
}
```

The `aMutex` variable is a global `sync.Mutex` variable defined outside the `main()` function that is accessible from everywhere in the program. Although this is not required, having such a global variable can save you from having to pass it to your functions all of the time.

Go Concurrency – Advanced Topics

Saving the new version of `raceC.go` as `noRaceC.go` and executing it will generate the expected output:

```
$ go run noRaceC.go 10
k = map[int]int{1:1, 0:0, 5:5, 3:3, 6:6, 9:9, 2:10, 4:4, 7:7, 8:8}
```

Processing `noRaceC.go` with the Go race detector will generate the following output:

```
$ go run -race noRaceC.go 10
k = map[int]int{5:5, 7:7, 9:9, 1:1, 0:0, 4:4, 6:6, 8:8, 2:10, 3:3}
```

Note that you need to use a locking mechanism while accessing the `k` map. If you do not use such a mechanism and just change the implementation of the anonymous function that is executed as a goroutine, you will get the following output from `go run noRaceC.go`:

```
$ go run noRaceC.go 10
fatal error: concurrent map writes
goroutine 10 [running]:
runtime.throw(0x10ca0bd, 0x15)
        /usr/local/Cellar/go/1.9.3/libexec/src/runtime/panic.go:605 +0x95
fp=0xc420024738 sp=0xc420024718 pc=0x10276b5
runtime.mapassign_fast64(0x10ae680, 0xc420074180, 0x5, 0x0)
        /usr/local/Cellar/go/1.9.3/libexec/src/runtime/hashmap_fast.go:607
+0x3d2 fp=0xc420024798 sp=0xc420024738 pc=0x100b582
main.main.func1(0xc420010090, 0xc420074180, 0x5)
        /Users/mtsouk/Desktop/masterGo/ch/ch10/code/noRaceC.go:35 +0x6b
fp=0xc4200247c8 sp=0xc420024798 pc=0x1096f5b
runtime.goexit()
        /usr/local/Cellar/go/1.9.3/libexec/src/runtime/asm_amd64.s:2337 +0x1
fp=0xc4200247d0 sp=0xc4200247c8 pc=0x1050c21
created by main.main
        /Users/mtsouk/Desktop/masterGo/ch/ch10/code/noRaceC.go:32 +0x15a
goroutine 1 [semacquire]:
sync.runtime_Semacquire(0xc42001009c)
        /usr/local/Cellar/go/1.9.3/libexec/src/runtime/sema.go:56 +0x39
sync.(*WaitGroup).Wait(0xc420010090)
        /usr/local/Cellar/go/1.9.3/libexec/src/sync/waitgroup.go:131 +0x72
main.main()
        /Users/mtsouk/Desktop/masterGo/ch/ch10/code/noRaceC.go:40 +0x17a
goroutine 12 [runnable]:
sync.(*WaitGroup).Done(0xc420010090)
        /usr/local/Cellar/go/1.9.3/libexec/src/sync/waitgroup.go:99 +0x43
main.main.func1(0xc420010090, 0xc420074180, 0x7)
        /Users/mtsouk/Desktop/masterGo/ch/ch10/code/noRaceC.go:37 +0x79
created by main.main
        /Users/mtsouk/Desktop/masterGo/ch/ch10/code/noRaceC.go:32 +0x15a
exit status 2
```

The root of the problem is pretty obvious: `concurrent map writes`.

The context package

The main purpose of the `context` package is to define the `Context` type and support **cancellation**! Yes, you heard that right; the main purpose of the `context` package is supporting cancellation because there are times that, for some reason, you want to abandon what you are doing. However, it would be very helpful to be able to include some extra information about your cancelling decisions. The `context` package allows you to do exactly that!

If you take a look at the source code of the `context` package, you will realize that its implementation is pretty simple—even the implementation of the `Context` type is pretty simple, yet the `context` package is very important.

> The `context` package existed for a while as an external Go package, but it first appeared as a standard Go package in Go version 1.7. So, if you have an older Go version, you will not be able to follow this section without first downloading the `context` package.

The `Context` type is an interface with four methods, named `Deadline()`, `Done()`, `Err()`, and `Value()`. The good news is that you do not need to implement all of these functions of the `Context` interface—you just need to modify a `Context` variable using functions such as `context.WithCancel()` and `context.WithTimeout()`.

The following is a simple use of the `context` package using the Go code of the `simpleContext.go` file, which will be presented in six parts.

The first code segment of `simpleContext.go` contains the following code:

```
package main

import (
    "context"
    "fmt"
    "os"
    "strconv"
    "time"
)
```

The second part of `simpleContext.go` follows next:

```
func f1(t int) {
    c1 := context.Background()
    c1, cancel := context.WithCancel(c1)
    defer cancel()

    go func() {
        time.Sleep(4 * time.Second)
        cancel()
    }()
```

The `f1()` function requires just one parameter, which is the time delay, because everything else is defined inside the function. Note that the type of the `cancel` variable is `context.CancelFunc`.

You need to call the `context.Background()` function in order to initialize an empty `Context` parameter. The `context.WithCancel()` function uses an existing `Context` and creates a child of it with cancellation. The `context.WithCancel()` function also creates a `Done` channel that can be closed, either when the `cancel()` function is called, as shown in the preceding code, or when the `Done` channel of the parent context is closed.

The third code portion of `simpleContext.go` contains the rest of the code of the `f1()` function:

```
    select {
    case <-c1.Done():
        fmt.Println("f1():", c1.Err())
        return
    case r := <-time.After(time.Duration(t) * time.Second):
        fmt.Println("f1():", r)
    }
    return
}
```

Here you see the use of the `Done()` function of a `Context` variable. When this function is called, you have a cancellation. The return value of `Context.Done()` is a channel, because otherwise you would have not been able to use it in a `select` statement.

The fourth part of `simpleContext.go` contains the following Go code:

```go
func f2(t int) {
    c2 := context.Background()
    c2, cancel := context.WithTimeout(c2, time.Duration(t)*time.Second)
    defer cancel()

    go func() {
        time.Sleep(4 * time.Second)
        cancel()
    }()

    select {
    case <-c2.Done():
        fmt.Println("f2():", c2.Err())
        return
    case r := <-time.After(time.Duration(t) * time.Second):
        fmt.Println("f2():", r)
    }
    return
}
```

This part showcases the use of the `context.WithTimeout()` function, which requires two parameters: a `Context` parameter and a `time.Duration` parameter. When the timeout period expires, the `cancel()` function is automatically called.

The fifth part of `simpleContext.go` is as follows:

```go
func f3(t int) {
    c3 := context.Background()
    deadline := time.Now().Add(time.Duration(2*t) * time.Second)
    c3, cancel := context.WithDeadline(c3, deadline)
    defer cancel()

    go func() {
        time.Sleep(4 * time.Second)
        cancel()
    }()

    select {
    case <-c3.Done():
        fmt.Println("f3():", c3.Err())
        return
    case r := <-time.After(time.Duration(t) * time.Second):
        fmt.Println("f3():", r)
```

```
        }
        return
}
```

The preceding Go code illustrates the use of the `context.WithDeadline()` function that requires two parameters: a `Context` variable and a time in the future that signifies the deadline of the operation. When the deadline passes, the `cancel()` function is automatically called.

The remaining Go code of `simpleContext.go` is as follows:

```
func main() {
    if len(os.Args) != 2 {
        fmt.Println("Need a delay!")
        return
    }

    delay, err := strconv.Atoi(os.Args[1])
    if err != nil {
        fmt.Println(err)
        return
    }
    fmt.Println("Delay:", delay)

    f1(delay)
    f2(delay)
    f3(delay)
}
```

Executing `simpleContext.go` will generate the following output:

```
$ go run simpleContext.go 4
Delay: 4
f1(): 2018-02-13 23:30:00.271587 +0200 EET m=+4.003435078
f2(): 2018-02-13 23:30:04.272678 +0200 EET m=+8.004706996
f3(): 2018-02-13 23:30:08.273738 +0200 EET m=+12.005937567
$ go run simpleContext.go 10
Delay: 10
f1(): context canceled
f2(): context canceled
f3(): context canceled
```

The long lines of the output are the return values of the `time.After()` function calls. They denote normal operation of the program. The point here is that the operation of the program is cancelled when there are delays in its execution.

This is as simple as it gets for the use of the `context` package, as the code presented did not do any serious work with the `Context` interface. However, the Go code included in the next section will present a more realistic example.

An advanced example of the context package

The functionality of the `context` package will be illustrated much better and in greater depth using the Go code of the `useContext.go` program, which is presented in five parts.

In this example, you will create an HTTP client that does not want to wait too long for a server response, which is not an unusual scenario. In fact, as almost all HTTP clients support such functionality, you will study another technique for timing out an HTTP request in Chapter 12, *The Foundations of Network Programming in Go*.

The `useContext.go` program requires two command-line arguments: the URL of the server to which it is going to connect and the delay for which it should wait. If the program has only one command-line argument, then the delay would be 5 seconds.

The first code segment of `useContext.go` follows next:

```
package main

import (
    "context"
    "fmt"
    "io/ioutil"
    "net/http"
    "os"
    "strconv"
    "sync"
    "time"
)

var (
    myUrl string
    delay int = 5
    w     sync.WaitGroup
)

type myData struct {
```

```
        r   *http.Response
        err error
}
```

Both `myURL` and `delay` are global variables, so they can be accessed from anything in the code. Additionally, there is a `sync.WaitGroup` variable named `w`, which also has a global scope, and the definition of a structure named `myData` for keeping together the response of the web server, along with an `error` variable in case there is an error somewhere.

The second part of `useContext.go` is shown in the following Go code:

```
func connect(c context.Context) error {
    defer w.Done()
    data := make(chan myData, 1)

    tr := &http.Transport{}
    httpClient := &http.Client{Transport: tr}

    req, _ := http.NewRequest("GET", myUrl, nil)
```

The preceding Go code deals with the HTTP connection.

 You will learn more about developing web servers and clients in Go in Chapter 12, *The Foundations of Network Programming in Go*.

The third code portion of `useContext.go` contains the following Go code:

```
go func() {
    response, err := httpClient.Do(req)
    if err != nil {
        fmt.Println(err)
        data <- myData{nil, err}
            return
    } else {
            pack := myData{response, err}
            data <- pack
    }
}()
```

The fourth code segment of `useContext.go` is shown in the following Go code:

```go
    select {
    case <-c.Done():
        tr.CancelRequest(req)
        <-data
        fmt.Println("The request was cancelled!")
        return c.Err()
    case ok := <-data:
        err := ok.err
        resp := ok.r
        if err != nil {
            fmt.Println("Error select:", err)
            return err
        }
        defer resp.Body.Close()

        realHTTPData, err := ioutil.ReadAll(resp.Body)
        if err != nil {
            fmt.Println("Error select:", err)
            return err
        }
        fmt.Printf("Server Response: %s\n", realHTTPData)

    }
    return nil
}
```

The remaining code of `useContext.go`, which is the implementation of the `main()` function, is as follows:

```go
func main() {
    if len(os.Args) == 1 {
        fmt.Println("Need a URL and a delay!")
        return
    }

    myUrl = os.Args[1]
    if len(os.Args) == 3 {
        t, err := strconv.Atoi(os.Args[2])
        if err != nil {
            fmt.Println(err)
            return
        }
        delay = t
    }
```

```go
        fmt.Println("Delay:", delay)
        c := context.Background()
        c, cancel := context.WithTimeout(c, time.Duration(delay)*time.Second)
        defer cancel()

        fmt.Printf("Connecting to %s \n", myUrl)
        w.Add(1)
        go connect(c)
        w.Wait()
        fmt.Println("Exiting...")
}
```

The timeout period is defined by the `context.WithTimeout()` method. The `connect()` function that is executed as a goroutine will either terminate normally or when the `cancel()` function is executed.

Although it is not necessary to know about the server part of the operation, it is good to see how a Go version of a web server can be slow in a random way; that is, a random number generator decides how slow your web server will be. The name of the source file is `slowWWW.go`, and its contents are as follows:

```go
package main

import (
    "fmt"
    "math/rand"
    "net/http"
    "os"
    "time"
)

func random(min, max int) int {
    return rand.Intn(max-min) + min
}

func myHandler(w http.ResponseWriter, r *http.Request) {
    delay := random(0, 15)
    time.Sleep(time.Duration(delay) * time.Second)

    fmt.Fprintf(w, "Serving: %s\n", r.URL.Path)
    fmt.Fprintf(w, "Delay: %d\n", delay)
    fmt.Printf("Served: %s\n", r.Host)
}

func main() {
    seed := time.Now().Unix()
    rand.Seed(seed)
```

```
        PORT := ":8001"
        arguments := os.Args
        if len(arguments) == 1 {
            fmt.Println("Using default port number: ", PORT)
        } else {
            PORT = ":" + arguments[1]
        }

        http.HandleFunc("/", myHandler)
        err := http.ListenAndServe(PORT, nil)
        if err != nil {
            fmt.Println(err)
            os.Exit(10)
        }
    }
```

As you can see, you do not need to use the `context` package in the `slowWWW.go` file because it is the job of the web client to decide how much time it can wait for a response!

The code of the `myHandler()` function is responsible for the slowness of the web server program. The delay can be from 0 seconds to 14 seconds, as introduced by the `random(0, 15)` function call.

If you try to use the `slowWWW.go` web server using a tool such as `wget(1)`, you will receive the following output:

```
$ wget -qO- http://localhost:8001/
Serving: /
Delay: 4
$ wget -qO- http://localhost:8001/
Serving: /
Delay: 13
```

This happens because the default timeout value of `wget(1)` is larger.

Executing `useContext.go` while `slowWWW.go` is already running in another Unix shell will create the following output when processed with the handy `time(1)` utility:

```
$ time go run useContext.go http://localhost:8001/ 1
Delay: 1
Connecting to http://localhost:8001/
Get http://localhost:8001/: net/http: request canceled
The request was cancelled!
Exiting...
real    0m1.374s
user    0m0.304s
sys     0m0.117s
```

```
$ time go run useContext.go http://localhost:8001/ 10
Delay: 10
Connecting to http://localhost:8001/
Get http://localhost:8001/: net/http: request canceled
The request was cancelled!
Exiting...
real    0m10.381s
user    0m0.314s
sys     0m0.125s
$ time go run useContext.go http://localhost:8001/ 15
Delay: 15
Connecting to http://localhost:8001/
Server Response: Serving: /
Delay: 13
Exiting...
real    0m13.379s
user    0m0.309s
sys     0m0.118s
```

The output shows that only the third command actually got an answer from the HTTP server—the first two commands timed out!

Worker pools

Generally speaking, a **Worker pool** is a set of threads that are about to process jobs assigned to them. More or less, the Apache web server works that way: the main process accepts all incoming requests that are forwarded to the worker processes for getting served. Once a worker process has finished its job, it is ready to serve a new client. Nevertheless, there is a central difference here because our worker pool will use goroutines instead of threads. Additionally, threads do not usually die after serving a request because the cost of ending a thread and creating a new one is too high, whereas goroutines do die after finishing their job.

As you will see shortly, worker pools in Go are implemented with the help of **buffered channels**, because they allow you to limit the number of goroutines running at the same time.

The next program, workerPool.go, will be presented in five parts. The program will implement a simple task: it will process integer numbers and print their square values using a single goroutine for serving each request. Despite the deceptive simplicity of workerPool.go, the Go code of the program can be easily used as a template for implementing much more difficult tasks.

This is an advanced technique that can help you create server processes in Go that can accept and serve multiple clients using goroutines!

The first part of workerPool.go follows next:

```go
package main

import (
    "fmt"
    "os"
    "strconv"
    "sync"
    "time"
)

type Client struct {
    id      int
    integer int
}

type Data struct {
    job    Client
    square int
}
```

Here you can see a technique that uses the Client structure for assigning a unique id to each request that will process. The Data structure is used for grouping the data of a Client with the actual results generated by the program. Put simply, the Client structure holds the input data of each request, whereas the Data structure holds the results of a request.

The second code portion of `workerPool.go` is shown in the following Go code:

```go
var (
    size    = 10
    clients = make(chan Client, size)
    data    = make(chan Data, size)
)

func worker(w *sync.WaitGroup) {
    for c := range clients {
        square := c.integer * c.integer
        output := Data{c, square}
        data <- output
        time.Sleep(time.Second)
    }
    w.Done()
}
```

The preceding code has two interesting parts. The first part creates three global variables. The `clients` and `data` buffered channels are used for getting new client requests and writing the results, respectively. If you want your program to run faster, you can increase the value of the `size` parameter.

The second part is the implementation of the `worker()` function, which reads the `clients` channel in order to get new requests to serve. Once the processing is complete, the result is written to the `data` channel. The delay that is introduced using the `time.Sleep(time.Second)` statement is not necessary, but it gives you a better sense of the way the generated output will be printed.

The third part of `workerPool.go` contains the following Go code:

```go
func makeWP(n int) {
    var w sync.WaitGroup
    for i := 0; i < n; i++ {
        w.Add(1)
        go worker(&w)
    }
    w.Wait()
    close(data)
}

func create(n int) {
    for i := 0; i < n; i++ {
        c := Client{i, i}
        clients <- c
    }
```

```
        close(clients)
    }
```

The preceding code implements two functions, named `makeWP()` and `create()`. The purpose of the `makeWP()` function is to generate the required number of `worker()` goroutines for processing all requests. Although the `w.Add(1)` function is called in `makeWP()`, the `w.Done()` is called in the `worker()` function once a worker has finished its job.

The purpose of the `create()` function is to create all requests properly using the `Client` type and then to write them to the `clients` channel for processing. Note that the `clients` channel is read by the `worker()` function.

The fourth code segment of `workerPool.go` is as follows:

```
func main() {
    fmt.Println("Capacity of clients:", cap(clients))
    fmt.Println("Capacity of data:", cap(data))

    if len(os.Args) != 3 {
        fmt.Println("Need #jobs and #workers!")
        os.Exit(1)
    }

    nJobs, err := strconv.Atoi(os.Args[1])
    if err != nil {
        fmt.Println(err)
        return
    }

    nWorkers, err := strconv.Atoi(os.Args[2])
    if err != nil {
        fmt.Println(err)
        return
    }
}
```

In the preceding code, you read your command-line parameters. First, however, you see that you can use the `cap()` function to find the capacity of a channel.

If the number of workers is larger than the size of the `clients` buffered channel, then the number of goroutines that is going to be created will be equal to the size of the `clients` channel. Similarly, if the number of jobs is larger than the number of workers, the jobs will be served in smaller sets.

The program allows you to define the number of workers and the number of jobs using its command-line arguments. However, in order to change the size of the `clients` and `data` channels, you will need to make changes to the source code of the program.

The remaining code of `workerPool.go` follows next:

```go
go create(nJobs)
finished := make(chan interface{})
go func() {
    for d := range data {
        fmt.Printf("Client ID: %d\tint: ", d.job.id)
        fmt.Printf("%dtsquare: %d\n", d.job.integer, d.square)
    }
        finished <- true
}()

    makeWP(nWorkers)
    fmt.Printf(": %v\n", <-finished)
}
```

First, you will call the `create()` function for mimicking the client requests that you will have to process. An anonymous goroutine is used for reading the `data` channel and printing the output to the screen. The `finished` channel is used for blocking the program until the anonymous goroutine is done reading the `data` channel. Therefore, the `finished` channel needs no particular type! Finally, you will call the `makeWP()` function for actually processing the requests. The `<-finished` statement in `fmt.Printf()` blocks, which means that it does not allow the program to end until somebody writes something to the `finished` channel. That somebody is the anonymous goroutine of the `main()` function. Additionally, although the anonymous function writes the value `true` to the `finished` channel, you can write `false` to it and have the same result, which is unblocking the `main()` function. Try it on your own!

Executing `workerPool.go` will generate the following output:

```
$ go run workerPool.go 15 5
Capacity of clients: 10
Capacity of data: 10
Client ID: 0      int: 0      square: 0
Client ID: 4      int: 4      square: 16
Client ID: 1      int: 1      square: 1
Client ID: 3      int: 3      square: 9
Client ID: 2      int: 2      square: 4
Client ID: 5      int: 5      square: 25
```

```
Client ID: 6      int: 6      square: 36
Client ID: 7      int: 7      square: 49
Client ID: 8      int: 8      square: 64
Client ID: 9      int: 9      square: 81
Client ID: 10     int: 10     square: 100
Client ID: 11     int: 11     square: 121
Client ID: 12     int: 12     square: 144
Client ID: 13     int: 13     square: 169
Client ID: 14     int: 14     square: 196
: true
```

When you want to serve each individual request without expecting an answer from it in the `main()` function, as happened with `workerPool.go`, you have fewer things to worry about. A simple way both to use goroutines for processing your requests and to get an answer from them in the `main()` function is using shared memory or a monitor process that will collect the data instead of just printing it on the screen.

Finally, the work of the `workerPool.go` program is much simpler because the `worker()` function cannot fail. This will not be the case when you have to work over computer networks or with other kinds of resources that can fail.

Additional resources

The following are very useful resources:

- Visit the documentation page of the `sync` package, which can be found at https://golang.org/pkg/sync/.
- Visit the documentation page of the `context` package, which can be found at https://golang.org/pkg/context/.
- You can learn more about the implementation of the Go scheduler by visiting https://golang.org/src/runtime/proc.go.
- You can view the design document of the Go scheduler at https://golang.org/s/go11sched.

Exercises

- Try to implement a concurrent version of `wc(1)` that uses a buffered channel.
- Next, try to implement a concurrent version of `wc(1)` that uses shared memory.
- Finally, try to implement a concurrent version of `wc(1)` that uses a monitor goroutine.
- Modify the Go code of `workerPool.go` in order to save the results in a file. Use a mutex and a critical section while dealing with the file, or a monitor goroutine that will keep writing your data on the disk.
- What will happen to the `workerPool.go` program when the value of the `size` global variable becomes 1? Why?
- Modify the Go code of `workerPool.go` in order to implement the functionality of the `wc(1)` command-line utility.
- Modify the Go code of `workerPool.go` so that the size of the `clients` and `data` buffered channels can be defined as command-line arguments.
- Try to write a concurrent version of the `find(1)` command-line utility that uses a monitor goroutine.
- Modify the code of `simpleContext.go` so that the anonymous function used in all `f1()`, `f2()`, and `f3()` functions becomes a separate function. What is the main challenge of this code change?
- Modify the Go code of `simpleContext.go` so that all the `f1()`, `f2()`, and `f3()` functions use an externally created `Context` variable instead of defining their own.
- Modify the Go code of `useContext.go` in order to use either `context.WithDeadline()` or `context.WithCancel()` instead of `context.WithTimeout()`.
- Finally, try to implement a concurrent version of the `find(1)` command-line utility using a `sync.Mutex` mutex.

Summary

This chapter addressed many important topics related to goroutines. Mainly, however, it clarified the power of the `select` statement. Additionally, it demonstrated the use of the `context` standard Go package.

Due to the capabilities of the `select` statement, channels are the preferred Go way for interconnecting the components of a Go program.

There are many rules in concurrent programming; however, the most important rule is that you should avoid sharing things unless you have a pretty good reason to do so. Shared data is the root of all nasty bugs in concurrent programming.

What you must remember from this chapter is that although shared memory used to be the only way for exchanging data over the threads of the same process, Go offers better ways for goroutines to communicate with each other. So think in Go terms before deciding to use shared memory in your Go code. Nonetheless, if you really have to use shared memory, you might want to use a monitor goroutine instead.

The primary subjects of the next chapter will be code testing, code optimization, and code profiling with Go. Apart from these topics, you will learn about benchmarking Go code, cross compilation, and finding unreachable Go code.

At the end of the next chapter, you will also learn how to document your Go code and how to generate HTML output using the `godoc` utility.

11
Code Testing, Optimization, and Profiling

The previous chapter discussed concurrency in Go and how the `select` statement allows you to use channels as glue for controlling goroutines and allowing goroutines to communicate.

The Go topics in this chapter are just moderately advanced, yet they are very practical and important, especially when you are interested in improving the performance of your Go programs and discovering bugs quickly. This chapter primarily addresses code optimization, code testing, code documentation, and code profiling.

Code optimization is a process where one or more developers try to make certain parts of a program run faster, more efficient, or use fewer resources. Put simply, code optimization is about eliminating the bottlenecks of a program.

Code testing is about making sure that your code does what you want it to do. In this chapter, you will experience the Go way of testing code. The best time to write code to test your program is during the development phase, as this can help reveal bugs in your code as early as possible.

Code profiling relates to measuring certain aspects of a program in order to get a detailed understanding of the way the code works. The results of code profiling may help you decide which parts of your code need to change.

I hope that you already recognize the importance of documenting your code in order to describe the decisions you made in developing the implementation of your program. In this chapter, you will see how Go can help you generate documentation for the modules that you implement.

Documentation is so important that some developers write the documentation first and the code afterwards!

In this chapter of *Mastering Go*, you will learn the following topics:

- Profiling Go code
- The `go tool pprof` utility
- Using the web interface of the Go profiler
- Testing Go code
- The `go test` command
- The `go tool trace` utility
- Benchmarking Go code
- Cross-compilation
- Generating documentation for your Go code
- Creating example functions
- Finding unreachable Go code in your programs

The Go version used in this chapter

Note that in this chapter, we will be using the latest Go version on a macOS machine. At the time of writing, this is Go version 1.10:

```
$ date
Fri Mar 9 23:17:49 EET 2018
$ go version
go version go1.10 darwin/amd64
```

The reason I mention this is that you will need at least Go version 1.10 in order to be able to use the web interface of the Go code profiler. However, the rest of the presented commands and code will work just fine with earlier Go versions.

Comparing Go version 1.10 with Go version 1.9

This section will list the most important differences between Go version 1.9 and Go version 1.10, just to give you an idea of the pace of Go changes.

Some of the changes in Go version 1.10 are as follows:

- There is a change in the way that the `GOROOT` and `GOTMPDIR` environment variables are defined.

- The `go test` command caches test results, which means that it might run a little faster.

- The `go doc` tool now includes output about functions that return slices of a type or a pointer to a type when you tell `go doc` to display data about that type.

- The `runtime/pprof` package that you will see later in this chapter now includes symbol information about the blocking and mutex profiles.

- The `go fix` tool replaces the `import` statements of `golang.org/x/net/context` with just `context`.

- The `go tool proff` includes a web interface. Although this is a major change, it has nothing to do with the Go programming language itself and therefore does not break any existing Go code. You will see that web interface in action later in this chapter.

- Go version 1.10 includes minor changes to some packages of the standard Go library, including `archive/tar` and `archive/zip`, `debug/macho`, `hash`, `html/template`, `math/big`, `net/url`, `os`, `time`, `unicode`, and `net/smtp`.

As you must realize, the Go programming language progresses without breaking things. This might look insignificant at first, but it is a very important feature of Go which guarantees that the existing Go code will continue to compile without changes while Go evolves. A side effect of this is that this book will be relevant for many years!

Installing a beta or RC version of Go

If you are eager to try the latest version of Go, even if it means that you will need to install a beta or a **Release Candidate** (**RC**) version of Go, there is a way!

Despite their name, Release Candidate versions of Go are pretty stable. Just do not install a RC version on a production Unix machine without having a very good reason to do so!

If you have a version of Go that is older than 1.10 and you want to try Go version 1.10 RC1, you can do that as follows:

```
$ go get golang.org/x/build/version/go1.10rc1
```

You should expect to execute an analogous command if you want to install a different version of Go. On a macOS High Sierra system that runs release 10.13.3, the new Go binary will be installed inside `~/go/bin`:

```
$ cd ~/go/bin
$ ls -l
-rwxr-xr-x 1 mtsouk staff 6051164 Jan 30 19:04 go1.10rc1
```

If you try to execute `go1.10rc1`, you will get the following output:

```
$ ./go1.10rc1
go1.10rc1: not downloaded. Run 'go1.10rc1 download' to install to
/Users/mtsouk/sdk/go1.10rc1
```

Next you will have to execute the remaining files of Go version 1.10 RC1 as follows:

```
$ ./go1.10rc1 download
$ ls -l ~/sdk/go1.10rc1/
total 248016
-rw-r--r--   1 mtsouk staff     47028 Jan 25 19:30 AUTHORS
-rw-r--r--   1 mtsouk staff      1576 Jan 25 19:30 CONTRIBUTING.md
-rw-r--r--   1 mtsouk staff     61940 Jan 25 19:30 CONTRIBUTORS
-rw-r--r--   1 mtsouk staff      1479 Jan 25 19:30 LICENSE
-rw-r--r--   1 mtsouk staff      1303 Jan 25 19:30 PATENTS
-rw-r--r--   1 mtsouk staff      1601 Jan 25 19:30 README.md
-rw-r--r--   1 mtsouk staff         9 Jan 25 19:30 VERSION
drwxr-xr-x  16 mtsouk staff       512 Jan 30 19:07 api
drwxr-xr-x   5 mtsouk staff       160 Jan 30 19:07 bin
drwxr-xr-x   4 mtsouk staff       128 Jan 30 19:07 blog
drwxr-xr-x  48 mtsouk staff      1536 Jan 30 19:07 doc
-rw-r--r--   1 mtsouk staff      5686 Jan 25 19:30 favicon.ico
-rw-r--r--   1 mtsouk staff 117832137 Jan 30 19:07 go1.10rc1.darwin-amd64.tar.gz
drwxr-xr-x   3 mtsouk staff        96 Jan 30 19:07 lib
drwxr-xr-x  16 mtsouk staff       512 Jan 30 19:07 misc
drwxr-xr-x   8 mtsouk staff       256 Jan 30 19:07 pkg
-rw-r--r--   1 mtsouk staff        26 Jan 25 19:30 robots.txt
drwxr-xr-x  68 mtsouk staff      2176 Jan 30 19:07 src
drwxr-xr-x 295 mtsouk staff      9440 Jan 30 19:07 test
```

You can now use Go version 1.10 RC1 as `go1.10rc1`:

```
$ ~/go/bin/go1.10rc1 version
go version go1.10rc1 darwin/amd64
```

About optimization

Code optimization is both an art and a science! This means that there is no deterministic way to help you optimize your Go code, or any other code in any programming language, and that you should use your brain and try many things if you want to make your code run faster.

You should make sure that you are optimizing code that does not contain any **bugs**, because there is no point in optimizing a bug. If you have any bugs in your program, you should debug it first!

If you are really into code optimization, you might want to read *Compilers: Principles, Techniques and Tools* by *Alfred V. Aho, Monica S. Lam, Ravi Sethi* and *Jeffrey D. Ullman, Pearson Education Limited, 2014*, which focuses on compiler construction. Additionally, all volumes in *The Art of Computer Programming* series by *Donald Knuth, Addison-Wesley Professional, 1998*, are a great resource for all aspects of programming.

Always remember what *Donald Knuth* said about optimization:

> *The real problem is that programmers have spent far too much time worrying about efficiency in the wrong places and at the wrong times; premature optimization is the root of all evil (or at least most of it) in programming.*

Optimizing Go code

Code optimization is the process where you try to discover the parts of your code that have a big influence on the performance of the entire program in order to make them run faster or use fewer resources.

The benchmarking section that appears later in this chapter will greatly help you understand what is going on with your code behind the scenes and which parameters of your program have the greatest impact on the performance of your program.

However, do not underestimate common sense. Put simply, if one of your functions is executed ten-thousand times more than the rest of the functions of a program, try to optimize that function first.

The general advice for optimization is that you must optimize bug-free code only. This means that you must optimize working code only. Therefore, you should first try to write correct code even if that code is slow. Finally, the single most frequent mistake that programmers make is trying to optimize the first version of their code, which is the root of most bugs!

Again, code optimization is both an art and a science, which means that it is a pretty difficult task. The next section about profiling Go code will definitely help you with code optimization because the main purpose of profiling is to find the bottlenecks in your code in order to optimize the right and most important parts of your program.

Profiling Go code

Profiling is a process of dynamic program analysis that measures various values related to program execution in order to give you a better understanding on the behavior of your program. In this section, you will learn how to profile Go code in order to understand your code better and improve its performance. Sometimes, code profiling can even reveal bugs!

First, we will use the command-line interface of the Go profiler. Next, we will use the brand new web interface of the Go profiler!

The single most important thing to remember is that if you want to profile Go code, you will need to import the `runtime/pprof` standard Go package, either directly or indirectly. You can find the help page of the `pprof` tool by executing the `go tool pprof -help` command, which will generate lots of output.

The net/http/pprof standard Go package

Although Go comes with the low-level `runtime/pprof` standard Go package, there is also a high-level `net/http/pprof` package that should be used when you want to profile a web application written in Go. As this chapter will not talk about creating HTTP servers in Go, you will learn more about the `net/http/pprof` package in Chapter 12, *The Foundations of Network Programming in Go*.

A simple profiling example

Go supports two kinds of profiling: **CPU profiling** and **memory profiling**. It is not recommended that you profile an application for both kinds at the same time, because these two different kinds of profiling do not work well with each other. The `profileMe.go` application is an exception, however, because it is used for illustrating the two techniques.

The Go code to be profiled is saved as `profileMe.go`, and it will be presented in five parts. The first part of `profileMe.go` is shown in the following Go code:

```go
package main
import (
    "fmt"
    "math"
    "os"
    "runtime"
    "runtime/pprof"
    "time"
)
func fibo1(n int) int64 {
    if n == 0 || n == 1 {
        return int64(n)
    }
    time.Sleep(time.Millisecond)
    return int64(fibo2(n-1)) + int64(fibo2(n-2))
}
```

Notice that it is compulsory to import the `runtime/pprof` package directly or indirectly for your program to create profiling data. The reason for calling `time.Sleep()` in the `fibo1()` function is to slow it down a bit. You will learn why, near the end of this section.

The second code segment of `profileMe.go` follows next:

```go
func fibo2(n int) int {
    fn := make(map[int]int)
    for i := 0; i <= n; i++ {
        var f int
        if i <= 2 {
            f = 1
        } else {
            f = fn[i-1] + fn[i-2]
        }
        fn[i] = f
    }
    time.Sleep(50 * time.Millisecond)
    return fn[n]
}
```

The preceding code contains the implementation of another Go function that uses a different algorithm for calculating the numbers of the Fibonacci sequence.

The third part of `profileMe.go` contains the following Go code:

```go
func N1(n int) bool {
    k := math.Floor(float64(n/2 + 1))
    for i := 2; i < int(k); i++ {
        if (n % i) == 0 {
            return false
        }
    }
    return true
}
func N2(n int) bool {
    for i := 2; i < n; i++ {
        if (n % i) == 0 {
            return false
        }
    }
    return true
}
```

Both the `N1()` and `N2()` functions are used for finding out whether a given integer is a prime number or not. The first function is optimized because its `for` loop iterates over approximately half the numbers used in the `for` loop of `N2()`. As both functions are relatively slow, there is no need for a call to `time.Sleep()` here.

The fourth code segment of profileMe.go is as follows:

```go
func main() {
    cpuFile, err := os.Create("/tmp/cpuProfile.out")
    if err != nil {
        fmt.Println(err)
        return
    }
    pprof.StartCPUProfile(cpuFile)
    defer pprof.StopCPUProfile()
    total := 0
    for i := 2; i < 100000; i++ {
        n := N1(i)
        if n {
            total = total + 1
        }
    }
    fmt.Println("Total primes:", total)
    total = 0
    for i := 2; i < 100000; i++ {
        n := N2(i)
        if n {
            total = total + 1
        }
    }
    fmt.Println("Total primes:", total)
    for i := 1; i < 90; i++ {
        n := fibo1(i)
        fmt.Print(n, " ")
    }
    fmt.Println()
    for i := 1; i < 90; i++ {
        n := fibo2(i)
        fmt.Print(n, " ")
    }
    fmt.Println()
    runtime.GC()
```

The call to os.Create() is used to have a file to which to write the profiling data. The pprof.StartCPUProfile() call begins the **CPU profiling** of the program, and the call to pprof.StopCPUProfile() stops it.

The last part of profileMe.go follows next:

```go
    // Memory profiling!
    memory, err := os.Create("/tmp/memoryProfile.out")
    if err != nil {
```

```
                fmt.Println(err)
                return
        }
        defer memory.Close()
        for i := 0; i < 10; i++ {
                s := make([]byte, 50000000)
                if s == nil {
                        fmt.Println("Operation failed!")
                }
                time.Sleep(50 * time.Millisecond)
        }
        err = pprof.WriteHeapProfile(memory)
        if err != nil {
                fmt.Println(err)
                return
        }
}
```

In the last part, you can see how the **memory profiling** technique works. It is pretty similar to CPU profiling, and once again you will need a file for writing out the profiling data.

Executing `profileMe.go` will generate the following output:

```
$ go run profileMe.go
Total primes: 9592
Total primes: 9592
1 1 2 3 5 8 13 21 34 55 89 144 233 377 610 987 1597 2584 4181 6765 10946
17711 28657 46368 75025 121393 196418 317811 514229 832040 1346269 2178309
3524578 5702887 9227465 14930352 24157817 39088169 63245986 102334155
165580141 267914296 433494437 701408733 1134903170 1836311903 2971215073
4807526976 7778742049 12586269025 20365011074 32951280099 53316291173
86267571272 139583862445 225851433717 365435296162 591286729879
956722026041 1548008755920 2504730781961 4052739537881 6557470319842
10610209857723 17167680177565 27777890035288 44945570212853 72723460248141
117669030460994 190392490709135 308061521170129 498454011879264
806515533049393 1304969544928657 2111485077978050 3416454622906707
5527939700884757 8944394323791464 14472334024676221 23416728348467685
37889062373143906 61305790721611591 99194853094755497 160500643816367088
259695496911122585 420196140727489673 679891637638612258
1100087778366101931 1779979416004714189
1 1 2 3 5 8 13 21 34 55 89 144 233 377 610 987 1597 2584 4181 6765 10946
17711 28657 46368 75025 121393 196418 317811 514229 832040 1346269 2178309
3524578 5702887 9227465 14930352 24157817 39088169 63245986 102334155
165580141 267914296 433494437 701408733 1134903170 1836311903 2971215073
4807526976 7778742049 12586269025 20365011074 32951280099 53316291173
86267571272 139583862445 225851433717 365435296162 591286729879
956722026041 1548008755920 2504730781961 4052739537881 6557470319842
```

```
10610209857723 17167680177565 27777890035288 44945570212853 72723460248141
117669030460994 190392490709135 308061521170129 498454011879264
806515533049393 1304969544928657 2111485077978050 3416454622906707
5527939700884757 8944394323791464 14472334024676221 23416728348467685
37889062373143906 61305790721611591 99194853094755497 160500643816367088
259695496911122585 420196140727489673 679891637638612258
1100087778366101931 1779979416004714189
```

Apart from the output, the program will also collect the profiling data in two files:

```
$ cd /tmp
$ ls -l *Profile*
-rw-r--r-- 1 mtsouk wheel 1965 Mar 8 16:53 cpuProfile.out
-rw-r--r-- 1 mtsouk wheel 484 Mar 8 16:53 memoryProfile.out
```

It is only after collecting the profiling data that you can start to inspect it. Thus, you can now start the command-line profiler to examine the CPU data as follows:

```
$ go tool pprof /tmp/cpuProfile.out
Main binary filename not available.
Type: cpu
Time: Mar 8, 2018 at 4:53pm (EET)
Duration: 19.12s, Total samples = 4.18s (21.86%)
Entering interactive mode (type "help" for commands, "o" for options)
(pprof)
```

Typing `help` while in the profiler shell will generate the following output:

```
(pprof) help
Commands:
callgrind Outputs a graph in callgrind format
comments Output all profile comments
disasm Output assembly listings annotated with samples
dot Outputs a graph in DOT format
eog Visualize graph through eog
evince Visualize graph through evince
gif Outputs a graph image in GIF format
gv Visualize graph through gv
kcachegrind Visualize report in KCachegrind
list Output annotated source for functions matching regexp
pdf Outputs a graph in PDF format
peek Output callers/callees of functions matching regexp
png Outputs a graph image in PNG format
proto Outputs the profile in compressed protobuf format
ps Outputs a graph in PS format
raw Outputs a text representation of the raw profile
svg Outputs a graph in SVG format
tags Outputs all tags in the profile
```

```
text             Outputs top entries in text form
top              Outputs top entries in text form
topproto         Outputs top entries in compressed protobuf format
traces           Outputs all profile samples in text form
tree             Outputs a text rendering of call graph
web              Visualize graph through web browser
weblist          Display annotated source in a web browser
o/options        List options and their current values
quit/exit/^D     Exit pprof
Options:
call_tree              Create a context-sensitive call tree
compact_labels         Show minimal headers
divide_by              Ratio to divide all samples before visualization
drop_negative          Ignore negative differences
edgefraction           Hide edges below <f>*total
focus                  Restricts to samples going through a node matching regexp
hide                   Skips nodes matching regexp
ignore                 Skips paths going through any nodes matching regexp
mean                   Average sample value over first value (count)
nodecount              Max number of nodes to show
nodefraction           Hide nodes below <f>*total
normalize              Scales profile based on the base profile.
output                 Output filename for file-based outputs
positive_percentages   Ignore negative samples when computing percentages
prune_from             Drops any functions below the matched frame.
relative_percentages   Show percentages relative to focused subgraph
sample_index           Sample value to report (0-based index or name)
show                   Only show nodes matching regexp
source_path            Search path for source files
tagfocus               Restricts to samples with tags in range or matched by regexp
taghide                Skip tags matching this regexp
tagignore              Discard samples with tags in range or matched by regexp
tagshow                Only consider tags matching this regexp
trim                   Honor nodefraction/edgefraction/nodecount defaults
unit                   Measurement units to display
Option groups (only set one per group):
cumulative
  cum           Sort entries based on cumulative weight
  flat          Sort entries based on own weight
granularity
  addresses         Aggregate at the function level.
  addressnoinlines  Aggregate at the function level, including functions'
                    addresses in the output.
  files             Aggregate at the file level.
  functions         Aggregate at the function level.
  lines             Aggregate at the source code line level.
  noinlines         Aggregate at the function level.
  :                 Clear focus/ignore/hide/tagfocus/tagignore
```

```
type "help <cmd|option>" for more information
(pprof)
```

 Do find the time to try out all of the commands of the `go tool pprof` utility and familiarize yourself with them.

The `top` command returns the top 10 entries in text form:

```
(pprof) top
Showing nodes accounting for 4120ms, 98.56% of 4180ms total
Dropped 26 nodes (cum <= 20.90ms)
Showing top 10 nodes out of 21
  flat  flat%   sum%        cum   cum%
1950ms 46.65% 46.65%     1950ms 46.65%  main.N2
1170ms 27.99% 74.64%     1170ms 27.99%  runtime.usleep
 910ms 21.77% 96.41%      910ms 21.77%  main.N1
  90ms  2.15% 98.56%       90ms  2.15%  runtime.mach_semaphore_signal
     0     0% 98.56%     2890ms 69.14%  main.main
     0     0% 98.56%       80ms  1.91%  runtime.goready
     0     0% 98.56%       80ms  1.91%  runtime.goready.func1
     0     0% 98.56%       80ms  1.91%  runtime.goroutineReady
     0     0% 98.56%       90ms  2.15%  runtime.mach_semrelease
     0     0% 98.56%     2890ms 69.14%  runtime.main
```

As the first line of the output states, the functions presented are responsible for `98.56%` of the total execution time of the program. The `main.N2` function in particular is responsible for the `46.65%` of the execution time of the program.

The `top10 --cum` command returns the cumulative time for each function:

```
(pprof) top10 --cum
Showing nodes accounting for 4030ms, 96.41% of 4180ms total
Dropped 26 nodes (cum <= 20.90ms)
Showing top 10 nodes out of 21
  flat  flat%   sum%        cum   cum%
     0     0%     0%     2890ms 69.14%  main.main
     0     0%     0%     2890ms 69.14%  runtime.main
1950ms 46.65% 46.65%     1950ms 46.65%  main.N2
     0     0% 46.65%     1180ms 28.23%  runtime.mstart
     0     0% 46.65%     1180ms 28.23%  runtime.mstart1
     0     0% 46.65%     1180ms 28.23%  runtime.sysmon
1170ms 27.99% 74.64%     1170ms 27.99%  runtime.usleep
 910ms 21.77% 96.41%      910ms 21.77%  main.N1
     0     0% 96.41%      130ms  3.11%  runtime.systemstack
     0     0% 96.41%      110ms  2.63%  runtime.timerproc
```

What if you want to find out what is happening with a particular function? You can use the `list` command followed by the function name, combined with the package name, and you'll get more detailed information about the performance of that function:

```
(pprof) list main.N1
Total: 4.18s
ROUTINE ======= main.N1 in /Users/mtsouk/ch11/code/profileMe.go
910ms  910ms  (flat, cum) 21.77% of Total
    .      .     32:        return fn[n]
    .      .     33:}
    .      .     34:
    .      .     35:func N1(n int) bool {
    .      .     36:       k := math.Floor(float64(n/2 + 1))
 60ms   60ms    37:       for i := 2; i < int(k); i++ {
850ms  850ms   38:               if (n % i) == 0 {
    .      .     39:                       return false
    .      .     40:               }
    .      .     41:       }
    .      .     42:       return true
    .      .     43:}
(pprof)
```

The output shows that the `for` loop of `main.N1` is responsible for almost all of the execution time of the entire function. Specifically, the `if (n % i) == 0` statement is responsible for `850ms` out of `910ms` of the execution time of the entire function.

You can also create a PDF output of the profiling data from the shell of the Go profiler, using the `pdf` command:

```
(pprof) pdf
Generating report in profile001.pdf
```

Note that you will need **Graphviz** in order to generate a PDF file that can be viewed using your favorite PDF reader.

Finally, a warning: if your program executes too quickly, then the profiler will not have enough time to get its required samples and you might see the `Total samples = 0` output when you load the data file. In that case, you will not be able to get any useful information from the profiling process. This is the reason for using the `time.Sleep()` function in some of the functions of `profileMe.go`:

```
$ go tool pprof /tmp/cpuProfile.out
Type: cpu
Time: Mar 7, 2018 at 10:25pm (EET)
Duration: 202.29ms, Total samples = 0
```

```
Entering interactive mode (type "help" for commands, "o" for options)
(pprof)
```

A convenient external package for profiling

In this subsection, you will see the use of an external Go package that sets up the profiling environment much more conveniently than when using the `runtime/pprof` standard Go package. This point is illustrated in `betterProfile.go`, which will be presented in three parts.

The first part of `betterProfile.go` is as follows:

```go
package main
import (
    "fmt"
    "github.com/pkg/profile"
)
var VARIABLE int
func N1(n int) bool {
    for i := 2; i < n; i++ {
        if (n % i) == 0 {
            return false
        }
    }
    return true
}
```

In the preceding code, you can see the use of an external Go package that can be found at, `github.com/pkg/profile`. You can download it with the help of the `go get` command, as follows:

```
$ go get github.com/pkg/profile
```

The second code segment of `betterProfile.go` contains the following Go code:

```go
func Multiply(a, b int) int {
    if a == 1 {
        return b
    }
    if a == 0 || b == 0 {
        return 0
    }
    if a < 0 {
        return -Multiply(-a, b)
    }
}
```

Code Testing, Optimization, and Profiling

```
        return b + Multiply(a-1, b)
    }
    func main() {
        defer profile.Start(profile.ProfilePath("/tmp")).Stop()
```

The `github.com/pkg/profile` package by Dave Cheney requires that you insert just a single statement in order to enable **CPU profiling** in your Go application. If you want to enable **memory profiling**, you should insert the following statement instead:

```
        defer profile.Start(profile.MemProfile).Stop()
```

The remaining Go code of the program is as follows:

```
    total := 0
    for i := 2; i < 200000; i++ {
        n := N1(i)
        if n {
            total++
        }
    }
    fmt.Println("Total: ", total)
    total = 0
    for i := 0; i < 5000; i++ {
        for j := 0; j < 400; j++ {
            k := Multiply(i, j)
            VARIABLE = k
            total++
        }
    }
    fmt.Println("Total: ", total)
}
```

Executing `betterProfile.go` generates the following output:

```
$ go run betterProfile.go
2018/03/08 17:01:28 profile: cpu profiling enabled, /tmp/cpu.pprof
Total:  17984
Total:  2000000
2018/03/08 17:01:56 profile: cpu profiling disabled, /tmp/cpu.pprof
```

The `github.com/pkg/profile` package will help you with the data capturing portion; the processing part is the same as before!

```
$ go tool pprof /tmp/cpu.pprof
Main binary filename not available.
Type: cpu
Time: Mar 8, 2018 at 5:01pm (EET)
Duration: 27.51s, Total samples = 23.62s (85.87%)
Entering interactive mode (type "help" for commands, "o" for options)
(pprof)
```

The web interface of the Go profiler

The good news is that with Go version 1.10, the `go tool pprof` command comes with a web user interface!

For the web user interface feature to work, you will need to have **Graphviz** installed and your web browser must support JavaScript. If you want to play it safe, use either **Chrome** or **Firefox**.

You can start the interactive Go profiler as follows:

```
$ go tool pprof -http=[host]:[port] aProfile
```

A profiling example that uses the web interface

We will use the data captured from the execution of `profileMe.go` to study the web interface of the Go profiler as there is no need for creating a specific code example for doing this. As you learned in the previous subsection, you will first need to execute the following command:

```
$ go tool pprof -http=localhost:8080 /tmp/cpuProfile.out
Main binary filename not available.
```

Code Testing, Optimization, and Profiling

The following figure shows the initial screen of the web user interface of the Go profiler after executing the preceding command:

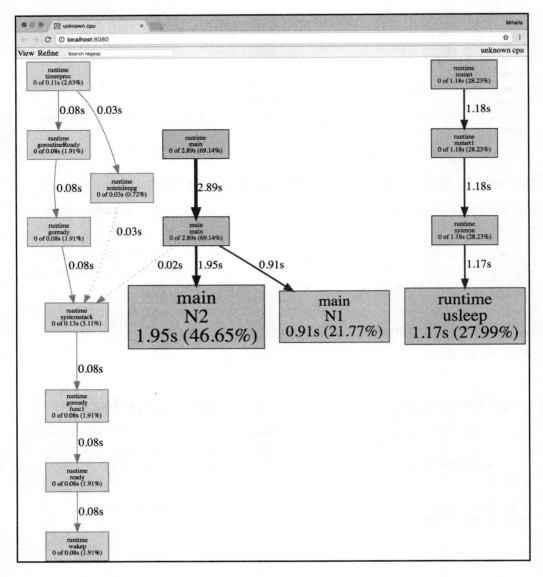

The web interface of the Go profiler in action

Similarly, the following figure shows the `http://localhost:8080/source` URL of the Go profiler, which displays analytical information for each function of the program:

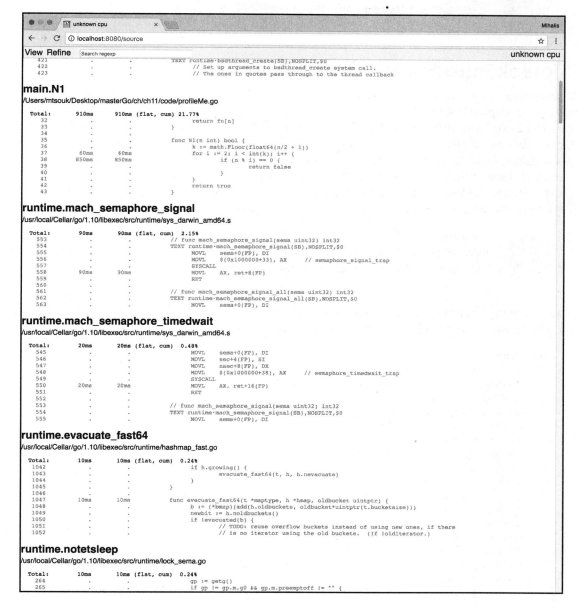

Using the /source URL of the Go profiler

> As I cannot possibly show every single page of the Go profiler web interface, you should begin to familiarize yourself with it on your own, as it is a great tool for examining the operation of your programs.

A quick introduction to Graphviz

Graphviz is a very handy compilation of utilities and a computer language that allows you to draw complex graphs. Strictly speaking, Graphviz is a collection of tools for manipulating both directed and undirected **graph** structures and generating graph layouts. Graphviz has its own language, named **DOT**, which is simple, elegant, and powerful. The good thing about Graphviz is that you can write its code using a simple plain text editor. A wonderful side effect of this feature is that you can easily develop scripts that generate Graphviz code! Also, most programming languages, including **Python**, **Ruby**, **C++**, and **Perl**, provide their own interface for creating Graphviz files using native code.

> You need not know all of these things about Graphviz in order to use the web interface of the Go profiler. It is just useful to know how Graphviz works and what its code looks like.

The following Graphviz code, which is saved as `graph.dot`, briefly illustrates the way that Graphviz works and the look of the Graphviz language:

```
digraph G
{
    graph [dpi = 300, bgcolor = "gray"];
    rankdir = LR;
    node [shape=record, width=.2, height=.2, color="white" ];
    node0 [label = "<p0>  |<p1>|<p2>|<p3>|<p4>|  | ", height = 3];
    node[ width=2 ];
    node1 [label = "{<e> r0  | 123  | <p> }", color="gray" ];
    node2 [label = "{<e> r10 | 13   | <p> }" ];
    node3 [label = "{<e> r11 | 23   | <p> }" ];
    node4 [label = "{<e> r12 | 326  | <p> }" ];
    node5 [label = "{<e> r13 | 1f3  | <p> }" ];
    node6 [label = "{<e> r20 | 143  | <p> }" ];
    node7 [label = "{<e> r40 | b23  | <p> }" ];
    node0:p0 -> node1:e [dir=both color="red:blue"];
    node0:p1 -> node2:e [dir=back arrowhead=diamond];
    node2:p -> node3:e;
    node3:p -> node4:e [dir=both arrowtail=box color="red"];
    node4:p -> node5:e [dir=forward];
    node0:p2 -> node6:e [dir=none color="orange"];
```

```
    node0:p4 -> node7:e;
}
```

The `color` attribute changes the color of a node, whereas the `shape` attribute changes the shape of a node. Additionally, the `dir` attribute, which can be applied to edges, defines whether an edge is going to have two arrows, one, or none. Furthermore, the style of the arrowhead can be specified using the `arrowhead` and `arrowtail` attributes.

Compiling the preceding code using one of the Graphviz command-line tools in order to create a PNG image requires the execution of the following command in your favorite Unix shell:

```
$ dot -T png graph.dot -o graph.png
$ ls -l graph.png
-rw-r--r--@ 1 mtsouk staff 94155 Mar 1 07:30 graph.png
```

The following diagram shows the graphics file generated from the execution of the preceding command:

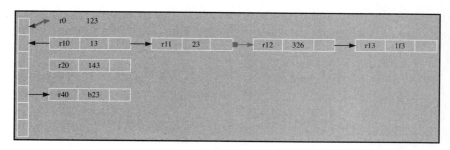

Using Graphviz to create graphs

Thus, if you want to visualize any kind of structure, you should definitely consider using Graphviz and its tools, especially if you want to automate things with your own scripts.

The go tool trace utility

The `go tool trace` utility is a tool for viewing trace files that can be generated with any one of the following three ways:

- Using the `runtime/trace` package
- Using the `net/http/pprof` package
- Executing the `go test -trace` command

This section will use the first technique only. The output of the following command will greatly help you understand what the **Go execution tracer** does:

```
$ go doc runtime/trace
package trace // import "runtime/trace"
Go execution tracer. The tracer captures a wide range of execution events
like goroutine creation/blocking/unblocking, syscall enter/exit/block, GC-
related events, changes of heap size, processor start/stop, etc and writes
them to an io.Writer in a compact form. A precise nanosecond-precision
timestamp and a stack trace is captured for most events. A trace can be
analyzed later with 'go tool trace' command.
func Start(w io.Writer) error
func Stop()
```

In Chapter 2, *Understanding Go Internals*, we talked about the **Go garbage collector** and presented a Go utility `gColl.go`, which allowed us to see some of the variables of the Go garbage collector. In this section, we will gather even more information about the operation of `gColl.go`, using the `go tool trace` utility.

First, let's examine the modified version of the `gColl.go` program, which tells Go to collect performance data. It is saved as `goGC.go`, and it will be presented in three parts.

The first part of `goGC.go` is as follows:

```
package main
import (
    "fmt"
    "os"
    "runtime"
    "runtime/trace"
    "time"
)
func printStats(mem runtime.MemStats) {
    runtime.ReadMemStats(&mem)
    fmt.Println("mem.Alloc:", mem.Alloc)
    fmt.Println("mem.TotalAlloc:", mem.TotalAlloc)
    fmt.Println("mem.HeapAlloc:", mem.HeapAlloc)
    fmt.Println("mem.NumGC:", mem.NumGC)
    fmt.Println("-----")
}
```

As you already know, you first need to import the `runtime/trace` standard Go package in order to collect data for the `go tool trace` utility.

The second code segment of goGC.go is shown in the following Go code:

```go
func main() {
    f, err := os.Create("/tmp/traceFile.out")
    if err != nil {
        panic(err)
    }
    defer f.Close()
    err = trace.Start(f)
    if err != nil {
        fmt.Println(err)
        return
    }
    defer trace.Stop()
```

This part is all about acquiring data for the `go tool trace` utility, and it has nothing to do with the functionality of the actual program. First, you will create a new file that will hold the tracing data for the `go tool trace` utility. Then you can start the tracing process, using `trace.Start()`. When you are done, you will call the `trace.Stop()` function. The `defer` call to this function means that you want to terminate tracing when your program ends.

Using the `go tool trace` utility is a process with two phases that requires extra Go code. First you collect the data, and then you display and process it.

The remaining Go code is as follows:

```go
    var mem runtime.MemStats
    printStats(mem)
    for i := 0; i < 3; i++ {
        s := make([]byte, 50000000)
        if s == nil {
            fmt.Println("Operation failed!")
        }
    }
    printStats(mem)
    for i := 0; i < 5; i++ {
        s := make([]byte, 100000000)
        if s == nil {
            fmt.Println("Operation failed!")
        }
        time.Sleep(time.Millisecond)
```

```
        }
        printStats(mem)
}
```

Executing `goGC.go` produces the following output as well as a new file named `/tmp/traceFile.out` with the tracing information:

```
$ go run goGC.go
mem.Alloc: 107264
mem.TotalAlloc: 107264
mem.HeapAlloc: 107264
mem.NumGC: 0
-----
mem.Alloc: 50117672
mem.TotalAlloc: 150129416
mem.HeapAlloc: 50117672
mem.NumGC: 3
-----
mem.Alloc: 117320
mem.TotalAlloc: 650174208
mem.HeapAlloc: 117320
mem.NumGC: 8
-----
$ cd /tmp
$ ls -l traceFile.out
-rw-r--r-- 1 mtsouk wheel 8275 Mar 7 08:37 traceFile.out
$ file /tmp/traceFile.out
/tmp/traceFile.out: data
```

The `go tool trace` utility uses a web interface that starts automatically when you execute the following command:

```
$ go tool trace /tmp/traceFile.out
2018/03/07 08:34:36 Parsing trace...
2018/03/07 08:34:36 Serializing trace...
2018/03/07 08:34:36 Splitting trace...
2018/03/07 08:34:36 Opening browser. Trace viewer is listening on
http://127.0.0.1:61428
```

The following screenshot shows the initial image of the web interface of the `go tool trace` utility when examining the `/tmp/traceFile.out` trace file:

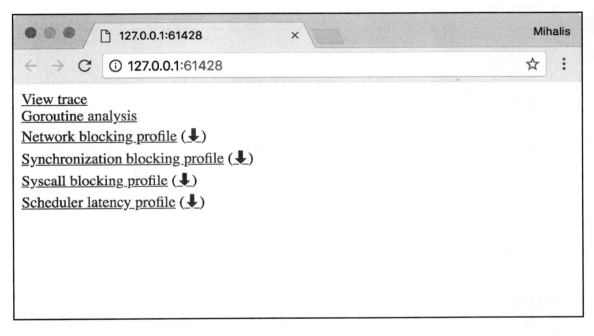

The initial screen of the web interface of the go trace tool

Code Testing, Optimization, and Profiling

You should now select the `View trace` link. This will take you to the next figure, which shows you another view of the web interface of the `go tool trace` utility that uses the data from `/tmp/traceFile.out`:

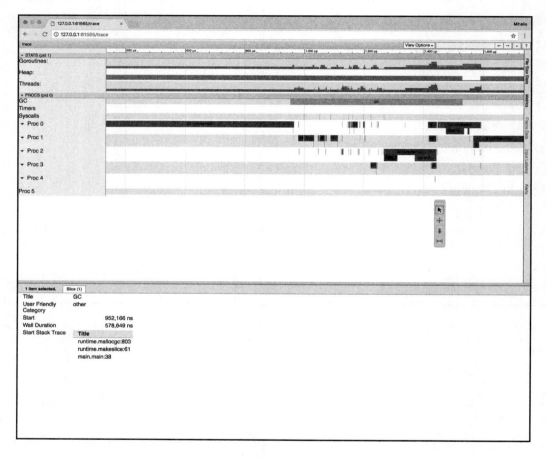

Examining the operation of the Go Garbage Collector using the go tool trace

In the preceding figure, you can see that the Go GC runs on its own goroutine, but it does not run all of the time. Additionally, you can see the number of goroutines used by the program. You can learn more about this by selecting certain parts of the interactive view. As we are interested in the operation of the GC, a useful piece of information that is displayed is how often and for how much long the Go GC runs.

Note that although `go tool trace` is a very handy and powerful utility, it cannot solve every kind of performance problem. There are times where `go tool pprof` is more appropriate, especially when you want to reveal where your program spends most of its time by individual function.

Testing Go code

Software testing is a very large subject, and it cannot be covered in a single section of a chapter in a book. So, this brief section will try to present as much practical information as possible.

Go allows you to write tests for your Go code in order to detect bugs. Strictly speaking, this section is about automated **testing**, which involves writing extra code to verify whether the real code; that is, the production code, works as expected or not. Thus, the result of a test function is either `PASS` or `FAIL`. You will see how this works shortly.

Although the Go approach to testing might look simple at first, especially if you compare it with the testing practices of other programming languages, it is very efficient and effective because it does not require too much of the developer's time.

Go follows certain conventions regarding testing. First of all, testing functions should be included in Go source files that end with `_test.go`. So, if you have a package named `aGoPackage.go`, then your tests should be placed in a file named `aGoPackage_test.go`. A test function begins with `Test`, and it checks the correctness of the behavior of a function of the production package.

Finally, you will need to import the `testing` standard Go package for the `go test` subcommand to work correctly. As you will soon see, this import requirement also applies to two additional cases.

Once the testing code is correct, the `go test` subcommand does all the dirty work for you, which includes scanning all the `*_test.go` files for special functions, generating a temporary `main` package properly, calling these special functions, getting the results, and generating the final output.

Always put the testing code in a different source file. There is no need to create a huge source file that is hard to read and maintain.

Writing tests for existing Go code

In this section, you will learn how to write tests for an existing Go application that includes two functions: one for calculating the numbers of the Fibonacci sequence and one for finding out the length of a string. The main reason for using these two functions that implement such relatively trivial tasks is simplicity. The tricky point here is that each function will have two different implementations: one that works fine and another one that has some issues.

The Go package in this example is named `testMe`, and it is saved as `testMe.go`. The code of this package will be presented in three parts.

The first part of `testMe.go` has the following Go code:

```
package testMe
func f1(n int) int {
    if n == 0 {
        return 0
    }
    if n == 1 {
        return 1
    }
    return f1(n-1) + f1(n-2)
}
```

In the preceding code, you can see the definition of a function named `f1()` that calculates natural numbers of the Fibonacci sequence.

The second part of `testMe.go` is shown in the following Go code:

```
func f2(n int) int {
    if n == 0 {
        return 0
    }
    if n == 1 {
        return 2
    }
    return f2(n-1) + f2(n-2)
}
```

In this code, you can see the implementation of another function that calculates the numbers of the Fibonacci sequence named `f2()`. However, this function contains a bug because it does not return 1 when the value of n is 1, which destroys the entire functionality of the function.

The remaining code of `testMe.go` is shown in the following Go code:

```go
func s1(s string) int {
    if s == "" {
        return 0
    }
    n := 1
    for range s {
        n++
    }
    return n
}

func s2(s string) int {
    return len(s)
}
```

In this part, we implement two functions named `s1()` and `s2()`, which work on strings. Both of these functions find the length of a string. However, the implementation of `s1()` is wrong because the initial value of n is 1 instead of 0.

It is now time to start thinking about tests and test cases. First of all, you should create a `testMe_test.go` file that will be used for storing your testing functions. Next, it is important to realize that you do not need to make any code changes to `testMe.go`. Finally, remember that you should try to write as many tests as required to cover all potential inputs and outputs.

The first part of `testMe_test.go` is shown in the following Go code:

```go
package testMe
import "testing"
func TestS1(t *testing.T) {
    if s1("123456789") != 9 {
        t.Error(`s1("123456789") != 9`)
    }
    if s1("") != 0 {
        t.Error(`s1("") != 0`)
    }
}
```

The preceding function performs two tests on the `s1()` function: one using `"123456789"` as input and another one using `""` as input.

The second part of testMe_test.go is as follows:

```go
func TestS2(t *testing.T) {
    if s2("123456789") != 9 {
        t.Error(`s2("123456789") != 9`)
    }
    if s2("") != 0 {
        t.Error(`s2("") != 0`)
    }
}
```

The preceding testing code performs the same two tests on the s2() function.

The remaining code of testMe_test.go comes next:

```go
func TestF1(t *testing.T) {
    if f1(0) != 0 {
        t.Error(`f1(0) != 0`)
    }
    if f1(1) != 1 {
        t.Error(`f1(1) != 1`)
    }
    if f1(2) != 1 {
        t.Error(`f1(2) != 1`)
    }
    if f1(10) != 55 {
        t.Error(`f1(10) != 55`)
    }
}

func TestF2(t *testing.T) {
    if f2(0) != 0 {
        t.Error(`f2(0) != 0`)
    }
    if f2(1) != 1 {
        t.Error(`f2(1) != 1`)
    }
    if f2(2) != 1 {
        t.Error(`f2(2) != 1`)
    }
    if f2(10) != 55 {
        t.Error(`f2(10) != 55`)
    }
}
```

The previous code tests the operation of the f1() and the f2() functions.

Executing the tests will generate the following output:

```
$ go test testMe.go testMe_test.go -v
=== RUN   TestS1
--- FAIL: TestS1 (0.00s)
    testMe_test.go:7: s1("123456789") != 9
=== RUN   TestS2
--- PASS: TestS2 (0.00s)
=== RUN   TestF1
--- PASS: TestF1 (0.00s)
=== RUN   TestF2
--- FAIL: TestF2 (0.00s)
    testMe_test.go:50: f2(1) != 1
    testMe_test.go:54: f2(2) != 1
    testMe_test.go:58: f2(10) != 55
FAIL
FAIL    command-line-arguments    0.005s
```

If you do not include the -v parameter, which produces richer output, you will get the following output:

```
$ go test testMe.go testMe_test.go
--- FAIL: TestS1 (0.00s)
    testMe_test.go:7: s1("123456789") != 9
--- FAIL: TestF2 (0.00s)
    testMe_test.go:50: f2(1) != 1
    testMe_test.go:54: f2(2) != 1
    testMe_test.go:58: f2(10) != 55
FAIL
FAIL    command-line-arguments    0.005s
```

Should you wish to execute certain tests, you should use to -run command-line option that accepts a regular expression and executes all tests that have a function name which matches the given regular expression:

```
$ go test testMe.go testMe_test.go -run='F2' -v
=== RUN   TestF2
--- FAIL: TestF2 (0.00s)
    testMe_test.go:50: f2(1) != 1
    testMe_test.go:54: f2(2) != 1
    testMe_test.go:58: f2(10) != 55
FAIL
FAIL    command-line-arguments    0.005s
$ go test testMe.go testMe_test.go -run='F1'
ok      command-line-arguments    (cached)
```

The last command verifies that the go test command used caching.

[451]

Software testing can only show the presence of one or more bugs, not the absence of bugs! This means that you can never be absolutely sure that your code has no bugs.

Benchmarking Go code

Benchmarking can give you information about the performance of a function or a program in order to understand better how much faster or slower a function is compared to another function, or compared to the rest of the application. Using that information, you can easily reveal the part of the Go code that needs to be rewritten in order to improve its performance.

Never benchmark your Go code on a busy Unix machine that is currently being used for other, more important purposes unless you have a very good reason to do so.

Go follows certain conventions regarding benchmarking. The most important convention is that the name of a benchmark function must begin with `Benchmark`.

Once again, the `go test` subcommand is responsible for benchmarking a program. As a result, you still need to import the `testing` standard Go package and include benchmarking functions in Go files that end with `_test.go`.

A simple benchmarking example

In this section, I will show you a basic benchmarking example that will measure the performance of three algorithms that generate numbers belonging to the **Fibonacci sequence**. The good news is that such algorithms require lots of mathematical calculations, which make them perfect candidates for benchmarking.

For the purposes of this section, I will create a new `main` package, which will be saved as `benchmarkMe.go` and presented in three parts.

The first part of `benchmarkMe.go` is as follows:

```
package main
import (
    "fmt"
```

```go
)

func fibo1(n int) int {
    if n == 0 {
        return 0
    } else if n == 1 {
        return 1
    } else {
        return fibo1(n-1) + fibo1(n-2)
    }
}
```

The preceding code contains the implementation of the `fibo1()` function that uses recursion in order to calculate the numbers of the Fibonacci sequence. Although the algorithm works fine, this is a relatively simple and somehow slow approach.

The second code segment of `benchmarkMe.go` is shown in the following Go code:

```go
func fibo2(n int) int {
    if n == 0 || n == 1 {
        return n
    }
    return fibo2(n-1) + fibo2(n-2)
}
```

In this part, you see the implementation of the `fibo2()` function that is almost identical to the `fibo1()` function that you saw earlier. However, it will be interesting to see whether a small code change—a single `if` statement as opposed to an `if else if` block—has any impact to the performance of the function.

The third code portion of `benchmarkMe.go` contains yet another implementation of a function that calculates numbers that belong to the Fibonacci sequence:

```go
func fibo3(n int) int {
    fn := make(map[int]int)
    for i := 0; i <= n; i++ {
        var f int
        if i <= 2 {
            f = 1
        } else {
            f = fn[i-1] + fn[i-2]
        }
        fn[i] = f
    }
    return fn[n]
}
```

The `fibo3()` function presented here uses a totally new approach that requires a Go **map** and has a `for` loop. It remains to be seen whether this approach is indeed faster than the other two implementations. The algorithm presented in `fibo3()` will be also used in Chapter 13, *Network Programming – Building Servers and Clients,* where it will be explained in greater detail. As you will see in a while, choosing an efficient algorithm can save you from a lot of trouble!

The remaining code of `benchmarkMe.go` follows next:

```
func main() {
    fmt.Println(fibo1(40))
    fmt.Println(fibo2(40))
    fmt.Println(fibo3(40))
}
```

Executing `benchmarkMe.go` will generate the following output:

```
$ go run benchmarkMe.go
102334155
102334155
102334155
```

The good news is that all three implementations returned the same number. Now it is time to add some benchmarks to `benchmarkMe.go` in order to understand the efficiency of the each one of the three algorithms.

As the Go rules require, the version of `benchmarkMe.go` containing the benchmark functions will be saved as `benchmarkMe_test.go`. This program is presented in five parts.

The first code segment of `benchmarkMe_test.go` contains the following Go code:

```
package main
import (
    "testing"
)

var result int
func benchmarkfibo1(b *testing.B, n int) {
    var r int
    for i := 0; i < b.N; i++ {
        r = fibo1(n)
    }
    result = r
}
```

In the preceding code, you can see the implementation of a function with a name that begins with the benchmark string instead of the Benchmark string. As a result, this function will not run automatically because it begins with a lowercase b instead of an uppercase B.

The reason for storing the result of fibo1(n) in a variable named r and using another global variable named result afterwards is tricky. This technique is used for preventing the compiler from performing any optimizations that will exclude the function that you want to measure because its result are never used! The same technique will be applied in functions benchmarkfibo2() and benchmarkfibo3(), which will be presented next.

The second part of benchmarkMe_test.go is shown in the following Go code:

```go
func benchmarkfibo2(b *testing.B, n int) {
    var r int
    for i := 0; i < b.N; i++ {
        r = fibo2(n)
    }
    result = r
}

func benchmarkfibo3(b *testing.B, n int) {
    var r int
    for i := 0; i < b.N; i++ {
        r = fibo3(n)
    }
    result = r
}
```

The preceding code defines two more benchmark functions that will not run automatically because they begin with a lowercase b instead of an uppercase B.

Now I will tell you a big secret: even if these three functions were named BenchmarkFibo1(), BenchmarkFibo2(), and BenchmarkFibo3(), they would not have been invoked automatically by the go test command because their signature is not func(*testing.B). So, the reason for naming them with a lowercase b is this! However, there is nothing that prevents you from invoking them from other benchmark functions afterwards, as you will see shortly.

The third part of benchmarkMe_test.go follows next:

```go
func Benchmark30fibo1(b *testing.B) {
    benchmarkfibo1(b, 30)
}
```

This is a correct benchmark function with the correct name and the correct signature, which means that it will get executed by `go tool`.

Note that although `Benchmark30fibo1()` is a valid benchmark function name, `BenchmarkfiboIII()` is not because there is no uppercase letter or a number after the `Benchmark` string. This is very important because a benchmark function with an incorrect name will not get executed automatically.

The fourth code segment of `benchmarkMe_test.go` contains the following Go code:

```
func Benchmark30fibo2(b *testing.B) {
    benchmarkfibo2(b, 30)
}

func Benchmark30fibo3(b *testing.B) {
    benchmarkfibo3(b, 30)
}
```

Both the `Benchmark30fibo2()` and `Benchmark30fibo3()` benchmark functions are similar to `Benchmark30fibo1()`.

The last part of `benchmarkMe_test.go` is as follows:

```
func Benchmark50fibo1(b *testing.B) {
    benchmarkfibo1(b, 50)
}

func Benchmark50fibo2(b *testing.B) {
    benchmarkfibo2(b, 50)
}

func Benchmark50fibo3(b *testing.B) {
    benchmarkfibo3(b, 50)
}
```

In this part, you see three additional benchmark functions that calculate the 50th number in the Fibonacci sequence.

Remember that each benchmark is executed for at least 1 second by default. If the benchmark function returns in a time that is less than 1 second, the value of `b.N` is increased and the function is run again. The first time the value of `b.N` is 1, then it becomes 2, 5, 10, 20, 50, and so on. This happens because the faster the function, the more times you need to run it to get accurate results.

Executing `benchmarkMe_test.go` will generate the following output:

```
$ go test -bench=. benchmarkMe.go benchmarkMe_test.go
goos: darwin
goarch: amd64
Benchmark30fibo1-8        300      4494213 ns/op
Benchmark30fibo2-8        300      4463607 ns/op
Benchmark30fibo3-8     500000         2829 ns/op
Benchmark50fibo1-8          1  67272089954 ns/op
Benchmark50fibo2-8          1  67300080137 ns/op
Benchmark50fibo3-8     300000         4138 ns/op
PASS
ok      command-line-arguments  145.827s
```

There are two important points here: first, the value of the -bench parameter specifies the benchmark functions that will be executed. The . value used is a regular expression that matches all valid benchmark functions. The second point is that if you omit the -bench parameter, no benchmark function will be executed!

So, what does this output tell us? First of all, -8 at the end of each benchmark function (Benchmark10fibo1-8) signifies the number of goroutines used during its execution, which is essentially the value of the GOMAXPROCS environment variable. You will recall that we talked about the GOMAXPROCS environment variable back in Chapter 10, *Go Concurrency – Advanced Topics*. Similarly, you can see the values of GOOS and GOARCH, which show the operating system and the architecture of your machine.

The second column in the output displays the number of times that the relevant function was executed. Faster functions are executed more times than slower functions. As an example, function Benchmark30fibo3() was executed 500,000 times, while function Benchmark50fibo2() was executed only once! The third column in the output shows the average time of each run.

As you can see, the fibo1() and fibo2() functions are really slow compared to the fibo3() function. Should you wish to include memory allocation statistics in the output, you can execute the following command:

```
$ go test -benchmem -bench=. benchmarkMe.go benchmarkMe_test.go
goos: darwin
goarch: amd64
Benchmark30fibo1-8 300      4413791 ns/op         0 B/op      0 allocs/op
Benchmark30fibo2-8 300      4430097 ns/op         0 B/op      0 allocs/op
Benchmark30fibo3-8 500000      2774 ns/op      2236 B/op      6 allocs/op
Benchmark50fibo1-8 1     71534648696 ns/op         0 B/op      0 allocs/op
Benchmark50fibo2-8 1     72551120174 ns/op         0 B/op      0 allocs/op
Benchmark50fibo3-8 300000      4612 ns/op      2481 B/op     10 allocs/op
```

```
PASS
ok      command-line-arguments    150.500s
```

The preceding output is similar to the one without the -benchmem command-line parameter, but it includes two more columns in its output. The fourth column shows the amount of memory that was allocated on average in each execution of the benchmark function. The fifth column shows the number of allocations used for allocating the memory value of the fourth column. So, Benchmark50fibo3() allocated 2481 bytes in 10 allocations on average.

As you can understand, functions fibo1() and fibo2() do not need any special kind of memory apart from the expected one, because neither of them is using any kind of data structure, which is not the case with fibo3() which uses a map variable; hence the larger than zero values in both the fourth and fifth columns of the output of Benchmark10fibo3-8.

A wrong benchmark function

Look at the Go code of the following benchmark function:

```
func BenchmarkFiboI(b *testing.B) {
    for i := 0; i < b.N; i++ {
        _ = fibo1(i)
    }
}
```

The BenchmarkFibo() function has a valid name and the correct signature. The bad news, however, is that this benchmark function is wrong, and you will not get any output from it after executing the go test command!

The reason for this is that as the b.N value grows according to the way described in a previous section, the run time of the benchmark function will also increase because of the for loop. This fact prevents BenchmarkFiboI() from converging to a stable number, which prevents the function from completing and therefore returning.

For analogous reasons, the next benchmark function is also wrongly implemented:

```
func BenchmarkfiboII(b *testing.B) {
    for i := 0; i < b.N; i++ {
        _ = fibo2(b.N)
    }
}
```

Conversely, there is nothing wrong with the implementation of the following two benchmark functions:

```
func BenchmarkFiboIV(b *testing.B) {
    for i := 0; i < b.N; i++ {
        _ = fibo3(10)
    }
}

func BenchmarkFiboIII(b *testing.B) {
    _ = fibo3(b.N)
}
```

Benchmarking buffered writing

In this section, we will explore how the size of the write buffer affects the performance of the entire writing operation using the Go code of writingBU.go, which will be presented in five parts.

The writingBU.go program generates dummy files with randomly generated data. The variables of the program are the size of the buffer and the size of the output file.

The first part of writingBU.go is as follows:

```
package main
import (
    "fmt"
    "math/rand"
    "os"
    "strconv"
)

var BUFFERSIZE int
var FILESIZE int

func random(min, max int) int {
    return rand.Intn(max-min) + min
}
```

The second code portion of writingBU.go contains the following Go code:

```
func createBuffer(buf *[]byte, count int) {
    *buf = make([]byte, count)
    if count == 0 {
        return
```

Code Testing, Optimization, and Profiling

```
    }
    for i := 0; i < count; i++ {
        intByte := byte(random(0, 100))
        if len(*buf) > count {
            return
        }
        *buf = append(*buf, intByte)
    }
}
```

The third part of writingBU.go is shown in the following Go code:

```
func Create(dst string, b, f int) error {
    _, err := os.Stat(dst)
    if err == nil {
        return fmt.Errorf("File %s already exists.", dst)
    }
    destination, err := os.Create(dst)
    if err != nil {
        return err
    }
    defer destination.Close()
    if err != nil {
        panic(err)
    }
    buf := make([]byte, 0)
    for {
        createBuffer(&buf, b)
        buf = buf[:b]
        if _, err := destination.Write(buf); err != nil {
            return err
        }
        if f < 0 {
            break
        }
        f = f - len(buf)
    }
    return err
}
```

The Create() function does all of the work in the program, and it is the function that needs to be benchmarked.

Note that if the buffer size and the file size were not part of the signature of the Create() function, you would a have problem writing a benchmark function for Create() because you would be required to use the BUFFERSIZE and FILESIZE global variables which are both initialized in the main() function of writingBU.go.

This would be difficult to do in the `writingBU_test.go` file. This means that in order to create a benchmark for a function, you should also think about it when you are writing your code.

The fourth code segment of `writingBU.go` is as follows:

```
func main() {
    if len(os.Args) != 3 {
        fmt.Println("Need BUFFERSIZE FILESIZE!")
        return
    }
    output := "/tmp/randomFile"
    BUFFERSIZE, _ = strconv.Atoi(os.Args[1])
    FILESIZE, _ = strconv.Atoi(os.Args[2])
    err := Create(output, BUFFERSIZE, FILESIZE)
    if err != nil {
        fmt.Println(err)
    }
```

The remaining Go code of `writingBU.go` follows next:

```
    err = os.Remove(output)
    if err != nil {
        fmt.Println(err)
    }
}
```

Although the `os.Remove()` call that deletes the temporary file is inside of the `main()` function, which is not called by the benchmark functions, it is easy to call `os.Remove()` from the benchmark functions, so there is no problem here.

Executing `writingBU.go` twice on a macOS High Sierra machine with a SSD hard disk using the `time(1)` utility to check the speed of the program generates the following output:

```
$ time go run writingBU.go 1 100000
real    0m1.193s
user    0m0.349s
sys     0m0.809s
$ time go run writingBU.go 10 100000
real    0m0.283s
user    0m0.195s
sys     0m0.228s
```

So, although it is obvious that the size of the write buffer plays a key role in the performance of the program, we need to be much more specific and accurate. Therefore, we will write benchmark functions that will be stored in `writingBU_test.go`.

The first part of `writingBU_test.go` is shown in the following Go code:

```go
package main
import (
    "fmt"
    "os"
    "testing"
)
var ERR error

func benchmarkCreate(b *testing.B, buffer, filesize int) {
    var err error
    for i := 0; i < b.N; i++ {
        err = Create("/tmp/random", buffer, filesize)
    }
    ERR = err
    err = os.Remove("/tmp/random")
    if err != nil {
        fmt.Println(err)
    }
}
```

As you will recall, this is not a valid benchmark function.

The second code segment of `writingBU_test.go` is as follows:

```go
func Benchmark1Create(b *testing.B) {
    benchmarkCreate(b, 1, 1000000)
}

func Benchmark2Create(b *testing.B) {
    benchmarkCreate(b, 2, 1000000)
}
```

The remaining code of `writingBU_test.go` follows next:

```go
func Benchmark4Create(b *testing.B) {
    benchmarkCreate(b, 4, 1000000)
}

func Benchmark10Create(b *testing.B) {
    benchmarkCreate(b, 10, 1000000)
}

func Benchmark1000Create(b *testing.B) {
    benchmarkCreate(b, 1000, 1000000)
}
```

Here we created five benchmark functions that will check the performance of the `benchmarkCreate()` function, which checks the performance of the `Create()` function for various write buffer sizes.

Executing `go test` on both `writingBU.go` and `writingBU_test.go` files will generate the following output:

```
$ go test -bench=. writingBU.go writingBU_test.go
goos: darwin
goarch: amd64
Benchmark1Create-8             1     6231302525 ns/op
Benchmark2Create-8             1     2956333326 ns/op
Benchmark4Create-8             1     1498240817 ns/op
Benchmark10Create-8       100000          10644 ns/op
Benchmark1000Create-8     300000           4653 ns/op
PASS
ok      command-line-arguments  23.774s
```

The following output also checks the memory allocations of the benchmark functions:

```
$ go test -bench=. writingBU.go writingBU_test.go -benchmem
goos: darwin
goarch: amd64
Benchmark1Create-8 1     6209493161 ns/op    16000840 B/op      2000017 allocs/op
Benchmark2Create-8 1     3177139645 ns/op     8000584 B/op      1000013 allocs/op
Benchmark4Create   1     1632772604 ns/op     4000424 B/op       500011 allocs/op
Benchmark10Create-8 100000     11238 ns/op        336 B/op            7 allocs/op
Benchmark1000Create-8 200000    5122 ns/op        303 B/op            5 allocs/op
PASS
ok      command-line-arguments  24.031s
```

It is now time to interpret the output of the two `go test` commands.

It is extremely obvious that using a write buffer with a size of 1 byte is totally inefficient and slows everything down. Additionally, such a buffer size requires too many memory operations, which slows down the program even more!

Using a write buffer with 2 bytes makes the entire program twice as fast, which is a good thing. However, it is still very slow. The same applies to a write buffer with a size of 4 bytes.

Where things get faster is when we decide to use a write buffer with a size of 10 bytes. Finally, the results show that using a write buffer with a size of 1,000 bytes does not make things 100 faster than when using a buffer size of 10 bytes, which means that the sweet spot between speed and write buffer size is between these two values.

Finding unreachable Go code

Go code that cannot be executed is a logical error, and therefore it is pretty difficult to be revealed by developers or a normal execution of the Go compiler. Put simply, there is nothing wrong with unreachable code apart from the fact that there is no way for this code to get executed!

Take a look at the following Go code, which is saved as `cannotReach.go`:

```
package main
import (
    "fmt"
)

func f1() int {
    fmt.Println("Entering f1()")
    return -10
    fmt.Println("Exiting f1()")
    return -1
}

func f2() int {
    if true {
        return 10
    }
    fmt.Println("Exiting f2()")
    return 0
}

func main() {
    fmt.Println(f1())
    fmt.Println("Exiting program...")
}
```

There is nothing syntactically incorrect with the Go code of `cannotReach.go`. As a result, you can execute `cannotReach.go` without getting any error messages from the compiler:

```
$ go run cannotReach.go
Entering f1()
```

```
-1
Exiting program...
```

Note that the `f2()` is never used in the program. However, it is easy to guess that the following Go code of `f2()` never gets executed because the condition in the preceding `if` statement is always `true`:

```
fmt.Println("Exiting f2()")
return 0
```

So, what can you do about it? You can execute `go tool vet` as follows:

```
$ go tool vet cannotReach.go
cannotReach.go:10: unreachable code
```

The output tells us that there is unreachable code at line 10 of the program. Now let's remove the `return -10` statement from function `f1()` and rerun `go tool vet`:

```
$ go tool vet cannotReach.go
```

Here there are no new error messages despite the fact the there is still unreachable code in the `f2()` function. This means that `go tool vet` cannot catch every possible type of logical error.

Cross-compilation

Cross-compilation is the process of generating a binary executable file for a different architecture than the one on which you are working.

The main benefit that you receive from cross-compilation is that you do not need a second or a third machine for creating executable files for different architectures. This means that you basically need just a single machine for your development. Fortunately, Go has built-in support for cross-compilation!

For the purpose of this section, we will use the Go code of `xCompile.go` for illustrating the cross-compilation process. The Go code of `xCompile.go` is as follows:

```
package main
import (
    "fmt"
    "runtime"
)

func main() {
```

```
        fmt.Print("You are using ", runtime.Compiler, " ")
        fmt.Println("on a", runtime.GOARCH, "machine")
        fmt.Println("with Go version", runtime.Version())
}
```

Running `xCompile.go` on a macOS High Sierra machine generates the following output:

```
$ go run xCompile.go
You are using gc on a amd64 machine
with Go version go1.10
```

In order to cross compile a Go source file, you will need to set the GOOS and GOARCH environment variables to the target operating system and architecture, respectively, which is not as difficult as it sounds.

So, the cross-compilation process goes like this:

```
$ env GOOS=linux GOARCH=arm go build xCompile.go
$ file xCompile
xCompile: ELF 32-bit LSB executable, ARM, EABI5 version 1 (SYSV),
statically linked, with debug_info, not stripped
$ ./xCompile
-bash: ./xCompile: cannot execute binary file
```

The first command generates a binary file that works on Linux machines which use the ARM architecture, while the output of `file(1)` verifies that the generated binary file is indeed for a different architecture.

As the Debian Linux machine that will be used for this example has an Intel processor, we will have to execute the `go build` command once more using the correct GOARCH value:

```
$ env GOOS=linux GOARCH=386 go build xCompile.go
$ file xCompile
xCompile: ELF 32-bit LSB executable, Intel 80386, version 1 (SYSV),
statically linked, with debug_info, not stripped
```

Executing the generated binary executable file on a Linux machine will produce the following expected output:

```
$ ./xCompile
You are using gc on a 386 machine
with Go version go1.10
$ go version
go version go1.3.3 linux/amd64
$ go run xCompile.go
You are using gc on a amd64 machine
with Go version go1.3.3
```

One thing to notice here is that the cross-compiled binary file of xCompile.go prints the Go version of the machine used for compiling it. The second thing to notice is that the architecture of the Linux machine is actually amd64 instead of 386, which was used in the cross-compilation process.

 You can find the list of the available values for the GOOS and GOARCH environment variables at https://golang.org/doc/install/source. Keep in mind, however, that not all the GOOS and GOARCH pairs are valid.

Creating example functions

Part of the documentation process is generating example code that showcases the use of some or all of the functions and types of a package.

Example functions have many benefits including the fact that they are executable tests which are executed by go test! Therefore, if an example function contains an // Output: line, the go test tool will check whether the calculated output matches the values found after the // Output: line.

Additionally, examples are really useful when seen in the documentation of the package, which is the subject of the next section. Finally, example functions that are presented on the Go documentation server (https://golang.org/pkg/io/#example_Copy) allow the reader of the documentation to experiment with the example code. The Go playground at https://play.golang.org/ also supports this functionality.

As the go test subcommand is responsible for the examples of a program, you will need to import the testing standard Go package and include example functions in Go files that end with _test.go. Moreover, the name of each example function must begin with Example. Lastly, **example functions** take no input parameters and return no results!

Now let's create some example functions for the following package, which is saved as ex.go:

```
package ex
func F1(n int) int {
    if n == 0 {
        return 0
    }
    if n == 1 || n == 2 {
        return 1
    }
```

```
        return F1(n-1) + F1(n-2)
}

func S1(s string) int {
    return len(s)
}
```

The ex.go source file contains the implementation of two functions named F1() and S1(). Note that ex.go does not need to import the fmt package.

As you know, the example functions will be included in the ex_test.go file, which will be presented in three parts.

The first part of ex_test.go is as follows:

```
package ex
import (
    "fmt"
)
```

The second code portion of ex_test.go is shown in the following Go code:

```
func ExampleF1() {
    fmt.Println(F1(10))
    fmt.Println(F1(2))
    // Output:
    // 55
    // 1
}
```

The remaining Go code of ex_test.go follows next:

```
func ExampleS1() {
    fmt.Println(S1("123456789"))
    fmt.Println(S1(""))
    // Output:
    // 8
    // 0
}
```

Executing the go test command on the ex.go package will generate the following output:

```
$ go test ex.go ex_test.go -v
=== RUN   ExampleF1
--- PASS: ExampleF1 (0.00s)
=== RUN   ExampleS1
--- FAIL: ExampleS1 (0.00s)
```

```
got:
9
0
want:
8
0
FAIL
FAIL    command-line-arguments    0.006s
```

You will observe that the preceding output tells us that there is something wrong with the S1() function based on the data after the // Output: comment.

Generating documentation

Go offers the godoc tool, which allows you to view the documentation of your packages—provided that you have included some extra information in your files.

> My general advice is that you should try to document everything, but leave out obvious things. Put simply, do not say, *Here I create a new* int *variable*. It would be better to state the use of that int variable! Nevertheless, really good code does not usually need documentation!

The rule about writing documentation in Go is pretty simple and straightforward. In order to document something, you have to put one or more regular comment lines that start with // just before its declaration. This convention can be used for documenting functions, variables, constants, or even packages!

Additionally, you'll notice that the first line of the documentation of a package of any size will appear in the package list of godoc, as it happens in https://golang.org/pkg/. This means that it should be pretty descriptive and complete.

Keep in mind that comments which begin with BUG(something) will appear in the Bugs section of the documentation of a package, even if they do not precede a declaration. If you are looking for such an example, you can visit the source code and the documentation page of the bytes package, which can be found at https://golang.org/src/bytes/bytes.go and https://golang.org/pkg/bytes/, respectively.

Lastly, all comments that are not related to a top-level declaration are omitted from the output that is generated by the godoc utility.

Take a look at the following Go code, which is saved as `documentMe.go`:

```go
// This package is for showcasing the documentation capabilities of Go
// It is a naive package!
package documentMe

// Pie is a global variable
// This is a silly comment!
const Pie = 3.1415912

// The S1() function finds the length of a string
// It iterates over the string using range
func S1(s string) int {
    if s == "" {
        return 0
    }
    n := 0
    for range s {
        n++
    }
    return n
}

// The F1() function returns the double value of its input integer
// A better function name would have been Double()!
func F1(n int) int {
    return 2 * n
}
```

As discussed in the previous section, we will need to create a `documentMe_test.go` file in order to develop example functions for it. The contents of `documentMe_test.go` will follow next:

```go
package documentMe
import (
    "fmt"
)
func ExampleS1() {
    fmt.Println(S1("123456789"))
    fmt.Println(S1(""))
    // Output:
    // 9
    // 0
}

func ExampleF1() {
    fmt.Println(F1(10))
```

```
        fmt.Println(F1(2))
        // Output:
        // 1
        // 55
}
```

In order to be able to see the documentation of documentMe.go, you will need to install the package on your local machine as you learned back in Chapter 6, *What You Might Not Know About Go Packages*. This will require the execution of the following commands from your favorite Unix shell:

```
$ mkdir ~/go/src/documentMe
$ cp documentMe* ~/go/src/documentMe/
$ ls -l ~/go/src/documentMe/
total 16
-rw-r--r--@ 1 mtsouk  staff  542 Mar  6 21:11 documentMe.go
-rw-r--r--@ 1 mtsouk  staff  223 Mar  6 21:11 documentMe_test.go
$ go install documentMe
$ cd ~/go/pkg/darwin_amd64
$ ls -l documentMe.a
-rw-r--r-- 1 mtsouk  staff  1626 Mar  6 21:11 documentMe.a
```

Next, you should execute the godoc utility as follows:

```
$ godoc -http=":8080"
```

Note that you can use you any port number you want, provided that the port is not already in use by another process. If that is the case, you will see an error message similar to the following:

```
$ godoc -http=":22"
2018/03/06 21:03:05 ListenAndServe :22: listen tcp :22: bind: permission denied
```

After taking care of that, you will be able to browse the HTML documentation created using your favorite web browser. The URL that will take you to the documentation is http://localhost:8080/pkg/.

Code Testing, Optimization, and Profiling

The following screenshot shows the root directory of the `godoc` server that we just started. There you can see the `documentMe` package that you created in `documentMe.go` among the other Go packages:

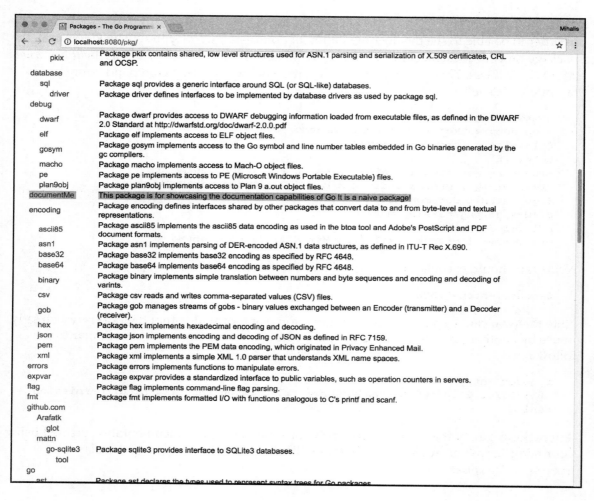

The root directory of the godoc server

The following screenshot shows the root directory of the documentation of the `documentMe` package implemented in the `documentMe.go` source file:

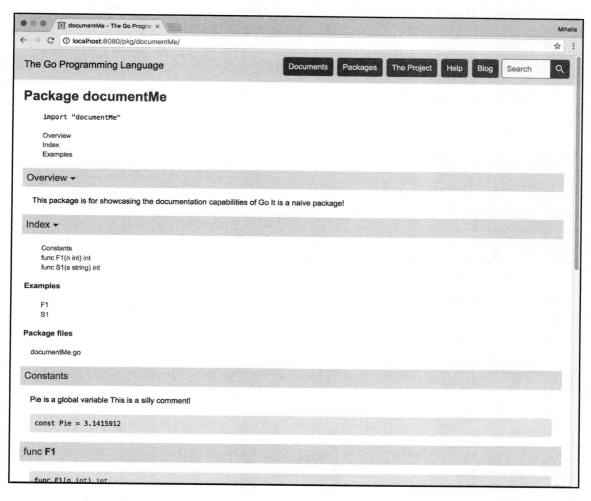

The root page of the documentMe.go file

Code Testing, Optimization, and Profiling

Similarly, the following screenshot shows the documentation of the `S1()` function of the `documentMe.go` package in greater detail, which also includes the example code. The example code is not dynamic in this case, but you can see the code and its output:

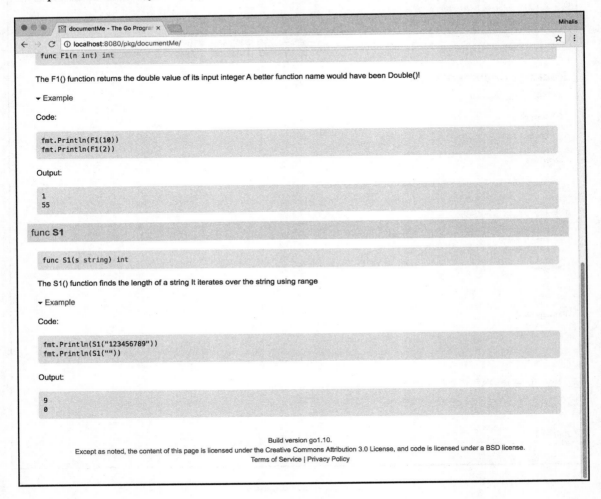

The documentation page and the example of the S1() function

Executing the `go test` command will generate the following output, which might reveal potential problems and bugs in our code:

```
$ go test -v documentMe*
=== RUN ExampleS1
--- PASS: ExampleS1 (0.00s)
=== RUN ExampleF1
--- FAIL: ExampleF1 (0.00s)
got:
20
4
want:
1
55
FAIL
FAIL    command-line-arguments   0.005s
```

Additional resources

Visit the following web links:

- Visit the website of Graphviz at `http://graphviz.org`

- Visit the documentation page of the `testing` package, which can be found at `https://golang.org/pkg/testing/`

- You can learn more about the `godoc` utility by visiting its documentation page at `https://godoc.org/golang.org/x/tools/cmd/godoc`

- Visit the documentation page of the `runtime/pprof` standard Go package that can be found at `https://golang.org/pkg/runtime/pprof/`

- You can view the Go code of the `net/http/pprof` package by visiting `https://golang.org/src/net/http/pprof/pprof.go`

- You can find the documentation page of the `net/http/pprof` package at `https://golang.org/pkg/net/http/pprof/`

- For the complete list of changes between Go version 1.10 and Go version 1.9, you should visit `https://golang.org/doc/go1.10`

- You can learn more about the `pprof` tool by visiting its development page at https://github.com/google/pprof

- Watch the Advanced Testing with Go video presented in GopherCon 2017 by Mitchell Hashimoto at https://www.youtube.com/watch?v=8hQG7QlcLBk

- You can find the source code of the `testing` package at https://golang.org/src/testing/testing.go

- You can learn more about the `profile` package by visiting its web page at https://github.com/pkg/profile

- You can learn more about the `go fix` tool by visiting its web page at https://golang.org/cmd/fix/

Exercises

- Write testing functions for the `byWord.go` program that we developed in `Chapter 8`, *Telling a Unix System What to Do*.

- Write benchmark functions for the `readSize.go` program that we developed in `Chapter 8`, *Telling a Unix System What to Do*.

- Try to fix the problems in the Go code of `documentMe.go` and `documentMe_test.go`.

- Use the text interface of the `go tool pprof` utility to examine the `memoryProfile.out` file generated by `profileMe.go`.

- Next, use the web interface of the `go tool pprof` utility to examine the `memoryProfile.out` file generated by `profileMe.go`.

Summary

In this chapter, we talked about code testing, code optimization, and code profiling. Toward the end of the chapter, you learned how to find unreachable code and how to cross-compile Go code. The `go test` command is used for testing and benchmarking Go code, as well as for offering extra documentation with the use of example functions.

Although the discussion of the Go profiler and `go tool trace` is far from complete, you should understand that with topics such as profiling and code tracing, nothing can replace experimenting and trying new techniques on your own!

In the next chapter, we will start talking about network programming in Go, which involves programming applications that work over TCP/IP computer networks, which also includes the internet. Some of the subjects in the next chapter are the `net/http` package, creating web clients and web servers in Go, the `http.Response` and `http.Request` structures, profiling HTTP servers, and timing out network connections.

Additionally, the next chapter will discuss the **IPv4** and **IPv6** protocols as well as the Wireshark and tshark tools, which are used for capturing and analyzing network traffic.

12
The Foundations of Network Programming in Go

In the previous chapter, we discussed benchmarking Go code using benchmark functions, testing in Go, example functions, cross-compilation and profiling Go code, as well as generating documentation in Go.

This chapter is all about network programming, which means that you will learn how to create web applications that work over computer networks and the internet. However, you will have to wait for the next chapter to learn how to develop TCP and UDP applications.

Note that in order to follow this chapter and the next one successfully, you'll need to know some basic information about HTTP, networking, and how networks work.

In this chapter of *Mastering Go*, you will learn the following topics:

- What TCP/IP is and why it is important
- The IPv4 and IPv6 protocols
- The **netcat** command-line utility
- Performing **DNS** lookups in Go
- The `net/http` package
- The `http.Response`, `http.Request`, and `http.Transport` structures
- Creating web servers in Go
- Programming web clients in Go
- Creating websites in Go
- The `http.NewServeMux` type
- **Wireshark** and **tshark**
- Timing out HTTP connections that take too long to finish either on the server or client end

Feel free to skip some of the low-level information presented in this chapter—you can always revisit it at another time.

About net/http, net, and http.RoundTripper

The star of this chapter will be the `net/http` package, which offers functions that allow you to develop powerful web servers and web clients. The `http.Get()` and `http.Get()` methods can be used for making HTTP and HTTPS requests, whereas the `http.ListenAndServe()` function can be used for creating web servers by specifying the IP address and the TCP port to which the server will listen, as well as the functions that will handle incoming requests.

Apart from `net/http`, we will use the `net` package in some of the programs presented in this chapter. The functionality of the `net` package, however, is going to be explained in greater detail in Chapter 13, *Network Programming – Building Servers and Clients*.

Finally, if you know that `http.RoundTripper` is an **interface** that is used for making sure that a Go element is capable of executing an HTTP transaction, it can be handy. Put simply, this means that a Go element can get `http.Response` for given `http.Request`. You will learn about `http.Response` and `http.Request` in a while.

The http.Response type

The definition of the `http.Response` structure, which can be found in the https://golang.org/src/net/http/response.go file, is as follows:

```
type Response struct {
    Status     string // e.g. "200 OK"
    StatusCode int    // e.g. 200
    Proto      string // e.g. "HTTP/1.0"
    ProtoMajor int    // e.g. 1
    ProtoMinor int    // e.g. 0
    Header Header
    Body io.ReadCloser
    ContentLength int64
    TransferEncoding []string
    Close bool
    Uncompressed bool
    Trailer Header
```

```
        Request *Request
        TLS *tls.ConnectionState
}
```

The goal of this pretty complex `http.Response` type is to represent the response of an HTTP request. The source file contains more information about the purpose of each field of the structure, which is the case with most `struct` types found in the standard Go library.

The http.Request type

The purpose of the `http.Request` type is to represent an HTTP request as received by a server, or as it is about to be sent to a server by an HTTP client.

The `http.Request` structure type as defined in https://golang.org/src/net/http/request.go is as follows:

```
type Request struct {
    Method string
    URL *url.URL
    Proto          string // "HTTP/1.0"
    ProtoMajor int     // 1
    ProtoMinor int     // 0
    Header Header
    Body io.ReadCloser
    GetBody func() (io.ReadCloser, error)
    ContentLength int64
    TransferEncoding []string
    Close bool
    Host string
    Form url.Values
    PostForm url.Values
    MultipartForm *multipart.Form
    Trailer Header
    RemoteAddr string
    RequestURI string
    TLS *tls.ConnectionState
    Cancel <-chan struct{}
    Response *Response
    ctx context.Context
}
```

The http.Transport type

The definition of the `http.Transport` structure, which can be found in https://golang.org/src/net/http/transport.go, is as follows:

```
type Transport struct {
    idleMu     sync.Mutex
    wantIdle   bool
    idleConn   map[connectMethodKey][]*persistConn
    idleConnCh map[connectMethodKey]chan *persistConn
    idleLRU    connLRU
    reqMu      sync.Mutex
    reqCanceler map[*Request]func(error)
    altMu      sync.Mutex
    altProto   atomic.Value
    Proxy func(*Request) (*url.URL, error)
    DialContext func(ctx context.Context, network, addr string) (net.Conn, error)
    Dial func(network, addr string) (net.Conn, error)
    DialTLS func(network, addr string) (net.Conn, error)
    TLSClientConfig *tls.Config
    TLSHandshakeTimeout time.Duration
    DisableKeepAlives bool
    DisableCompression bool
    MaxIdleConns int
    MaxIdleConnsPerHost int
    IdleConnTimeout time.Duration
    ResponseHeaderTimeout time.Duration
    ExpectContinueTimeout time.Duration
    TLSNextProto map[string]func(authority string, c *tls.Conn) RoundTripper
    ProxyConnectHeader Header
    MaxResponseHeaderBytes int64
    nextProtoOnce sync.Once
    h2transport   *http2Transport
}
```

As you can see, the `http.Transport` structure is pretty complex and contains a very large number of fields. The good news is that you will not need to use the `http.Transport` structure in all of your programs and that you are not required to deal with all of its fields each time that you use it.

The `http.Transport` structure implements the `http.RoundTripper` interface and supports HTTP, HTTPS, and HTTP proxies. Note that `http.Transport` is pretty low level, whereas the `http.Client` structure that is also used in this chapter implements a high-level HTTP client.

About TCP/IP

TCP/IP is a family of protocols that help the internet operate. Its name came from its two most well-known protocols—**TCP** and **IP**.

TCP stands for **Transmission Control Protocol**. TCP software transmits data between machines using segments, which are also called TCP **packets**. The main characteristic of TCP is that it is a reliable protocol, which means that it makes sure that a packet was delivered without needing any extra code from the programmer. If there is no proof of a packet delivery, TCP resends that particular packet. Among other things, a TCP packet can be used for establishing connections, transferring data, sending acknowledgements, and closing connections.

When a TCP connection is established between two machines, a full duplex virtual circuit, similar to the telephone call, is created between those two machines. The two machines constantly communicate to make sure that data are sent and received correctly. If the connection fails for some reason, the two machines try to find the problem and report to the relevant application.

IP stands for **Internet Protocol**. The main characteristic of IP is that it is not a reliable protocol by nature. IP encapsulates the data that travels over a TCP/IP network because it is responsible for delivering packets from the source host to the destination host according to the IP addresses. IP has to find an addressing method to send the packet to its destination effectively. Although there are dedicated devices called routers that perform IP routing, every TCP/IP device has to perform some basic routing.

The **UDP (User Datagram Protocol)** protocol is based on IP, which means that it is also unreliable. Generally speaking, the UDP protocol is simpler than the TCP protocol mainly because UDP is not reliable by design. As a result, UDP messages can be lost, duplicated, or arrive out of order. Furthermore, packets can arrive faster than the recipient can process them. So, UDP is used when speed is more important than reliability.

About IPv4 and IPv6

The first version of the **IP protocol** is now called **IPv4** in order to differentiate it from the latest version of the IP protocol, which is called **IPv6**.

The main problem with IPv4 is that it is at the verge of running out of IP addresses, which is the main reason for creating the IPv6 protocol. This happens because an IPv4 address is represented using 32 bits only, which allows a total number of 2^{32} (4,294,967,296) different IP addresses. On the other hand, IPv6 uses 128 bits for defining each one of its addresses.

The format of an IPv4 address is 10.20.32.245 (four parts separated by dots), while the format of an IPv6 address is 3fce:1706:4523:3:150:f8ff:fe21:56cf (eight parts separated by colons).

The nc(1) command-line utility

The `nc(1)` utility, which is also called `netcat(1)`, comes in very handy when you want to test TCP/IP servers and clients. This section will present some of its more common uses.

You can use `nc(1)` as a client for a TCP service that runs on a machine with the 10.10.1.123 IP address and listens to the port number 1234, as follows:

```
$ nc 10.10.1.123 1234
```

By default, `nc(1)` uses the TCP protocol. However, if you execute `nc(1)` with the -u flag, then `nc(1)` will use the UDP protocol.

The -l option tells `netcat(1)` to act as a server, which means that `netcat(1)` will start listening for connections at the given port number.

Finally, the -v and -vv options tell `netcat(1)` to generate verbose output, which can come in handy when you want to troubleshoot network connections.

Although `netcat(1)` can help you test HTTP applications, it will become niftier in Chapter 13, *Network Programming – Building Servers and Clients*, when we will develop TCP and UDP clients and servers. As a result, the `netcat(1)` utility will be used only once in this chapter.

Reading the configuration of network interfaces

There are four core elements in a network configuration: the IP address of the interface, the network mask of the interface, the DNS servers of the machine, and the default gateway or default router of the machine. However, there is a problem here: you cannot find every piece of information using native and portable Go code. This means that there is no portable way to discover the DNS configuration and the default gateway information of a Unix machine.

As a result, in this section, you will learn how to read the configuration of the network interfaces of a Unix machine with Go. For this purpose, I will present two portable utilities that allow you to find out information about your network interfaces.

The source code of the first utility, netConfig.go, is presented in three parts.

The first part of netConfig.go is shown in the following Go code:

```go
package main

import (
    "fmt"
    "net"
)

func main() {
    interfaces, err := net.Interfaces()
    if err != nil {
        fmt.Println(err)
        return
    }
```

The net.Interfaces() function returns all of the interfaces of the current machine as a slice with elements of the net.Interface type. This slice will be used shortly to acquire the desired information.

The second code portion of `netConfig.go` contains the following Go code:

```go
for _, i := range interfaces {
    fmt.Printf("Interface: %v\n", i.Name)
        byName, err := net.InterfaceByName(i.Name)
        if err != nil {
            fmt.Println(err)
        }
```

In the preceding code, you visit each element of the slice with the `net.Interface` elements to retrieve the desired information.

The remaining Go code of `netConfig.go` is as follows:

```go
        addresses, err := byName.Addrs()
        for k, v := range addresses {
            fmt.Printf("Interface Address #%v: %v\n", k, v.String())
        }
        fmt.Println()
    }
}
```

Executing `netConfig.go` on a macOS High Sierra machine with Go version 1.10 generates the following output:

```
$ go run netConfig.go
Interface: lo0
Interface Address #0: 127.0.0.1/8
Interface Address #1: ::1/128
Interface Address #2: fe80::1/64
Interface: gif0
Interface: stf0
Interface: XHC20
Interface: en0
Interface Address #0: fe80::18fa:901a:ea9:eb5f/64
Interface Address #1: 192.168.1.200/24
Interface Address #2: 2a02:587:3006:b800:1cb8:bf1b:b154:4d0c/64
Interface Address #3: 2a02:587:3006:b800:d84a:f0c:c932:35d1/64
Interface: en1
Interface: p2p0
Interface: awdl0
Interface: en2
Interface: en3
Interface: bridge0
Interface: utun0
Interface Address #0: fe80::2514:c3a3:ca83:e1c6/64
Interface: utun1
```

```
Interface Address #0: fe80::4e0b:a9a6:9abe:81a4/64
Interface: en5
Interface Address #0: fe80::1cb4:a29e:97bc:6fb5/64
Interface Address #1: 169.254.72.59/16
```

As you can see, the `netConfig.go` utility returns rather large output because today's computers tend to have lots of network interfaces and because the program supports both the IPv4 and IPv6 protocols.

Executing `netConfig.go` on a Debian Linux machine with Go version 1.7.4 will generate the following output:

```
$ go run netConfig.go
Interface: lo
Interface Address #0: 127.0.0.1/8
Interface Address #1: ::1/128
Interface: dummy0
Interface: eth0
Interface Address #0: 10.74.193.253/24
Interface Address #1: 2a01:7e00::f03c:91ff:fe69:1381/64
Interface Address #2: fe80::f03c:91ff:fe69:1381/64
Interface: teql0
Interface: tunl0
Interface: gre0
Interface: gretap0
Interface: erspan0
Interface: ip_vti0
Interface: ip6_vti0
Interface: sit0
Interface: ip6tnl0
Interface: ip6gre0
```

Note that the main reason why a network interface may not have a network address is that it is down, which essentially means that it is not currently configured.

Not all of the listed interfaces have a real hardware network device attached to them. The more representative example is the `lo0` interface, which is the loopback device. The **loopback device** is a special, virtual network interface that is used by your computer in order to communicate with itself.

The Go code for the next utility, `netCapabilities.go`, is also presented in three parts. The purpose of the `netCapabilities.go` utility is to reveal the capabilities of each network interface found on your Unix system.

The `netCapabilities.go` utility uses the fields of the `net.Interface` structure, which is defined as follows:

```
type Interface struct {
    Index        int
    MTU          int
    Name         string
    HardwareAddr HardwareAddr
    Flags        Flags
}
```

The first part from the Go code of `netCapabilities.go` follows next:

```
package main

import (
    "fmt"
    "net"
)
```

The second code portion of `netCapabilities.go` contains the following Go code:

```
func main() {
    interfaces, err := net.Interfaces()

    if err != nil {
        fmt.Print(err)
        return
    }
```

The last part of `netCapabilities.go` comes in the following Go code:

```
    for _, i := range interfaces {
        fmt.Printf("Name: %v\n", i.Name)
        fmt.Println("Interface Flags:", i.Flags.String())
        fmt.Println("Interface MTU:", i.MTU)
        fmt.Println("Interface Hardware Address:", i.HardwareAddr)
        fmt.Println()
    }
}
```

Running `netCapabilities.go` on a macOS High Sierra machine will generate the following output:

```
$ go run netCapabilities.go
Name: lo0
Interface Flags: up|loopback|multicast
Interface MTU: 16384
Interface Hardware Address:
Name: gif0
Interface Flags: pointtopoint|multicast
Interface MTU: 1280
Interface Hardware Address:
Name: stf0
Interface Flags: 0
Interface MTU: 1280
Interface Hardware Address:
Name: XHC20
Interface Flags: 0
Interface MTU: 0
Interface Hardware Address:
Name: en0
Interface Flags: up|broadcast|multicast
Interface MTU: 1500
Interface Hardware Address: 98:5a:eb:d7:84:cd
Name: en1
Interface Flags: up|broadcast|multicast
Interface MTU: 1500
Interface Hardware Address: d0:03:4b:cf:84:d3
Name: p2p0
Interface Flags: broadcast|multicast
Interface MTU: 2304
Interface Hardware Address: 02:03:4b:cf:84:d3
Name: awdl0
Interface Flags: broadcast|multicast
Interface MTU: 1484
Interface Hardware Address: 02:ac:d4:3b:d9:29
Name: en2
Interface Flags: up|broadcast|multicast
Interface MTU: 1500
Interface Hardware Address: 0a:00:00:a5:32:b0
Name: en3
Interface Flags: up|broadcast|multicast
Interface MTU: 1500
Interface Hardware Address: 0a:00:00:a5:32:b1
Name: bridge0
Interface Flags: up|broadcast|multicast
Interface MTU: 1500
```

```
Interface Hardware Address: 0a:00:00:a5:32:b0
Name: utun0
Interface Flags: up|pointtopoint|multicast
Interface MTU: 2000
Interface Hardware Address:
Name: utun1
Interface Flags: up|pointtopoint|multicast
Interface MTU: 1380
Interface Hardware Address:
Name: en5
Interface Flags: up|broadcast|multicast
Interface MTU: 1500
Interface Hardware Address: 6e:72:e7:1b:cd:5f
```

Executing `netCapabilities.go` on a Debian Linux machine will generate similar output.

Finally, if you are really interested in finding out the default gateway of the machine, you can execute the `netstat -nr` command either externally or using `exec.Command()`, taking its output using a pipe or `exec.CombinedOutput()`, and processing it as text using Go. This solution, however, is neither elegant nor perfect.

Performing DNS lookups

DNS stands for **Domain Name System**, which relates to the way an IP address is translated into a name such as `packt.com` and vice versa. The logic behind the `DNS.go` utility, which will be developed in this section, is pretty simple: if the given command-line argument is a valid IP address, the program will process it as an IP address; otherwise, it will assume that it is dealing with a hostname that needs to be translated into one or more IP addresses.

The code for the `DNS.go` utility will be presented in three parts. The first part of the program contains the following Go code:

```
package main

import (
    "fmt"
    "net"
    "os"
)

func lookIP(address string) ([]string, error) {
    hosts, err := net.LookupAddr(address)
    if err != nil {
        return nil, err
```

```
        }
        return hosts, nil
}

func lookHostname(hostname string) ([]string, error) {
        IPs, err := net.LookupHost(hostname)
        if err != nil {
                return nil, err
        }
        return IPs, nil
}
```

The `lookIP()` function gets an IP address as input and returns the list of names that match that IP address with the help of the `net.LookupAddr()` function.

On the other hand, the `lookHostname()` function gets a host name as input and returns a list with the associated IP addresses using the `net.LookupHost()` function.

The second part of DNS.go is the following Go code:

```
func main() {
        arguments := os.Args
        if len(arguments) == 1 {
                fmt.Println("Please provide an argument!")
                return
        }

        input := arguments[1]
        IPaddress := net.ParseIP(input)
```

The `net.ParseIP()` function parses a string as an IPv4 or an IPv6 address. If the IP address is not valid, `net.ParseIP()` returns `nil`.

The remaining Go code of the DNS.go utility follows next:

```
        if IPaddress == nil {
                IPs, err := lookHostname(input)
                if err == nil {
                        for _, singleIP := range IPs {
                                fmt.Println(singleIP)
                        }
                }
        } else {
                hosts, err := lookIP(input)
                if err == nil {
                        for _, hostname := range hosts {
                                fmt.Println(hostname)
```

```
            }
          }
        }
      }
    }
```

Executing `DNS.go` with various kinds of input will generate the following output:

```
$ go run DNS.go 127.0.0.1
localhost
$ go run DNS.go 192.168.1.1
cisco
$ go run DNS.go packtpub.com
83.166.169.231
$ go run DNS.go google.com
2a00:1450:4001:816::200e
216.58.210.14
$ go run DNS.go www.google.com
2a00:1450:4001:816::2004
216.58.214.36
$ go run DNS.go cnn.com
2a04:4e42::323
2a04:4e42:600::323
2a04:4e42:400::323
2a04:4e42:200::323
151.101.193.67
151.101.1.67
151.101.129.67
151.101.65.67
```

Note that the output of the `go run DNS.go 192.168.1.1` command is taken from my `/etc/hosts` file, because the `cisco` hostname is an alias for the `192.168.1.1` IP address in my `/etc/hosts` file.

The output of the last command shows that sometimes a single hostname (`cnn.com`) might have multiple public IP addresses. Pay special attention to the word *public* here, because although `www.google.com` has multiple IP addresses, it uses just a single public IP address (`216.58.214.36`).

Getting the NS records of a domain

A very popular DNS request has to do with finding out the **name servers** of a domain, which are stored in the **NS records** of that domain. This functionality is will be illustrated in the code of `NSrecords.go`.

The code of `NSrecords.go` will be presented in two parts. The first part of `NSrecords.go` follows next:

```
package main

import (
    "fmt"
    "net"
    "os"
)

func main() {
    arguments := os.Args
    if len(arguments) == 1 {
        fmt.Println("Need a domain name!")
        return
    }
```

In this part, you check whether you have at least one command-line argument in order to have something with which you can work.

The remaining Go code of `NSrecords.go` is as follows:

```
    domain := arguments[1]
    NSs, err := net.LookupNS(domain)
    if err != nil {
        fmt.Println(err)
        return
    }

    for _, NS := range NSs {
        fmt.Println(NS.Host)
    }
}
```

All the work is done by the `net.LookupNS()` function, which returns the **NS records** of a domain as a slice variable of the `net.NS` type. This is the reason for printing the `Host` field of each `net.NS` element of the slice.

Executing `NSrecords.go` will generate the following output:

```
$ go run NSrecords.go mtsoukalos.eu
ns5.linode.com.
ns4.linode.com.
ns1.linode.com.
ns2.linode.com.
ns3.linode.com.
$ go run NSrecords.go www.mtsoukalos.eu
lookup www.mtsoukalos.eu on 8.8.8.8:53: no such host
```

You can verify the correctness of the preceding output with the help of the `host(1)` utility:

```
$ host -t ns www.mtsoukalos.eu
www.mtsoukalos.eu has no NS record
$ host -t ns mtsoukalos.eu
mtsoukalos.eu name server ns3.linode.com.
mtsoukalos.eu name server ns1.linode.com.
mtsoukalos.eu name server ns4.linode.com.
mtsoukalos.eu name server ns2.linode.com.
mtsoukalos.eu name server ns5.linode.com.
```

Getting the MX records of a domain

Another very popular DNS request has to do with getting the **MX records** of a domain. The MX records specify the mail servers of a domain. The code of the `MXrecords.go` utility will perform this task with Go.

The first part of the `MXrecords.go` utility is shown in the following Go code:

```go
package main

import (
    "fmt"
    "net"
    "os"
)

func main() {
    arguments := os.Args
    if len(arguments) == 1 {
        fmt.Println("Need a domain name!")
        return
    }
```

The second part of `MXrecords.go` contains the following Go code:

```go
    domain := arguments[1]
    MXs, err := net.LookupMX(domain)
    if err != nil {
        fmt.Println(err)
        return
    }

    for _, MX := range MXs {
        fmt.Println(MX.Host)
    }
}
```

The code of `MXrecords.go` works in a similar way to the code of `NXrecords.go` presented in the previous section. The only difference is that you use the `net.LookupMX()` function instead of the `net.LookupNS()` function.

Executing `MXrecords.go` will generate the following output:

```
$ go run MXrecords.go golang.com
aspmx.l.google.com.
alt3.aspmx.l.google.com.
alt1.aspmx.l.google.com.
alt2.aspmx.l.google.com.
$ go run MXrecords.go www.mtsoukalos.eu
lookup www.mtsoukalos.eu on 8.8.8.8:53: no such host
```

Once again, you can verify the validity of the preceding output with the help of the `host(1)` utility:

```
$ host -t mx golang.com
golang.com mail is handled by 2 alt3.aspmx.l.google.com.
golang.com mail is handled by 1 aspmx.l.google.com.
golang.com mail is handled by 2 alt1.aspmx.l.google.com.
golang.com mail is handled by 2 alt2.aspmx.l.google.com.
$ host -t mx www.mtsoukalos.eu
www.mtsoukalos.eu has no MX record
```

Creating a web server in Go

Go allows you to create a web server on your own, using some of the functions of its standard library. The first time that you saw a web server application programmed in Go in this book was back in Chapter 10, *Go Concurrency – Advanced Topics*, when we talked about the context package.

Although a web server programmed in Go can do many things efficiently and securely, if what you really need is a powerful web server that will support modules, multiple websites, and virtual hosts, then you would be better off using a web server such as **Apache** or **Nginx**.

The name of the Go program for this example will be www.go, and it will be presented in five parts. The first part of www.go contains the expected import statements:

```
package main

import (
    "fmt"
    "net/http"
    "os"
    "time"
)
```

The time package is not necessary for a web server to operate. However, it is needed in this case because the server will send the time and the date to its clients.

The second code segment of www.go is shown in the following Go code:

```
func myHandler(w http.ResponseWriter, r *http.Request) {
    fmt.Fprintf(w, "Serving: %s\n", r.URL.Path)
    fmt.Printf("Served: %s\n", r.Host)
}
```

This is the implementation of the first handler function of the program. A **handler function** serves one or more URLs depending on the configuration used.

The third part from www.go contains the following Go code:

```
func timeHandler(w http.ResponseWriter, r *http.Request) {
    t := time.Now().Format(time.RFC1123)
    Body := "The current time is:"
    fmt.Fprintf(w, "<h1 align=\"center\">%s</h1>", Body)
    fmt.Fprintf(w, "<h2 align=\"center\">%s</h2>\n", t)
    fmt.Fprintf(w, "Serving: %s\n", r.URL.Path)
    fmt.Printf("Served time for: %s\n", r.Host)
}
```

In the preceding Go code, you can see the implementation of the second handler function of the program. This function generates dynamic content.

The fourth section of code of our web server deals with the command-line arguments and the definition of the supported URLs:

```
func main() {
    PORT := ":8001"
    arguments := os.Args
    if len(arguments) == 1 {
        fmt.Println("Using default port number: ", PORT)
    } else {
        PORT = ":" + arguments[1]
    }

    http.HandleFunc("/time", timeHandler)
    http.HandleFunc("/", myHandler)
```

The `http.HandleFunc()` function associates a URL with a handler function.

The last part of the www.go program is as follows:

```
    err := http.ListenAndServe(PORT, nil)
    if err != nil {
        fmt.Println(err)
        return
    }
}
```

You should start the web server with the help of the `http.ListenAndServe()` function using the desired port number.

Executing www.go and connecting to its web server will generate the following output:

```
$ go run www.go
Using default port number:   :8001
Served: localhost:8001
Served: localhost:8001
Served time for: localhost:8001
Served: localhost:8001
Served time for: localhost:8001
Served: localhost:8001
Served time for: localhost:8001
Served: localhost:8001
Served: localhost:8001
Served: localhost:8001
```

Although the output of the program provides some handy material, I think that you would prefer to see the real output of the program using your favorite web browser. The following screenshot shows the output of the myHandler() function of our web server as displayed in **Google Chrome**:

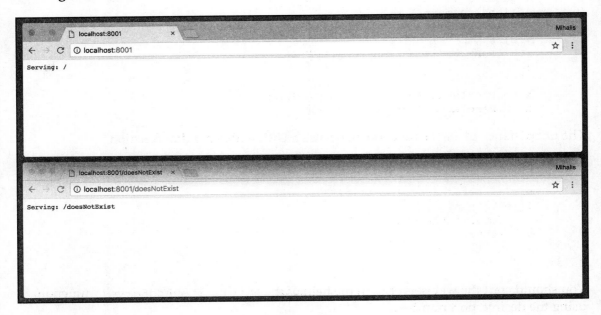

The home page of the www.go web server

The following screenshot shows that `www.go` can generate dynamic pages as well. In this case, it is a web page registered in `/time`, which displays the current time and date:

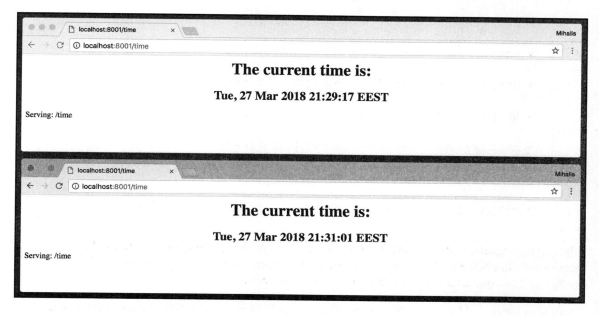

Getting the current time and date from the www.go web server

What is really interesting here is that all URLs apart from `/time` are served by the `myHandler()` function because its first argument, which is `/`, matches every URL that is not matched by another handler!

Profiling an HTTP server

As you learned in Chapter 11, *Code Testing, Optimization, and Profiling*, there is a standard Go package named `net/http/pprof` that should be used when you want to profile a Go application with its own HTTP server. Therefore, importing `net/http/pprof` will install various handlers under the `/debug/pprof/` URL. You will see more of this in a short while. For now, it is enough that you remember that the `net/http/pprof` package should be used for profiling web applications with an HTTP server, whereas the `runtime/pprof` standard Go package should be used for profiling all other kinds of applications.

Note that if your profiler works using the `http://localhost:8080` address, you will get support for the following web links:

- `http://localhost:8080/debug/pprof/goroutine`
- `http://localhost:8080/debug/pprof/heap`
- `http://localhost:8080/debug/pprof/threadcreate`
- `http://localhost:8080/debug/pprof/block`
- `http://localhost:8080/debug/pprof/mutex`
- `http://localhost:8080/debug/pprof/profile`
- `http://localhost:8080/debug/pprof/trace?seconds=5`

The next program to be presented will use `www.go` as its starting point and add the necessary Go code to allow you to profile it. The name of the new program is `wwwProfile.go`, and it will be presented in four parts.

Note that `wwwProfile.go` will use an `http.NewServeMux` variable for registering the supported paths of the program. The main reason for doing so is that the use of `http.NewServeMux` requires defining the HTTP endpoints manually. Also note that you are allowed to define a subset of the supported HTTP endpoints. If you decide not to use `http.NewServeMux`, then the HTTP endpoints are registered automatically, which also means that you will have to import the `net/http/pprof` package using the _ character in front of it.

The first part of `wwwProfile.go` contains the following Go code:

```
package main

import (
    "fmt"
    "net/http"
    "net/http/pprof"
    "os"
    "time"
)

func myHandler(w http.ResponseWriter, r *http.Request) {
    fmt.Fprintf(w, "Serving: %s\n", r.URL.Path)
```

```
        fmt.Printf("Served: %s\n", r.Host)
}

func timeHandler(w http.ResponseWriter, r *http.Request) {
    t := time.Now().Format(time.RFC1123)
    Body := "The current time is:"
    fmt.Fprintf(w, "<h1 align=\"center\">%s</h1>", Body)
    fmt.Fprintf(w, "<h2 align=\"center\">%s</h2>\n", t)
    fmt.Fprintf(w, "Serving: %s\n", r.URL.Path)
    fmt.Printf("Served time for: %s\n", r.Host)
}
```

The implementations of these two handler functions are exactly the same as before.

The second code segment from `wwwProfile.go` is as follows:

```
func main() {
    PORT := ":8001"
    arguments := os.Args
    if len(arguments) == 1 {
        fmt.Println("Using default port number: ", PORT)
    } else {
        PORT = ":" + arguments[1]
        fmt.Println("Using port number: ", PORT)
    }

    r := http.NewServeMux()
    r.HandleFunc("/time", timeHandler)
    r.HandleFunc("/", myHandler)
```

In the preceding Go code, you define the URLs that will be supported by your web server using `http.NewServeMux()` and `HandleFunc()`.

The third code portion from `wwwProfile.go` is shown in the following Go code:

```
    r.HandleFunc("/debug/pprof/", pprof.Index)
    r.HandleFunc("/debug/pprof/cmdline", pprof.Cmdline)
    r.HandleFunc("/debug/pprof/profile", pprof.Profile)
    r.HandleFunc("/debug/pprof/symbol", pprof.Symbol)
    r.HandleFunc("/debug/pprof/trace", pprof.Trace)
```

The preceding Go code defines the HTTP endpoints related to profiling. Without them, you will not be able to profile your web application.

The remaining Go code is as follows:

```
err := http.ListenAndServe(PORT, r)
    if err != nil {
        fmt.Println(err)
        return
    }
}
```

This code begins the Go web server, and this allows it to serve connections from HTTP clients. You'll notice that the second parameter to `http.ListenAndServe()` is no longer `nil`.

As you can see, `wwwProfile.go` does not define the `/debug/pprof/goroutine` HTTP endpoint, which makes perfect sense as `wwwProfile.go` does not use any goroutines!

Executing `wwwProfile.go` will generate the following output:

```
$ go run wwwProfile.go 1234
Using port number:   :1234
Served time for: localhost:1234
```

Using the Go profiler to get data is a pretty simple task that requires the execution of the next command, which will take you to the shell of the Go profiler automatically:

```
$ go tool pprof http://localhost:1234/debug/pprof/profile
Fetching profile over HTTP from http://localhost:1234/debug/pprof/profile
Saved profile in /Users/mtsouk/pprof/pprof.samples.cpu.003.pb.gz
Type: cpu
Time: Mar 27, 2018 at 10:04pm (EEST)
Duration: 30s, Total samples = 21.04s (70.13%)
Entering interactive mode (type "help" for commands, "o" for options)
(pprof) top
Showing nodes accounting for 19.94s, 94.77% of 21.04s total
Dropped 159 nodes (cum <= 0.11s)
Showing top 10 nodes out of 75
      flat   flat%   sum%       cum   cum%
     13.73s 65.26% 65.26%    13.73s 65.26%  syscall.Syscall
      1.58s  7.51% 72.77%     1.58s  7.51%  runtime.kevent
      1.36s  6.46% 79.23%     1.36s  6.46%  runtime.mach_semaphore_signal
      1.02s  4.85% 84.08%     1.02s  4.85%  runtime.usleep
      0.80s  3.80% 87.88%     0.80s  3.80%  runtime.mach_semaphore_wait
      0.53s  2.52% 90.40%     2.11s 10.03%  runtime.netpoll
      0.44s  2.09% 92.49%     0.44s  2.09%  internal/poll.convertErr
      0.26s  1.24% 93.73%     0.26s  1.24%  net.(*TCPConn).Read
```

```
       0.18s  0.86% 94.58%     0.18s  0.86%  runtime.freedefer
       0.04s  0.19% 94.77%     1.05s  4.99%  runtime.runqsteal
(pprof)
```

You can now use the profiling data and analyze it using `go tool pprof`, as you learned in Chapter 11, *Code Testing, Optimization, and Profiling*.

You can visit `http://HOSTNAME:PORTNUMBER/debug/pprof/` and see the profiling results from there. When the value of `HOSTNAME` is `localhost` and that of `PORTNUMBER` is `1234`, you should visit `http://localhost:1234/debug/pprof/`.

Should you wish to test the performance of your web server application, you can use the `ab(1)` utility, which is more widely known as the **Apache HTTP server benchmarking tool**, in order to create some traffic and benchmark `wwwProfile.go`. This will also allow `go tool pprof` to collect more accurate data, as follows:

```
$ ab -k -c 10 -n 100000 "http://127.0.0.1:1234/time"
This is ApacheBench, Version 2.3 <$Revision: 1807734 $>
Copyright 1996 Adam Twiss, Zeus Technology Ltd, http://www.zeustech.net/
Licensed to The Apache Software Foundation, http://www.apache.org/
Benchmarking 127.0.0.1 (be patient)
Completed 10000 requests
Completed 20000 requests
Completed 30000 requests
Completed 40000 requests
Completed 50000 requests
Completed 60000 requests
Completed 70000 requests
Completed 80000 requests
Completed 90000 requests
Completed 100000 requests
Finished 100000 requests
Server Software:
Server Hostname:        127.0.0.1
Server Port:            1234
Document Path:          /time
Document Length:        114 bytes
Concurrency Level:      10
Time taken for tests:   2.114 seconds
Complete requests:      100000
Failed requests:        0
Keep-Alive requests:    100000
Total transferred:      25500000 bytes
HTML transferred:       11400000 bytes
Requests per second:    47295.75 [#/sec] (mean)
```

```
Time per request:        0.211 [ms] (mean)
Time per request:        0.021 [ms] (mean, across all concurrent requests)
Transfer rate:           11777.75 [Kbytes/sec] received
Connection Times (ms)
              min   mean[+/-sd] median   max
Connect:       0     0    0.0      0      0
Processing:    0     0    0.7      0     13
Waiting:       0     0    0.7      0     13
Total:         0     0    0.7      0     13
Percentage of the requests served within a certain time (ms)
  50%      0
  66%      0
  75%      0
  80%      0
  90%      0
  95%      0
  98%      0
  99%      0
 100%     13 (longest request)
```

Can you use the net/http/pprof package for profiling command-line applications? Yes! However, the net/http/pprof package is particularly useful when you want to profile a running web application and capture live data, which is the main reason for presenting it in this chapter.

Creating a website in Go

Do you remember the keyValue.go application from Chapter 4, *The Uses of Composite Types*, and kvSaveLoad.go from Chapter 8, *Telling a Unix System What to Do*? In this section, you will learn how to create a web interface for them, using the capabilities of the standard Go library. The name of the new Go source code file is kvWeb.go, and it will be presented in six parts.

The first difference between the Go code of kvWeb.go and www.go, developed earlier in this chapter, is that kvWeb.go uses the http.NewServeMux type for dealing with HTTP requests because it is much more versatile for nontrivial web applications.

The first part of kvWeb.go is as follows:

```
package main

import (
    "encoding/gob"
    "fmt"
```

```go
        "html/template"
        "net/http"
        "os"
)

type myElement struct {
    Name     string
    Surname  string
    Id       string
}

var DATA = make(map[string]myElement)
var DATAFILE = "/tmp/dataFile.gob"
```

You have already seen this code in `kvSaveLoad.go` in Chapter 8, *Telling a Unix System What to Do*.

The second part of `kvWeb.go` is shown in the following Go code:

```go
func save() error {
    fmt.Println("Saving", DATAFILE)
    err := os.Remove(DATAFILE)
    if err != nil {
        fmt.Println(err)
    }

    saveTo, err := os.Create(DATAFILE)
    if err != nil {
        fmt.Println("Cannot create", DATAFILE)
        return err
    }
    defer saveTo.Close()

    encoder := gob.NewEncoder(saveTo)
    err = encoder.Encode(DATA)
    if err != nil {
        fmt.Println("Cannot save to", DATAFILE)
        return err
    }
    return nil
}

func load() error {
    fmt.Println("Loading", DATAFILE)
    loadFrom, err := os.Open(DATAFILE)
    defer loadFrom.Close()
    if err != nil {
```

```go
            fmt.Println("Empty key/value store!")
            return err
    }

    decoder := gob.NewDecoder(loadFrom)
    decoder.Decode(&DATA)
    return nil
}

func ADD(k string, n myElement) bool {
    if k == "" {
        return false
    }

    if LOOKUP(k) == nil {
        DATA[k] = n
        return true
    }
    return false
}

func DELETE(k string) bool {
    if LOOKUP(k) != nil {
        delete(DATA, k)
        return true
    }
    return false
}

func LOOKUP(k string) *myElement {
    _, ok := DATA[k]
    if ok {
        n := DATA[k]
        return &n
    } else {
        return nil
    }
}

func CHANGE(k string, n myElement) bool {
    DATA[k] = n
    return true
}

func PRINT() {
    for k, d := range DATA {
```

```
        fmt.Printf("key: %s value: %v\n", k, d)
    }
}
```

You should also be familiar with the preceding Go code, as it first appeared in kvSaveLoad.go in Chapter 8, *Telling a Unix System What to Do*.

The third code portion of kvWeb.go follows next:

```
func homePage(w http.ResponseWriter, r *http.Request) {
    fmt.Println("Serving", r.Host, "for", r.URL.Path)
    myT := template.Must(template.ParseGlob("home.gohtml"))
    myT.ExecuteTemplate(w, "home.gohtml", nil)
}

func listAll(w http.ResponseWriter, r *http.Request) {
    fmt.Println("Listing the contents of the KV store!")

    fmt.Fprintf(w, "<a href=\"/\" style=\"margin-right: 20px;\">Home sweet home!</a>")
    fmt.Fprintf(w, "<a href=\"/list\" style=\"margin-right: 20px;\">List all elements!</a>")
    fmt.Fprintf(w, "<a href\"=/change\" style=\"margin-right: 20px;\">Change an element!</a>")
    fmt.Fprintf(w, "<a href\"=/insert\" style=\"margin-right: 20px;\">Insert new element!</a>")

    fmt.Fprintf(w, "<h1>The contents of the KV store are:</h1>")
    fmt.Fprintf(w, "<ul>")
    for k, v := range DATA {
        fmt.Fprintf(w, "<li>")
        fmt.Fprintf(w, "<strong>%s</strong> with value: %v\n", k, v)
        fmt.Fprintf(w, "</li>")
    }

    fmt.Fprintf(w, "</ul>")
}
```

The listAll() function does not use any Go templates for generating its dynamic output. Instead, its output is generated on the fly, using Go. You may consider this an exception, as web applications usually work better with HTML templates and the html/template standard Go package.

The fourth part of kvWeb.go contains the following Go code:

```go
func changeElement(w http.ResponseWriter, r *http.Request) {
    fmt.Println("Changing an element of the KV store!")
    tmpl := template.Must(template.ParseFiles("update.gohtml"))
    if r.Method != http.MethodPost {
        tmpl.Execute(w, nil)
        return
    }

    key := r.FormValue("key")
    n := myElement{
        Name:    r.FormValue("name"),
        Surname: r.FormValue("surname"),
        Id:      r.FormValue("id"),
    }

    if !CHANGE(key, n) {
        fmt.Println("Update operation failed!")
    } else {
        err := save()
        if err != nil {
            fmt.Println(err)
            return
        }
        tmpl.Execute(w, struct{ Success bool }{true})
    }
}
```

In the preceding Go code, you can see how to read the values from the fields of an HTML form with the help of the FormValue() function. The template.Must() function is a helper function which makes sure that the template file provided contains no errors.

The fifth code segment of kvWeb.go is contained in the following Go code:

```go
func insertElement(w http.ResponseWriter, r *http.Request) {
    fmt.Println("Inserting an element to the KV store!")
    tmpl := template.Must(template.ParseFiles("insert.gohtml"))
    if r.Method != http.MethodPost {
        tmpl.Execute(w, nil)
        return
    }

    key := r.FormValue("key")
    n := myElement{
        Name:    r.FormValue("name"),
        Surname: r.FormValue("surname"),
```

```
            Id:        r.FormValue("id"),
    }

    if !ADD(key, n) {
        fmt.Println("Add operation failed!")
    } else {
        err := save()
        if err != nil {
            fmt.Println(err)
            return
        }
        tmpl.Execute(w, struct{ Success bool }{true})
    }
}
```

The remaining Go code is as follows:

```
func main() {
    err := load()
    if err != nil {
        fmt.Println(err)
    }

    PORT := ":8001"
    arguments := os.Args
    if len(arguments) == 1 {
        fmt.Println("Using default port number: ", PORT)
    } else {
        PORT = ":" + arguments[1]
    }

    http.HandleFunc("/", homePage)
    http.HandleFunc("/change", changeElement)
    http.HandleFunc("/list", listAll)
    http.HandleFunc("/insert", insertElement)
    err = http.ListenAndServe(PORT, nil)
    if err != nil {
        fmt.Println(err)
    }
}
```

The main() function of kvWeb.go is much simpler than the main() function of kvSaveLoad.go from Chapter 8, *Telling a Unix System What to Do*, because these two program have a totally different design.

It is now time to look at the gohtml files used for this project starting with home.gohtml, which is as follows:

```html
<!doctype html>
<html lang="en">
<head>
    <meta charset="UTF-8">
    <title>A Key Value Store!</title>
</head>
<body>

<a href="/" style="margin-right: 20px;">Home sweet home!</a>
<a href="/list" style="margin-right: 20px;">List all elements!</a>
<a href="/change" style="margin-right: 20px;">Change an element!</a>
<a href="/insert" style="margin-right: 20px;">Insert new element!</a>

<h2>Welcome to the Go KV store!</h2>

</body>
</html>
```

The `home.gohtml` file is static, which means that its contents do not change. However, the remaining `gohtml` files display information dynamically.

The contents of `update.gohtml` are as follows:

```html
<!doctype html>
<html lang="en">
<head>
    <meta charset="UTF-8">
    <title>A Key Value Store!</title>
</head>
<body>

<a href="/" style="margin-right: 20px;">Home sweet home!</a>
<a href="/list" style="margin-right: 20px;">List all elements!</a>
<a href="/change" style="margin-right: 20px;">Change an element!</a>
<a href="/insert" style="margin-right: 20px;">Insert new element!</a>

{{if .Success}}    <h1>Element updated!</h1>{{else}}
<h1>Please fill in the fields:</h1>
    <form method="POST">
        <label>Key:</label><br />
        <input type="text" name="key"><br />
        <label>Name:</label><br />
        <input type="text" name="name"><br />
        <label>Surname:</label><br />
        <input type="text" name="surname"><br />
        <label>Id:</label><br />
        <input type="text" name="id"><br />
```

```
            <input type="submit">
        </form>
{{end}}

</body>
</html>
```

The preceding is mainly HTML code. Its most interesting part is the `if` statement that specifies whether you should see the form or the `Element updated!` message.

Finally, the contents of `insert.gohtml` are as follows:

```
<!doctype html>
<html lang="en">
<head>
    <meta charset="UTF-8">
    <title>A Key Value Store!</title>
</head>
<body>

<a href="/" style="margin-right: 20px;">Home sweet home!</a>
<a href="/list" style="margin-right: 20px;">List all elements!</a>
<a href="/change" style="margin-right: 20px;">Change an element!</a>
<a href="/insert" style="margin-right: 20px;">Insert new element!</a>

{{if .Success}}
    <h1>Element inserted!</h1>
{{else}}
    <h1>Please fill in the fields:</h1>
    <form method="POST">
        <label>Key:</label><br />
        <input type="text" name="key"><br />
        <label>Name:</label><br />
        <input type="text" name="name"><br />
        <label>Surname:</label><br />
        <input type="text" name="surname"><br />
        <label>Id:</label><br />
        <input type="text" name="id"><br />
        <input type="submit">
    </form>
{{end}}

</body>
</html>
```

As you can see, `insert.gohtml` and `update.gohtml` are identical apart from the text in the `<title>` tag!

Executing `kvWeb.go` will generate the following output on a Unix shell:

```
$ go run kvWeb.go
Loading /tmp/dataFile.gob
Using default port number:   :8001
Serving localhost:8001 for /
Serving localhost:8001 for /favicon.ico
Listing the contents of the KV store!
Serving localhost:8001 for /favicon.ico
Inserting an element to the KV store!
Serving localhost:8001 for /favicon.ico
Inserting an element to the KV store!
Add operation failed!
Serving localhost:8001 for /favicon.ico
Inserting an element to the KV store!
Serving localhost:8001 for /favicon.ico
Inserting an element to the KV store!
Saving /tmp/dataFile.gob
Serving localhost:8001 for /favicon.ico
Inserting an element to the KV store!
Serving localhost:8001 for /favicon.ico
Changing an element of the KV store!
Serving localhost:8001 for /favicon.ico
```

Additionally, what is really interesting is how you can interact with `kvWeb.go` from a web browser. The home page of the website as defined in `home.gohtml` is shown in the following screenshot:

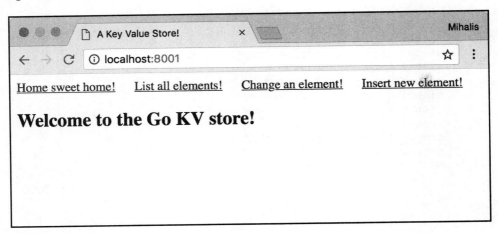

The static home page of our web application

The following screenshot presents the contents of the key-value store:

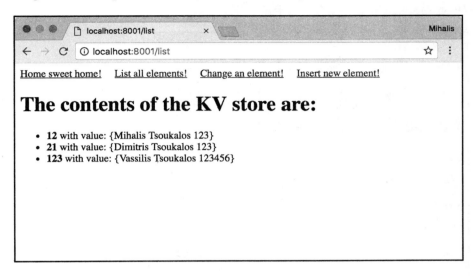

The contents of the key-value store

The following screenshot shows the appearance of the web page that allows you to add new data to the key-value store, using the web interface of the `kvWeb.go` web application:

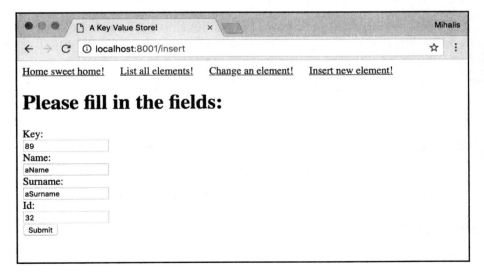

Adding a new entry to the key-value store

The following screenshot shows you how to update the value of an existing key, using the web interface of the kvWeb.go web application:

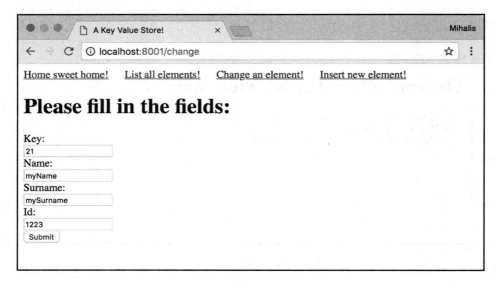

Updating the value of a key in the key-value store

The kvWeb.go web application is far from perfect, so feel free to improve it as an exercise!

 This section illustrated how you can develop entire websites and web applications in Go. Although your requirements will undoubtedly vary, the techniques are the same as the ones used in kvWeb.go. Note that custom-made sites are considered more secure than sites created using some popular content management system.

HTTP tracing

Go supports **HTTP tracing** with the help of the net/http/httptrace standard package! This package allows you to trace the phases of an HTTP request. So, the use of the net/http/httptrace standard Go package will be illustrated in httpTrace.go, which will be presented in five parts.

The first part of `httpTrace.go` is as follows:

```
package main

import (
    "fmt"
    "io"
    "net/http"
    "net/http/httptrace"
    "os"
)
```

As you might expect, you need to import the `net/http/httptrace` package in order to enable HTTP tracing.

The second part of `httpTrace.go` contains the following Go code:

```
func main() {
    if len(os.Args) != 2 {
        fmt.Printf("Usage: URL\n")
        return
    }

    URL := os.Args[1]
    client := http.Client{}
```

In this part, we read the command-line arguments and create a new `http.Client` variable. As this is the first time you are seeing an `http.Client` object in action, we will talk a little bit more about it. The `http.Client` object offers a way to send a request to a server and get a reply. Its `Transport` field permits you to set various HTTP details instead of using the default values.

Note that you should never use the default values of the `http.Client` object in production software because these values do not specify request timeouts, which can jeopardize the performance of your programs and your goroutines. Moreover, by design, `http.Client` objects can be safely used in concurrent programs.

The third code segment of `httpTrace.go` is contained in the following Go code:

```
    req, _ := http.NewRequest("GET", URL, nil)
    trace := &httptrace.ClientTrace{
        GotFirstResponseByte: func() {
            fmt.Println("First response byte!")
        },
        GotConn: func(connInfo httptrace.GotConnInfo) {
            fmt.Printf("Got Conn: %+v\n", connInfo)
```

```
        },
        DNSDone: func(dnsInfo httptrace.DNSDoneInfo) {
            fmt.Printf("DNS Info: %+v\n", dnsInfo)
        },
        ConnectStart: func(network, addr string) {
            fmt.Println("Dial start")
        },
        ConnectDone: func(network, addr string, err error) {
            fmt.Println("Dial done")
        },
        WroteHeaders: func() {
            fmt.Println("Wrote headers")
        },
    }
```

The preceding code is all about tracing HTTP requests. The `httptrace.ClientTrace` object defines the events that interest us. When such an event occurs, the relevant code is executed. You can find more information about the supported events and their purpose in the documentation of the `net/http/httptrace` package.

The fourth part of the `httpTrace.go` utility follows next:

```
req = req.WithContext(httptrace.WithClientTrace(req.Context(), trace))
fmt.Println("Requesting data from server!")
_, err := http.DefaultTransport.RoundTrip(req)
if err != nil {
    fmt.Println(err)
    return
}
```

The `httptrace.WithClientTrace()` function returns a new context, based on the given parent context; while the `http.DefaultTransport.RoundTrip()` method wraps `http.DefaultTransport.RoundTrip` in order to tell it to keep track of the current request. Note that Go HTTP tracing is designed so that it can trace the events of a single `http.Transport.RoundTrip`. However, as you may have multiple URL redirects when serving a single HTTP request, you need to be able to identify the current request.

The remaining Go code of `httpTrace.go` is as followss:

```
        response, err := client.Do(req)
        if err != nil {
            fmt.Println(err)
            return
```

```
        }
        io.Copy(os.Stdout, response.Body)
}
```

This last part is about performing the actual request to the web server using `Do()`, getting the HTTP data, and displaying it on the screen.

Executing `httpTrace.go` will generate the following very informative output:

```
$ go run httpTrace.go http://localhost:8001/
Requesting data from server!
DNS Info: {Addrs:[{IP:::1 Zone:} {IP:127.0.0.1 Zone:}] Err:<nil>
Coalesced:false}
Dial start
Dial done
Got Conn: {Conn:0xc420142000 Reused:false WasIdle:false IdleTime:0s}
Wrote headers
First response byte!
DNS Info: {Addrs:[{IP:::1 Zone:} {IP:127.0.0.1 Zone:}] Err:<nil>
Coalesced:false}
Dial start
Dial done
Got Conn: {Conn:0xc420142008 Reused:false WasIdle:false IdleTime:0s}
Wrote headers
First response byte!
Serving: /
```

Be aware that since `httpTrace.go` prints the full HTML response from the HTTP server, you might get lots of output when you test it on a real web server, which is the main reason for using the web server developed in `www.go` here.

If you have the time to look at the source code of the `net/http/httptrace` package at https://golang.org/src/net/http/httptrace/trace.go, you will immediately realize that `net/http/httptrace` is a pretty low-level package that uses the `context`, `reflect`, and `internal/nettrace` packages to implement its functionality.

Testing HTTP handlers

In this section, we will learn how to test HTTP handlers in Go, which is a special case of testing in Go. We will begin with the Go code of `www.go` and modify it where needed.

The new version of www.go is called testWWW.go, and it will be presented in three parts. The first code portion of testWWW.go is as follows:

```go
package main

import (
    "fmt"
    "net/http"
    "os"
)

func CheckStatusOK(w http.ResponseWriter, r *http.Request) {
    w.WriteHeader(http.StatusOK)
    fmt.Fprintf(w, `Fine!`)
}
```

The second part of testWWW.go is contained in the following Go code:

```go
func StatusNotFound(w http.ResponseWriter, r *http.Request) {
    w.WriteHeader(http.StatusNotFound)
}

func MyHandler(w http.ResponseWriter, r *http.Request) {
    fmt.Fprintf(w, "Serving: %s\n", r.URL.Path)
    fmt.Printf("Served: %s\n", r.Host)
}
```

The remaining Go code of testWWW.go follows next:

```go
func main() {
    PORT := ":8001"
    arguments := os.Args
    if len(arguments) == 1 {
        fmt.Println("Using default port number: ", PORT)
    } else {
        PORT = ":" + arguments[1]
    }

    http.HandleFunc("/CheckStatusOK", CheckStatusOK)
    http.HandleFunc("/StatusNotFound", StatusNotFound)
    http.HandleFunc("/", MyHandler)

    err := http.ListenAndServe(PORT, nil)
    if err != nil {
        fmt.Println(err)
```

```
        return
    }
}
```

We now need to start testing `testWWW.go`, which means that we should create a `testWWW_test.go` file. The contents of that file will be presented in four parts.

The first part of `testWWW_test.go` is contained in the following Go code:

```
package main

import (
    "fmt"
    "net/http"
    "net/http/httptest"
    "testing"
)
```

Notice that you need to import the `net/http/httptest` standard Go package in order to test web applications in Go.

The second code portion of `testWWW_test.go` contains the following code:

```
func TestCheckStatusOK(t *testing.T) {
    req, err := http.NewRequest("GET", "/CheckStatusOK", nil)
    if err != nil {
        fmt.Println(err)
        return
    }

    rr := httptest.NewRecorder()
    handler := http.HandlerFunc(CheckStatusOK)
    handler.ServeHTTP(rr, req)
```

The `httptest.NewRecorder()` function returns an `httptest.ResponseRecorder` object, and it is used for recording the HTTP response.

The third part of `testWWW_test.go` is as follows:

```
    status := rr.Code
    if status != http.StatusOK {
        t.Errorf("handler returned %v", status)
    }

    expect := `Fine!`
    if rr.Body.String() != expect {
```

```
            t.Errorf("handler returned %v", rr.Body.String())
        }
    }
```

You first check that the response code is as expected, and then you verify that the body of the response is also what you expected.

The remaining Go code of testWWW_test.go is as follows:

```
func TestStatusNotFound(t *testing.T) {
    req, err := http.NewRequest("GET", "/StatusNotFound", nil)
    if err != nil {
        fmt.Println(err)
        return
    }

    rr := httptest.NewRecorder()
    handler := http.HandlerFunc(StatusNotFound)
    handler.ServeHTTP(rr, req)

    status := rr.Code
    if status != http.StatusNotFound {
        t.Errorf("handler returned %v", status)
    }
}
```

This test function verifies that the StatusNotFound() function of the main package works as expected.

Executing the two test functions of testWWW_test.go will generate the following output:

```
$ go test   testWWW.go testWWW_test.go   -v
=== RUN     TestCheckStatusOK
--- PASS:  TestCheckStatusOK (0.00s)
=== RUN     TestStatusNotFound
--- PASS:  TestStatusNotFound (0.00s)
PASS
ok      command-line-arguments    (cached)
```

Creating a web client in Go

In this section, you will learn more about developing web clients in Go. The name of the web client utility is webClient.go, and it will be presented in four parts.

The first part of `webClient.go` contains the following Go code:

```go
package main

import (
    "fmt"
    "io"
    "net/http"
    "os"
    "path/filepath"
)
```

The second part of `webClient.go` is where you read the desired URL as a command-line argument:

```go
func main() {
    if len(os.Args) != 2 {
        fmt.Printf("Usage: %s URL\n", filepath.Base(os.Args[0]))
        return
    }

    URL := os.Args[1]
```

The third code portion of `webClient.go`, is where the real action takes place:

```go
data, err := http.Get(URL)

if err != nil {
    fmt.Println(err)
    return
```

All of the work is done by the `http.Get()` call, which is pretty convenient when you do not want to deal with parameters and options. However, this type of call gives you no flexibility over the process. Note that `http.Get()` returns an `http.Response` variable.

The remaining Go code is as follows:

```go
    } else {
        defer data.Body.Close()
        _, err := io.Copy(os.Stdout, data.Body)
        if err != nil {
            fmt.Println(err)
            return
        }
    }
}
```

What the previous code does is to copy the contents of the `Body` field of the `http.Response` structure to standard output.

Executing `webClient.go` will generate the following type of output:

 Only a small portion of the output is presented here.

```
$ go run webClient.go http://www.mtsoukalos.eu/ | head -20
<!DOCTYPE html PUBLIC "-//W3C//DTD XHTML+RDFa 1.0//EN"
"http://www.w3.org/MarkUp/DTD/xhtml-rdfa-1.dtd">
<html xmlns="http://www.w3.org/1999/xhtml" xml:lang="en"
version="XHTML+RDFa 1.0" dir="ltr"
xmlns:content="http://purl.org/rss/1.0/modules/content/"
xmlns:dc="http://purl.org/dc/terms/"
xmlns:foaf="http://xmlns.com/foaf/0.1/"
xmlns:og="http://ogp.me/ns#"
xmlns:rdfs="http://www.w3.org/2000/01/rdf-schema#"
xmlns:sioc="http://rdfs.org/sioc/ns#"
xmlns:sioct="http://rdfs.org/sioc/types#"
xmlns:skos="http://www.w3.org/2004/02/skos/core#"
      xmlns:xsd="http://www.w3.org/2001/XMLSchema#">
<head profile="http://www.w3.org/1999/xhtml/vocab">
    <meta http-equiv="Content-Type" content="text/html; charset=utf-8" />
    <meta name="viewport" content="width=device-width, initial-scale=1" />
    <link rel="shortcut icon" href="http://www.mtsoukalos.eu/misc/favicon.ico"
type="image/vnd.microsoft.icon" />
    <meta name="HandheldFriendly" content="true" />
    <meta name="MobileOptimized" content="width" />
    <meta name="Generator" content="Drupal 7 (http://drupal.org)" />
```

The main problem with `webClient.go` is that it gives you almost no control over the process—you either get the entire HTML output or nothing at all!

Making your Go web client more advanced

As the web client of the previous section is relatively simplistic and does not give you any flexibility, in this section you will learn how to read a URL more elegantly without using the `http.Get()` function and by having more options. The name of the utility is `advancedWebClient.go`, and it will be presented in five parts.

The first code segment of `advancedWebClient.go` contains the following Go code:

```go
package main

import (
    "fmt"
    "net/http"
    "net/http/httputil"
    "net/url"
    "os"
    "path/filepath"
    "strings"
    "time"
)
```

The second part of `advancedWebClient.go` is as follows:

```go
func main() {
    if len(os.Args) != 2 {
        fmt.Printf("Usage: %s URL\n", filepath.Base(os.Args[0]))
        return
    }

    URL, err := url.Parse(os.Args[1])
    if err != nil {
        fmt.Println("Error in parsing:", err)
        return
    }
```

The third code portion of `advancedWebClient.go` contains the following Go code:

```go
    c := &http.Client{
        Timeout: 15 * time.Second,
    }
    request, err := http.NewRequest("GET", URL.String(), nil)
    if err != nil {
        fmt.Println("Get:", err)
        return
    }

    httpData, err := c.Do(request)
    if err != nil {
        fmt.Println("Error in Do():", err)
        return
    }
```

The `http.NewRequest()` function returns an `http.Request` object given a method, a URL, and an optional body. The `http.Do()` function sends an HTTP request (`http.Request`) using an `http.Client` and gets an HTTP response (`http.Response`). So, `http.Do()` does the job of `http.Get()` in a more comprehensive way.

The `GET` string used in `http.NewRequest()` can be replaced with `http.MethodGet`.

The fourth part of `advancedWebClient.go` is contained in the following Go code:

```
fmt.Println("Status code:", httpData.Status)
header, _ := httputil.DumpResponse(httpData, false)
fmt.Print(string(header))

contentType := httpData.Header.Get("Content-Type")
characterSet := strings.SplitAfter(contentType, "charset=")
if len(characterSet) > 1 {
    fmt.Println("Character Set:", characterSet[1])
}

if httpData.ContentLength == -1 {
    fmt.Println("ContentLength is unknown!")
} else {
    fmt.Println("ContentLength:", httpData.ContentLength)
}
```

In the preceding code, you can see how to start searching the server response in order to find what you want.

The last part of the `advancedWebClient.go` utility is as follows:

```
length := 0
var buffer [1024]byte
r := httpData.Body
for {
    n, err := r.Read(buffer[0:])
    if err != nil {
            fmt.Println(err)
            break
    }
    length = length + n
}
fmt.Println("Calculated response data length:", length)
```

In the preceding code, you can see a technique for discovering the size of the server HTTP response. If you want to display the HTML output on your screen, you can print the contents of the `r` buffer variable.

Using `advancedWebClient.go` for visiting a web page will generate the following output, which is much richer than before:

```
$ go run advancedWebClient.go http://www.mtsoukalos.eu
Status code: 200 OK
HTTP/1.1 200 OK
Accept-Ranges: bytes
Age: 0
Cache-Control: no-cache, must-revalidate
Connection: keep-alive
Content-Language: en
Content-Type: text/html; charset=utf-8
Date: Sat, 24 Mar 2018 18:52:17 GMT
Expires: Sun, 19 Nov 1978 05:00:00 GMT
Server: Apache/2.4.25 (Debian) PHP/5.6.33-0+deb8u1 mod_wsgi/4.5.11
Python/2.7
Vary: Accept-Encoding
Via: 1.1 varnish (Varnish/5.0)
X-Content-Type-Options: nosniff
X-Frame-Options: SAMEORIGIN
X-Generator: Drupal 7 (http://drupal.org)
X-Powered-By: PHP/5.6.33-0+deb8u1
X-Varnish: 886025
Character Set: utf-8
ContentLength is unknown!
EOF
Calculated response data length: 50176
```

Executing `advancedWebClient.go` for visiting a different URL will return a slightly different output:

```
$ go run advancedWebClient.go http://www.google.com
Status code: 200 OK
HTTP/1.1 200 OK
Cache-Control: private, max-age=0
Content-Type: text/html; charset=ISO-8859-7
Date: Sat, 24 Mar 2018 18:52:38 GMT
Expires: -1
P3p: CP="This is not a P3P policy! See g.co/p3phelp for more info."
Server: gws
Set-Cookie: 1P_JAR=2018-03-24-18; expires=Mon, 23-Apr-2018 18:52:38 GMT;
path=/; domain=.google.gr
Set-Cookie:
```

```
NID=126=csX1_koD30SJcC_ljAfcM2V8kTfRkppmAamLjINLfclracMxuk6JGe4glc0Pjs8uD00
bqGaxkSW-J-ZNDJexG2ZX9pNB9E_dRc2y1KZ05V7pk0bOczE2FtS1zb50Uofl; expires=Sun,
23-Sep-2018 18:52:38 GMT; path=/; domain=.google.gr; HttpOnly
X-Frame-Options: SAMEORIGIN
X-Xss-Protection: 1; mode=block
Character Set: ISO-8859-7
ContentLength in unknown!
EOF
Calculated response data length: 10240
```

If you try to fetch an erroneous URL with `advancedWebClient.go`, you will get the following output:

```
$ go run advancedWebClient.go http://www.google
Error in Do(): Get http://www.google: dial tcp: lookup www.google: no such host
$ go run advancedWebClient.go www.google.com
Error in Do(): Get www.google.com: unsupported protocol scheme ""
```

Feel free to make any changes you want to `advancedWebClient.go` in order to make the output match your requirements!

Timing out HTTP connections

This section will present a technique for timing out network connections that take too long to finish. Remember that you already know such a technique from Chapter 10, *Go Concurrency – Advanced Topics*, when we talked about the `context` standard Go package. This technique was presented in the `useContext.go` source code file.

The method that will be presented in this section is much easier to implement. The relevant code is saved in `clientTimeOut.go`, and it will be presented in four parts. The utility accepts two command-line arguments, which are the URL and the timeout period in seconds. Note that the second parameter is optional.

The first part code portion of `clientTimeOut.go` is as follows:

```
package main

import (
    "fmt"
    "io"
    "net"
    "net/http"
    "os"
```

```
    "path/filepath"
    "strconv"
    "time"
)

var timeout = time.Duration(time.Second)
```

The second code segment of `clientTimeOut.go` contains the following Go code:

```
func Timeout(network, host string) (net.Conn, error) {
    conn, err := net.DialTimeout(network, host, timeout)
    if err != nil {
        return nil, err
    }
    conn.SetDeadline(time.Now().Add(timeout))
    return conn, nil
}
```

You will learn more about the functionality of `SetDeadline()` in the next subsection. The `Timeout()` function will be used by the `Dial` field of a `http.Transport` variable.

The third code portion of `clientTimeOut.go` is contained in the following code:

```
func main() {
    if len(os.Args) == 1 {
        fmt.Printf("Usage: %s URL TIMEOUT\n", filepath.Base(os.Args[0]))
        return
    }

    if len(os.Args) == 3 {
        temp, err := strconv.Atoi(os.Args[2])
        if err != nil {
            fmt.Println("Using Default Timeout!")
        } else {
            timeout = time.Duration(time.Duration(temp) * time.Second)
        }
    }

    URL := os.Args[1]
    t := http.Transport{
        Dial: Timeout,
    }
```

The remaining Go code of the `clientTimeOut.go` utility is as follows:

```
    client := http.Client{
        Transport: &t,
    }
```

```go
        data, err := client.Get(URL)
        if err != nil {
            fmt.Println(err)
            return
        } else {
            defer data.Body.Close()
            _, err := io.Copy(os.Stdout, data.Body)
            if err != nil {
                fmt.Println(err)
                return
            }
        }
    }
```

The `clientTimeOut.go` web client will be tested using the `slowWWW.go` web server developed in `Chapter 10`, *Go Concurrency – Advanced Topics*.

Executing `clientTimeOut.go` two times will generate the following output:

```
$ go run clientTimeOut.go http://localhost:8001
Serving: /
Delay: 0
$ go run clientTimeOut.go http://localhost:8001
Get http://localhost:8001: read tcp [::1]:57397->[::1]:8001: i/o timeout
```

As you can see from the generated output, the first request had no problem connecting to the desired web server. However, the second `http.Get()` request took longer than expected and therefore timed out!

More information about SetDeadline

The `SetDeadline()` function is used by the `net` package to set the read and write deadlines of a given network connection. Due to the way the `SetDeadline()` function works, you will need to call `SetDeadline()` before any read or write operations. Keep in mind that Go uses deadlines to implement timeouts, so you do not need to reset it every time your application receives or sends any data.

Setting the timeout period on the server side

In this subsection, you will learn how to time out a connection on the server side. You need to do this because there are times that clients take much longer than expected to end an HTTP connection. This usually happens for two reasons: the first reason is when there are bugs in the client software, and the second reason is when a server process is experiencing an attack!

The technique will be implemented in the `serverTimeOut.go` source code file, which will be presented in four parts.

The first part of `serverTimeOut.go` is as follows:

```
package main

import (
    "fmt"
    "net/http"
    "os"
    "time"
)

func myHandler(w http.ResponseWriter, r *http.Request) {
    fmt.Fprintf(w, "Serving: %s\n", r.URL.Path)
    fmt.Printf("Served: %s\n", r.Host)
}
```

The second code portion of `serverTimeOut.go` contains the following Go code:

```
func timeHandler(w http.ResponseWriter, r *http.Request) {
    t := time.Now().Format(time.RFC1123)
    Body := "The current time is:"
    fmt.Fprintf(w, "<h1 align=\"center\">%s</h1>", Body)
    fmt.Fprintf(w, "<h2 align=\"center\">%s</h2>n", t)
    fmt.Fprintf(w, "Serving: %s\n", r.URL.Path)
    fmt.Printf("Served time for: %s\n", r.Host)
}
```

The third code segment of `serverTimeOut.go` is shown in the following Go code:

```
func main() {
    PORT := ":8001"
    arguments := os.Args
    if len(arguments) == 1 {
        fmt.Printf("Listening on http://0.0.0.0%s\n", PORT)
    } else {
        PORT = ":" + arguments[1]
```

```
        fmt.Printf("Listening on http://0.0.0.0%s\n", PORT)
    }

    m := http.NewServeMux()
    srv := &http.Server{
        Addr:         PORT,
        Handler:      m,
        ReadTimeout:  3 * time.Second,
        WriteTimeout: 3 * time.Second,
    }
}
```

In this case, we are using an `http.Server` structure that supports two kinds of timeouts with its fields. The first one is called `ReadTimeout`, and the second one is called `WriteTimeout`. The value of the `ReadTimeout` field specifies the maximum duration for reading the entire request, including the body.

The value of the `WriteTimeout` field specifies the maximum duration before timing out writes of the response. Put simply, this is the time from the end of the request header read to the end of the response write.

The remaining Go code of `serverTimeOut.go` is as follows:

```
    m.HandleFunc("/time", timeHandler)
    m.HandleFunc("/", myHandler)

    err := srv.ListenAndServe()
    if err != nil {
        fmt.Println(err)
        return
    }
}
```

We are now going to execute `serverTimeOut.go` in order to interact with it, using nc(1):

```
$ go run serverTimeOut.go
Listening on http://0.0.0.0:8001
```

On the `nc(1)` part, which in this case acts as a dummy HTTP client, you should issue the next command to connect to `serverTimeOut.go`:

```
$ time nc localhost 8001
real    0m3.012s
user    0m0.001s
sys     0m0.002s
```

As we did not issue any commands, the HTTP server ended the connection. The output of the `time(1)` utility verifies the time it took the server to close the connection.

Yet another way to time out!

This subsection will present yet another way to timeout an HTTP connection from the client side. As you will see, this is the simplest form of timeout, because you will just need to define an `http.Client` object and set its `Timeout` field to the desired timeout value!

The name of the utility that showcases the last type of timeout is `anotherTimeOut.go`, and it will be presented in four parts.

The first part of `anotherTimeOut.go` follows next:

```
package main

import (
    "fmt"
    "io"
    "net/http"
    "os"
    "strconv"
    "time"
)

var timeout = time.Duration(time.Second)
```

The second part of `anotherTimeOut.go` is containing the following Go code:

```
func main() {
    if len(os.Args) == 1 {
        fmt.Println("Please provide a URL")
        return
    }

    if len(os.Args) == 3 {
        temp, err := strconv.Atoi(os.Args[2])
```

```
            if err != nil {
                fmt.Println("Using Default Timeout!")
            } else {
                timeout = time.Duration(time.Duration(temp) * time.Second)
            }
    }
    URL := os.Args[1]
```

The third code segment of `anotherTimeOut.go` contains the following Go code:

```
client := http.Client{
    Timeout: timeout,
}
client.Get(URL)
```

This is the place where the timeout period is defined using the `Timeout` field of the `http.Client` variable.

The last code portion of `anotherTimeOut.go` follows next:

```
    data, err := client.Get(URL)
    if err != nil {
        fmt.Println(err)
        return
    } else {
        defer data.Body.Close()
        _, err := io.Copy(os.Stdout, data.Body)
        if err != nil {
            fmt.Println(err)
            return
        }
    }
}
```

Executing `anotherTimeOut.go` and interacting with the `slowWWW.go` web server developed in `Chapter 10`, *Go Concurrency – Advanced Topics,* will generate the following output:

```
$ go run anotherTimeOut.go http://localhost:8001
Get http://localhost:8001: net/http: request canceled (Client.Timeout exceeded while awaiting headers)
$ go run anotherTimeOut.go http://localhost:8001 15
Serving: /
Delay: 8
```

Wireshark and tshark tools

This section will briefly mention the powerful Wireshark and tshark utilities. **Wireshark**, which is a graphical application, is the dominant tool for analyzing network traffic of almost any kind. Although Wireshark is very powerful, there may be times that you need something lighter that you can execute remotely without a graphical user interface. In such situations, you can use **tshark**, which is the command-line version of Wireshark.

Unfortunately, talking more about Wireshark and tshark is beyond the scope of this chapter.

Additional resources

The following resources will definitely broaden your horizons, so please find some time to visit them:

- The official web page of the Apache web server is located at http://httpd.apache.org/.

- The official web page of the Nginx web server is located at http://nginx.org/.

- Should you wish to learn more about the internet, TCP/IP, and its various services, you should start by reading **RFC** documents. One of the places that you can find such documents is at http://www.rfc-archive.org/.

- Visit the website of both Wireshark and tshark at https://www.wireshark.org/ to learn more about them.

- Visit the documentation page of the net standard Go package, which can be found at https://golang.org/pkg/net/. This is one of the largest documentation pages in the official Go documentation.

- Visit the documentation page of the net/http Go package at https://golang.org/pkg/net/http/.

- If you want to create a website without writing any Go code, you can try the **Hugo utility**, which is written in Go! You can learn more about the Hugo framework at https://gohugo.io/. However, what would really be interesting and educational for every Go programmer is to look at its Go code, which can be found at https://github.com/gohugoio/hugo.

- You can visit the documentation page of the `net/http/httptrace` package at https://golang.org/pkg/net/http/httptrace/.

- You can find the documentation page of the `net/http/pprof` package at https://golang.org/pkg/net/http/pprof/.

- Visit the manual page of the `nc(1)` command-line utility to learn more about its capabilities and its various command-line options.

- The `httpstat` utility that was developed by Dave Cheney can be found at https://github.com/davecheney/httpstat. It is a good example of the use of the `net/http/httptrace` Go package for HTTP tracing.

- You can find more information about `ab(1)` by visiting its manual page at https://httpd.apache.org/docs/2.4/programs/ab.html

Exercises

- Write a web client in Go on your own without looking at the code of this chapter

- Merge the code of `MXrecords.go` and `NSrecords.go` in order to create a single utility that does both jobs based on its command-line arguments

- Modify the code of `MXrecords.go` and `NSrecords.go` to also accept IP addresses as input

- Modify `advancedWebClient.go` in order to save the HTML output in an external file

- Try to create a simple version of `ab(1)` on your own using goroutines

- Modify `kvWeb.go` in order to support the `DELETE` and the `LOOKUP` operations found in the original version of the key-value store

- Modify `httpTrace.go` in order to have a flag that disables the execution of the `io.Copy(os.Stdout, response.Body)` statement

Summary

This chapter discussed Go code for programming web clients and web servers as well as creating a website in Go! You also learned about the `http.Response`, the `http.Request`, and the `http.Transport` structures that allow you to define the parameters of an HTTP connection.

Additionally, you learned how to get the network configuration of a Unix machine, using Go code, how to perform DNS lookups in a Go program, including getting the NS and MX records of a domain.

Finally, we talked about **Wireshark** and **tshark**, which are two very popular utilities that allow you to capture and, most importantly, analyze network traffic. At the beginning of this chapter, we also mentioned the `nc(1)` utility.

In the final chapter of this book, we will continue our discussion of network programming in Go. However, this time, we will present lower-level Go code that allows you to develop TCP clients and servers as well as UDP client and server processes. Additionally, you will learn about creating RCP clients and servers. I hope you'll enjoy it!

13
Network Programming – Building Servers and Clients

The previous chapter discussed topics related to network programming; including developing web clients, web servers, and web applications; performing DNS lookups; and timing out HTTP connections.

This chapter will take you to the next level by showing you how to program your own TCP clients and servers as well as your own UDP clients and servers. Additionally, it will showcase how you can program a concurrent TCP server using two examples. The first example will be relatively simple, as the concurrent TCP server will just return numbers of the Fibonacci sequence. However, the second example will use the code of the keyValue.go application from Chapter 4, *The Uses of Composite Types*, as its foundation and convert the **key-value store** into a concurrent TCP application that can operate without the need for a web browser!

In this chapter of *Mastering Go*, you will learn the following topics:

- The net standard Go package
- Developing TCP clients and TCP servers
- Developing a concurrent TCP server
- Developing UDP clients and UDP servers
- Modifying kvSaveLoad.go from Chapter 8, *Telling a Unix System What to Do*, so that it can serve requests through TCP connections
- Creating an RCP client and RCP server

The net standard Go package

You cannot create a TCP or UDP client or server in Go without using the functionality offered by the `net` package.

The `net.Dial()` function is used for connecting to a network as a client, whereas the `net.Listen()` function is used for telling a Go program to accept network connections and thus act as a server. The first parameter of both functions is the network type, but this is where their similarities end. The return value of both the `net.Dial()` and `net.Listen()` functions is of the `net.Conn` type, which implements the `io.Reader` and `io.Writer` interfaces.

A TCP client

As you already know from the previous chapter, **TCP** stands for **Transmission Control Protocol**, and its principal characteristic is that it is a reliable protocol. The TCP header of each packet includes **source port** and **destination port** fields. These two fields, plus the source and destination IP addresses, are combined to identify uniquely a single TCP connection. The name of the TCP client that will be developed in this section is `TCPclient.go`, and it will be presented in four parts. The first part of `TCPclient.go` is shown in the following Go code:

```
package main

import (
    "bufio"
    "fmt"
    "net"
    "os"
    "strings"
)
```

The second code segment of `TCPclient.go` follows next:

```
func main() {
    arguments := os.Args
    if len(arguments) == 1 {
        fmt.Println("Please provide host:port.")
        return
    }

    CONNECT := arguments[1]
    c, err := net.Dial("tcp", CONNECT)
```

```
        if err != nil {
            fmt.Println(err)
            return
        }
```

The `net.Dial()` function is used for connecting to the remote server. The first parameter of the `net.Dial()` function defines the network that will be used, while the second parameter defines the server address, which must also include the port number. Valid values for the first parameter are `tcp`, `tcp4` (IPv4-only), `tcp6` (IPv6-only), `udp`, `udp4` (IPv4-only), `udp6` (IPv6-only), `ip`, `ip4` (IPv4-only), `ip6` (IPv6-only), `Unix` (Unix sockets), `Unixgram`, and `Unixpacket`.

The third part of `TCPclient.go` contains the following code:

```
    for {
        reader := bufio.NewReader(os.Stdin)
        fmt.Print(">> ")
        text, _ := reader.ReadString('\n')
        fmt.Fprintf(c, text+"\n")
```

The preceding code is for getting input from the user, which can be verified using the `os.Stdin` file for reading. Ignoring the `error` value returned by `reader.ReadString()` is not a good practice, but it saves some space here. Certainly, you should never do that on production software.

The last part of `TCPclient.go` follows next:

```
        message, _ := bufio.NewReader(c).ReadString('\n')
        fmt.Print("->: " + message)
        if strings.TrimSpace(string(text)) == "STOP" {
            fmt.Println("TCP client exiting...")
            return
        }
    }
}
```

For testing purposes, `TCPclient.go` will connect to a TCP server that is implemented using `netcat(1)`, and it will create the following output:

```
$ go run TCPclient.go 8001
dial tcp: address 8001: missing port in address
$ go run TCPclient.go localhost:8001
>> Hello from TCPclient.go!
->: Hi from nc!
```

```
>> STOP
->:
TCP client exiting...
```

The output of the first command illustrates what will happen if you do not include a host name in the command-line arguments of `TCPclient.go`. The output of the `netcat(1)` TCP server, which should be executed first, is as follows:

```
$ nc -l 127.0.0.1 8001
Hello from TCPclient.go!
Hi from nc!
STOP
```

Note that a client for a given protocol such as TCP and UDP can be reasonably generic in nature, which means that it can talk to many kinds of servers that support its protocol. As you will soon see, this is not the case with server applications that must implement a prearranged functionality using a prearranged protocol.

A slightly different version of the TCP client

Go offers a different family of functions that also allow you to develop TCP clients and servers. In this section, you will learn how to program a TCP client using these functions.

The name of the TCP client will be `otherTCPclient.go`, and it will be presented in four parts. The first code segment from `otherTCPclient.go` is as follows:

```
package main

import (
    "bufio"
    "fmt"
    "net"
    "os"
    "strings"
)
```

The second code portion from `otherTCPclient.go` contains the following code:

```
func main() {
    arguments := os.Args
    if len(arguments) == 1 {
        fmt.Println("Please provide a server:port string!")
        return
    }
```

```
CONNECT := arguments[1]

tcpAddr, err := net.ResolveTCPAddr("tcp4", CONNECT)
if err != nil {
    fmt.Println("ResolveTCPAddr:", err.Error())
    return
}
```

The `net.ResolveTCPAddr()` function returns the address of a TCP end point (type `TCPAddr`) and can only be used for TCP networks.

The third part of `otherTCPclient.go` contains the following code:

```
conn, err := net.DialTCP("tcp4", nil, tcpAddr)
if err != nil {
    fmt.Println("DialTCP:", err.Error())
    return
}
```

The `net.DialTCP()` function is equivalent to `net.Dial()` for TCP networks.

The remaining code of `otherTCPclient.go` follows next:

```
for {
    reader := bufio.NewReader(os.Stdin)
    fmt.Print(">> ")
    text, _ := reader.ReadString('\n')
    fmt.Fprintf(conn, text+"\n")

    message, _ := bufio.NewReader(conn).ReadString('\n')
    fmt.Print("->: " + message)
    if strings.TrimSpace(string(text)) == "STOP" {
        fmt.Println("TCP client exiting...")
        conn.Close()
        return
    }
}
```

Executing `otherTCPclient.go` and interacting with a TCP server will generate the following type of output:

```
$ go run otherTCPclient.go localhost:8001
>> Hello from otherTCPclient.go!
->: Hi from netcat!
>> STOP
->:
TCP client exiting...
```

For this example, the TCP server is supported by the `netcat(1)` utility, which is executed as follows:

```
$ nc -l 127.0.0.1 8001
Hello from otherTCPclient.go!
Hi from netcat!
STOP
```

A TCP server

The TCP server that is going to be developed in this section will return the current date and time to the client in a single network packet. In practice, this means that after accepting a client connection, the server will get the time and the date from the Unix system and send this data back to the client.

The name of the utility is `TCPserver.go`, and it will be presented in four parts.

The first part of `TCPserver.go` is as follows:

```
package main

import (
    "bufio"
    "fmt"
    "net"
    "os"
    "strings"
    "time"
)
```

The second code portion of `TCPserver.go` contains the following Go code:

```go
func main() {
    arguments := os.Args
    if len(arguments) == 1 {
        fmt.Println("Please provide port number")
        return
    }

    PORT := ":" + arguments[1]
    l, err := net.Listen("tcp", PORT)
    if err != nil {
        fmt.Println(err)
        return
    }
    defer l.Close()
```

The `net.Listen()` function listens for connections. If the second parameter does not contain an IP address, but only a port number, `net.Listen()` will listen on all available IP addresses of the local system.

The third segment from `TCPserver.go` is as follows:

```go
    c, err := l.Accept()
    if err != nil {
        fmt.Println(err)
        return
    }
```

The `Accept()` function waits for the next connection and returns a generic `Conn` variable. The only thing that is wrong with this particular TCP server is that it can only serve the first TCP client, which is going to connect to it because the `Accept()` call is outside the `for` loop that is coming next.

The remaining Go code of `TCPserver.go` is as follows:

```go
    for {
        netData, err := bufio.NewReader(c).ReadString('\n')
        if err != nil {
            fmt.Println(err)
            return
        }
        if strings.TrimSpace(string(netData)) == "STOP" {
            fmt.Println("Exiting TCP server!")
            return
        }
```

```
            fmt.Print("-> ", string(netData))
            t := time.Now()
            myTime := t.Format(time.RFC3339) + "\n"
            c.Write([]byte(myTime))
    }
}
```

Executing `TCPserver.go` and interacting with it using a TCP client application will generate the following type of output:

```
$ go run TCPserver.go 8001
-> HELLO
Exiting TCP server!
```

On the client side, you will see the following output:

```
$ nc 127.0.0.1 8001
HELLO
2018-04-07T14:40:05+03:00
STOP
```

If the `TCPserver.go` utility attempts to use a TCP port that is already in use by another Unix process, you will get the following error message:

```
$ go run TCPserver.go 9000
listen tcp :9000: bind: address already in use
```

Finally, if the `TCPserver.go` utility attempts to use a TCP port in the 1-1024 range that requires root privileges on Unix systems, you will get the following error message:

```
$ go run TCPserver.go 80
listen tcp :80: bind: permission denied
```

A slightly different version of the TCP server

In this section, you will see an alternative implementation of a TCP server in Go. This time, the TCP server implements the **Echo service**, which basically returns to the client, the data that the client sent. The program is called `otherTCPserver.go`, and it will be presented in four parts.

The first part of `otherTCPserver.go` is as follows:

```
package main

import (
    "fmt"
    "net"
    "os"
    "strings"
)
```

The second portion of `otherTCPserver.go` contains the following Go code:

```
func main() {
    arguments := os.Args
    if len(arguments) == 1 {
        fmt.Println("Please provide a port number!")
        return
    }

    SERVER := "localhost" + ":" + arguments[1]

    s, err := net.ResolveTCPAddr("tcp", SERVER)
    if err != nil {
        fmt.Println(err)
        return
    }

    l, err := net.ListenTCP("tcp", s)
    if err != nil {
        fmt.Println(err)
        return
    }
```

The `net.ListenTCP()` function is equivalent to `net.Listen()` for TCP networks.

The third segment of `otherTCPserver.go` is as follows:

```
    buffer := make([]byte, 1024)
    conn, err := l.Accept()
    if err != nil {
        fmt.Println(err)
        return
    }
```

The remaining code of `otherTCPserver.go` is as follows:

```
    for {
        n, err := conn.Read(buffer)
        if err != nil {
            fmt.Println(err)
            return
        }

        if strings.TrimSpace(string(buffer[0:n])) == "STOP" {
            fmt.Println("Exiting TCP server!")
            conn.Close()
            return
        }

        fmt.Print("> ", string(buffer[0:n-1]))
        _, err = conn.Write(buffer)
        if err != nil {
            fmt.Println(err)
            return
        }
    }
}
```

Executing `otherTCPserver.go` and using a client for interacting with it will generate the following output:

```
$ go run otherTCPserver.go 8001
> 1
> 2
> Hello!
> Exiting TCP server!
```

On the client side, which in this case will be `otherTCPclient.go`, you will see the following output:

```
$ go run otherTCPclient.go localhost:8001
>> 1
->: 1
>> 2
->: 2
>> Hello!
->: Hello!
>> ->:
>> STOP
->: TCP client exiting...
```

Finally, I will present a way of finding out the name of the process that listens to a given TCP or UDP port on a Unix machine. So, if you want to discover which process uses TCP port number 8001, you should execute the following command:

```
$ sudo lsof -n -i :8001
COMMAND     PID    USER   FD   TYPE             DEVICE SIZE/OFF NODE NAME
TCPserver 86775  mtsouk    3u  IPv6 0x98d55014e6c9360f      0t0  TCP *:vcom-
tunnel (LISTEN)
```

A UDP client

If you know how to develop a TCP client, then developing a UDP client should be much simpler for you due to the simplicity of the UDP protocol.

The biggest difference between UDP and TCP is that UDP is not reliable by design. This also means that in general, UDP is simpler than TCP because UDP does not need to keep the state of a UDP connection.

The name of the utility presented for this topic is UDPclient.go, and it will be presented in four code segments. The first part of UDPclient.go is as follows:

```
package main

import (
    "bufio"
    "fmt"
    "net"
    "os"
    "strings"
)
```

The second segment of UDPclient.go is as follows:

```
func main() {
    arguments := os.Args
    if len(arguments) == 1 {
        fmt.Println("Please provide a host:port string")
        return
    }
    CONNECT := arguments[1]

    s, err := net.ResolveUDPAddr("udp4", CONNECT)
    c, err := net.DialUDP("udp4", nil, s)
```

```
        if err != nil {
            fmt.Println(err)
            return
        }

        fmt.Printf("The UDP server is %s\n", c.RemoteAddr().String())
        defer c.Close()
```

The `net.ResolveUDPAddr()` function returns an address of a UDP end point as defined by its second parameter. The first parameter (`udp4`) specifies that the program will support the IPv4 protocol only.

The `net.DialUDP()` function used is like `net.Dial()` for UDP networks.

The third segment of `UDPclient.go` contains the following code:

```
        for {
            reader := bufio.NewReader(os.Stdin)
            fmt.Print(">> ")
            text, _ := reader.ReadString('\n')
            data := []byte(text + "\n")
            _, err = c.Write(data)
            if strings.TrimSpace(string(data)) == "STOP" {
                fmt.Println("Exiting UDP client!")
                return
            }
```

The preceding code requires that the user types some text, which is sent to the UDP server. The user text is read from standard input, using `bufio.NewReader(os.Stdin)`. The `Write(data)` method sends the data over the UDP network connection.

The rest of the Go code is as follows:

```
        if err != nil {
            fmt.Println(err)
            return
        }

        buffer := make([]byte, 1024)
        n, _, err := c.ReadFromUDP(buffer)
        if err != nil {
            fmt.Println(err)
            return
        }
        fmt.Printf("Reply: %s\n", string(buffer[0:n]))
    }
}
```

Once the client data is sent, you must wait for the data the UDP server has to send, which is read using `ReadFromUDP()`.

Executing `UDPclient.go` and interacting with the `netcat(1)` utility that acts as a UDP server will generate the following output:

```
$ go run UDPclient.go localhost:8001
The UDP server is 127.0.0.1:8001
>> Hello!
Reply: Hi there!
>> Have to leave - bye!
Reply: OK.
>> STOP
Exiting UDP client!
```

On the UDP server side, the output will be as follows:

```
$ nc -v -u -l 127.0.0.1 8001
Hello!
Hi there!
Have to leave - bye!
OK.
STOP
^C
```

The reason for typing *Control* + *C* in order to terminate `nc(1)` is that `nc(1)` does not have any code that tells it to terminate when it receives the `STOP` string as input.

Developing a UDP server

The purpose of the UDP server that will be developed in this section is to return random numbers from 1 to 1,000 to its UDP clients. The name of the program is `UDPserver.go`, and it is presented in four segments.

The first part of `UDPserver.go` is as follows:

```
package main

import (
    "fmt"
    "math/rand"
    "net"
    "os"
    "strconv"
```

```go
    "strings"
    "time"
)

func random(min, max int) int {
    return rand.Intn(max-min) + min
}
```

The second segment of `UDPserver.go` is as follows:

```go
func main() {
    arguments := os.Args
    if len(arguments) == 1 {
        fmt.Println("Please provide a port number!")
        return
    }
    PORT := ":" + arguments[1]

    s, err := net.ResolveUDPAddr("udp4", PORT)
    if err != nil {
        fmt.Println(err)
        return
    }
```

The third segment of `UDPserver.go` contains the following code:

```go
    connection, err := net.ListenUDP("udp4", s)
    if err != nil {
        fmt.Println(err)
        return
    }

    defer connection.Close()
    buffer := make([]byte, 1024)
    rand.Seed(time.Now().Unix())
```

The `net.ListenUDP()` function acts like `net.ListenTCP()` for UDP networks.

The remaining Go code of `UDPserver.go` is as follows:

```go
    for {
        n, addr, err := connection.ReadFromUDP(buffer)
        fmt.Print("-> ", string(buffer[0:n-1]))

        if strings.TrimSpace(string(buffer[0:n])) == "STOP" {
            fmt.Println("Exiting UDP server!")
            return
        }
```

```go
            data := []byte(strconv.Itoa(random(1, 1001)))
            fmt.Printf("data: %s\n", string(data))
            _, err = connection.WriteToUDP(data, addr)
            if err != nil {
                fmt.Println(err)
                return
            }
        }
    }
}
```

The `ReadFromUDP()` function allows you to read data from a UDP connection using a buffer, which as expected is a **byte slice**.

Executing `UDPserver.go` and connecting to it with `UDPclient.go` will generate the following output:

```
$ go run UDPserver.go 8001
-> Hello!
data: 156
-> Another random number please :)
data: 944
-> Leaving...
data: 491
-> STOP
Exiting UDP server!
```

On the client side, the output will be as follows:

```
$ go run UDPclient.go localhost:8001
The UDP server is 127.0.0.1:8001
>> Hello!
Reply: 156
>> Another random number please :)
Reply: 944
>> Leaving...
Reply: 491
>> STOP
Exiting UDP client!
```

A concurrent TCP server

In this section, you will learn how to develop a **concurrent TCP server**, using goroutines. For each incoming connection to the TCP server, the TCP server will start a new goroutine to handle that request. This allows it to accept more requests, which means that a concurrent TCP server can serve multiple clients simultaneously.

The job of the TCP concurrent server is to accept a positive integer and return a natural number from the Fibonacci sequence. If there is an error in the input, the return value will be -1. As the calculation of numbers of the Fibonacci sequence can be slow, we will use an algorithm that was first presented in Chapter 11, *Code Testing, Optimization and Profiling*, and was included in benchmarkMe.go. Additionally, this time the algorithm used will be explained a bit further.

The name of the program is fiboTCP.go, and its code is presented in five parts. As it is considered good practice to be able to define the port number of a web service as a command-line parameter, fiboTCP.go will do exactly that.

The first part of fiboTCP.go contains the following Go code:

```go
package main

import (
    "bufio"
    "fmt"
    "net"
    "os"
    "strconv"
    "strings"
    "time"
)
```

The second code portion of fiboTCP.go contains the following Go code:

```go
func f(n int) int {
    fn := make(map[int]int)
    for i := 0; i <= n; i++ {
        var f int
        if i <= 2 {
            f = 1
        } else {
            f = fn[i-1] + fn[i-2]
        }
        fn[i] = f
    }
    return fn[n]
}
```

In the preceding code, you can see the implementation of the `f()` function that generates natural numbers that belong to the Fibonacci sequence. The algorithm used is difficult to understand at first, but it is very efficient and therefore fast. First, the `f()` function uses a Go map named `fn`, which is pretty unusual when calculating numbers of the Fibonacci sequence. Second, the `f()` function uses a `for` loop, which is also fairly unusual. Finally, the `f()` function does not use recursion, which is the main reason for the speed of its operation.

The idea behind the algorithm used in `f()`, which uses a **Dynamic Programming** technique, is that whenever a Fibonacci number is computed, it is put into the `fn` map so that it will not be computed again. This simple idea saves a lot of time, especially when large Fibonacci numbers need to be calculated because you do not have to calculate the same Fibonacci number multiple times.

The third code segment of `fiboTCP.go` is as follows:

```
func handleConnection(c net.Conn) {
    for {
        netData, err := bufio.NewReader(c).ReadString('\n')
        if err != nil {
            fmt.Println(err)
            os.Exit(100)
        }

        temp := strings.TrimSpace(string(netData))
        if temp == "STOP" {
            break
        }

        fibo := "-1\n"
        n, err := strconv.Atoi(temp)
        if err == nil {
            fibo = strconv.Itoa(f(n)) + "\n"
        }
        c.Write([]byte(string(fibo)))
    }
    time.Sleep(5 * time.Second)
    c.Close()
}
```

The `handleConnection()` function deals with each client of the concurrent TCP server.

The fourth part of `fiboTCP.go` follows next:

```
func main() {
    arguments := os.Args
    if len(arguments) == 1 {
        fmt.Println("Please provide a port number!")
        return
    }

    PORT := ":" + arguments[1]
    l, err := net.Listen("tcp4", PORT)
    if err != nil {
        fmt.Println(err)
        return
    }
    defer l.Close()
```

The remaining Go code of `fiboTCP.go` is as follows:

```
    for {
        c, err := l.Accept()
        if err != nil {
            fmt.Println(err)
            return
        }
        go handleConnection(c)
    }
}
```

The concurrency of the program is implemented by the `go handleConnection(c)` statement, which begins a new goroutine each time a new TCP client comes. The goroutine is executed concurrently, which gives the server the opportunity to serve even more clients!

Executing `fiboTCP.go` and interacting with it using both `netcat(1)` and `TCPclient.go` on two different Terminal windows will generate the following output:

```
$ go run fiboTCP.go 9000
n: 10
fibo: 55
n: 0
fibo: 1
n: -1
fibo: 0
n: 100
fibo: 3736710778780434371
n: 12
fibo: 144
```

```
n: 12
fibo: 144
```

The output will be as follows on the `TCPclient.go` side:

```
$ go run TCPclient.go localhost:9000
>> 12
->: 144
>> a
->: -1
>> STOP
->: TCP client exiting...
```

On the `netcat(1)` side, the output will be as follows:

```
$ nc localhost 9000
10
55
0
1
-1
0
100
37367107787804343 71
ads
-1
STOP
```

When you send the `STOP` string to the server process, the goroutine that serves that particular TCP client will terminate, which will cause the connection to end.

Finally, the impressive thing here is that both clients are served at the same time, which can be verified by the output of the following command:

```
$ netstat -anp TCP | grep 9000
tcp4       0  0  127.0.0.1.9000    127.0.0.1.57309    ESTABLISHED
tcp4       0  0  127.0.0.1.57309   127.0.0.1.9000     ESTABLISHED
tcp4       0  0  127.0.0.1.9000    127.0.0.1.57305    ESTABLISHED
tcp4       0  0  127.0.0.1.57305   127.0.0.1.9000     ESTABLISHED
tcp4       0  0  *.9000            *.*                LISTEN
```

The last line of the output of the preceding command tells us that there is a process that listens to port `9000`, which means that you can still connect to port `9000`. The first two lines of the output say that there is a client that uses port `57309` to talk to the server process. The third and fourth lines of the preceding output verify that there is another client that communicates with the server that listens to port `9000`. This client uses the TCP port `57305`.

A handy concurrent TCP server

Although the concurrent TCP server from the previous section works fine, it does not serve a practical application. Therefore, in this section, you will learn how to convert the `keyValue.go` application from Chapter 4, *The Uses of Composite Types,* into a fully featured concurrent TCP application.

We will create our own kind of TCP protocol in order to be able to interact with the key-value store from a network. You will need a keyword for each one of the functions of the key-value store. For reasons of simplicity, each keyword will be followed by the relevant data. The result of most commands will be a success or a failure message.

 Designing your own TCP or UDP protocols is not an easy job. This means that you should be particularly specific and careful when designing a new protocol. The key here is to document everything before you start writing production code.

The name of the utility shown in this topic is `kvTCP.go`, and it is presented in six parts.

The first part of `kvTCP.go` is as follows:

```
package main

import (
    "bufio"
    "encoding/gob"
    "fmt"
    "net"
    "os"
    "strings"
)

type myElement struct {
    Name    string
    Surname string
    Id      string
}

const welcome = "Welcome to the Key-value store!\n"

var DATA = make(map[string]myElement)
var DATAFILE = "/tmp/dataFile.gob"
```

The second part of kvTCP.go contains the following Go code:

```go
func handleConnection(c net.Conn) {
    c.Write([]byte(welcome))
    for {
        netData, err := bufio.NewReader(c).ReadString('\n')
        if err != nil {
            fmt.Println(err)
            return
        }

        command := strings.TrimSpace(string(netData))
        tokens  := strings.Fields(command)
        switch len(tokens) {
        case 0:
                continue
        case 1:
            tokens = append(tokens, "")
            tokens = append(tokens, "")
            tokens = append(tokens, "")
            tokens = append(tokens, "")
        case 2:
            tokens = append(tokens, "")
            tokens = append(tokens, "")
            tokens = append(tokens, "")
        case 3:
            tokens = append(tokens, "")
            tokens = append(tokens, "")
        case 4:
            tokens = append(tokens, "")
        }

        switch tokens[0] {
        case "STOP":
            err = save()
            if err != nil {
                    fmt.Println(err)
            }
            c.Close()
            return
        case "PRINT":
                PRINT(c)
        case "DELETE":
                if !DELETE(tokens[1]) {
                    netData := "Delete operation failed!\n"
                    c.Write([]byte(netData))
                } else {
```

```
                    netData := "Delete operation successful!\n"
                    c.Write([]byte(netData))
                }
            case "ADD":
                n := myElement{tokens[2], tokens[3], tokens[4]}
                if !ADD(tokens[1], n) {
                    netData := "Add operation failed!\n"
                    c.Write([]byte(netData))
                } else {
                    netData := "Add operation successful!\n"
                    c.Write([]byte(netData))
                }
                err = save()
                if err != nil {
                    fmt.Println(err)
                }
            case "LOOKUP":
                n := LOOKUP(tokens[1])
                if n != nil {
                    netData := fmt.Sprintf("%v\n", *n)
                    c.Write([]byte(netData))
                } else {
                    netData := "Did not find key!\n"
                    c.Write([]byte(netData))
                }
            case "CHANGE":
                n := myElement{tokens[2], tokens[3], tokens[4]}
                if !CHANGE(tokens[1], n) {
                    netData := "Update operation failed!\n"
                    c.Write([]byte(netData))
                } else {
                    netData := "Update operation successful!\n"
                    c.Write([]byte(netData))
                }
                err = save()
                if err != nil {
                    fmt.Println(err)
                }
            default:
                netData := "Unknown command - please try again!\n"
                c.Write([]byte(netData))
            }
        }
    }
```

The `handleConnection()` function communicates with each TCP client and interprets the client input.

The third segment of kvTCP.go contains the following Go code:

```go
func save() error {
    fmt.Println("Saving", DATAFILE)
    err := os.Remove(DATAFILE)
    if err != nil {
            fmt.Println(err)
    }

    saveTo, err := os.Create(DATAFILE)
    if err != nil {
            fmt.Println("Cannot create", DATAFILE)
            return err
    }
    defer saveTo.Close()

    encoder := gob.NewEncoder(saveTo)
    err = encoder.Encode(DATA)
    if err != nil {
            fmt.Println("Cannot save to", DATAFILE)
            return err
    }
    return nil
}

func load() error {
    fmt.Println("Loading", DATAFILE)
    loadFrom, err := os.Open(DATAFILE)
    defer loadFrom.Close()
    if err != nil {
            fmt.Println("Empty key/value store!")
            return err
    }

    decoder := gob.NewDecoder(loadFrom)
    decoder.Decode(&DATA)
    return nil
}
```

The fourth segment of kvTCP.go is as follows:

```go
func ADD(k string, n myElement) bool {
    if k == "" {
            return false
    }

    if LOOKUP(k) == nil {
            DATA[k] = n
```

```
            return true
    }
    return false
}

func DELETE(k string) bool {
    if LOOKUP(k) != nil {
            delete(DATA, k)
            return true
    }
    return false
}

func LOOKUP(k string) *myElement {
    _, ok := DATA[k]
    if ok {
            n := DATA[k]
            return &n
    } else {
            return nil
    }
}

func CHANGE(k string, n myElement) bool {
    DATA[k] = n
    return true
}
```

The implementation of the preceding functions is the same as in `keyValue.go`. None of them talks directly to a TCP client.

The fifth part of `kvTCP.go` contains the following Go code:

```
func PRINT(c net.Conn) {
    for k, d := range DATA {
            netData := fmt.Sprintf("key: %s value: %v\n", k, d)
            c.Write([]byte(netData))
    }
}
```

The `PRINT()` function sends data to a TCP client directly, one line at a time.

The remaining Go code of the program is as follows:

```
func main() {
    arguments := os.Args
    if len(arguments) == 1 {
            fmt.Println("Please provide a port number!")
```

```
        return
    }

    PORT := ":" + arguments[1]
    l, err := net.Listen("tcp", PORT)
    if err != nil {
        fmt.Println(err)
        return
    }
    defer l.Close()

    err = load()
    if err != nil {
        fmt.Println(err)
    }

    for {
        c, err := l.Accept()
         if err != nil {
                fmt.Println(err)
                os.Exit(100)
        }
        go handleConnection(c)
    }
}
```

Executing kvTCP.go will generate the following output:

```
$ go run kvTCP.go 9000
Loading /tmp/dataFile.gob
Empty key/value store!
open /tmp/dataFile.gob: no such file or directory
Saving /tmp/dataFile.gob
remove /tmp/dataFile.gob: no such file or directory
Saving /tmp/dataFile.gob
Saving /tmp/dataFile.gob
```

For the purposes of this section, the netcat(1) utility will act as the TCP client of kvTCP.go:

```
$ nc localhost 9000
Welcome to the Key-value store!
PRINT
LOOKUP 1
Did not find key!
ADD 1 2 3 4
Add operation successful!
LOOKUP 1
```

```
{2 3 4}
ADD 4 -1 -2 -3
Add operation successful!
PRINT
key: 1 value: {2 3 4}
key: 4 value: {-1 -2 -3}
STOP
```

The `kvTCP.go` file is a concurrent application that uses goroutines, and it can serve multiple TCP clients simultaneously. However, all of these TCP clients will share the same data!

Remote Procedure Call (RPC)

Remote Procedure Call (RPC) is a client-server mechanism for Interprocess communication that uses TCP/IP. Both the RPC client and the RPC server to be developed will use the following package, which is named `sharedRPC.go`:

```
package sharedRPC

type MyFloats struct {
    A1, A2 float64
}

type MyInterface interface {
    Multiply(arguments *MyFloats, reply *float64) error
    Power(arguments *MyFloats, reply *float64) error
}
```

The `sharedRPC` package defines one interface called `MyInterface` and one structure called `MyFloats`, which will be used by both the client and the server. However, only the RPC server will need to implement that interface.

After this, you will need to install the `sharedRPC.go` package by executing the following commands:

```
$ mkdir -p ~/go/src/sharedRPC
$ cp sharedRPC.go ~/go/src/sharedRPC/
$ go install sharedRPC
```

The RPC client

In this section, you will see the Go code of the RPC client, which will be saved as `RPCclient.go` and presented in four parts.

The first part of `RPCclient.go` is as follows:

```
package main

import (
    "fmt"
    "net/rpc"
    "os"
    "sharedRPC"
)
```

The second segment of `RPCclient.go` contains the following Go code:

```
func main() {
    arguments := os.Args
    if len(arguments) == 1 {
        fmt.Println("Please provide a host:port string!")
        return
    }

    CONNECT := arguments[1]
    c, err := rpc.Dial("tcp", CONNECT)
    if err != nil {
        fmt.Println(err)
        return
    }
```

Note the use of the `rpc.Dial()` function for connecting to the RPC server instead of the `net.Dial()` function, even though the RCP server uses TCP.

The third segment of `RPCclient.go` is as follows:

```
    args := sharedRPC.MyFloats{16, -0.5}
    var reply float64

    err = c.Call("MyInterface.Multiply", args, &reply)
    if err != nil {
        fmt.Println(err)
        return
    }
    fmt.Printf("Reply (Multiply): %f\n", reply)
```

What is being exchanged between the RPC client and the RPC server with the help of the `Call()` function are function names, their arguments, and the results of the function calls, as the RPC client knows nothing about the actual implementation of the functions.

The remaining Go code of `RPCclient.go` is as follows:

```
    err = c.Call("MyInterface.Power", args, &reply)
    if err != nil {
        fmt.Println(err)
        return
    }
    fmt.Printf("Reply (Power): %f\n", reply)
}
```

If you try to execute `RPCclient.go` without having a RPC server running, you will get the following type of error message:

```
$ go run RPCclient.go localhost:1234
dial tcp [::1]:1234: connect: connection refused
```

The RPC server

The RPC server will be saved as `RPCserver.go`, and it will be presented in five parts.

The first part of `RPCserver.go` is as follows:

```
package main

import (
    "fmt"
    "math"
    "net"
    "net/rpc"
    "os"
    "sharedRPC"
)
```

The second segment of `RPCserver.go` contains the following Go code:

```
type MyInterface struct{}

func Power(x, y float64) float64 {
    return math.Pow(x, y)
}
```

```go
func (t *MyInterface) Multiply(arguments *sharedRPC.MyFloats, reply
*float64) error {
    *reply = arguments.A1 * arguments.A2
    return nil
}

func (t *MyInterface) Power(arguments *sharedRPC.MyFloats, reply *float64)
error {
    *reply = Power(arguments.A1, arguments.A2)
    return nil
}
```

In the preceding Go code, the RPC server implements the desired interface as well as a helper function named Power().

The third segment of RPCserver.go is as follows:

```go
func main() {
    PORT := ":1234"
    arguments := os.Args
    if len(arguments) != 1 {
        PORT = ":" + arguments[1]
    }
```

The fourth part of RPCserver.go contains the following code:

```go
    myInterface := new(MyInterface)
    rpc.Register(myInterface)
    t, err := net.ResolveTCPAddr("tcp4", PORT)
    if err != nil {
        fmt.Println(err)
        return
    }
    l, err := net.ListenTCP("tcp4", t)
    if err != nil {
        fmt.Println(err)
        return
    }
```

What makes this program an RPC server is the use of the rpc.Register() function. However, as the RPC server uses TCP, it still needs to make function calls to net.ResolveTCPAddr() and net.ListenTCP().

The rest of the Go code of `RPCclient.go` is as follows:

```
    for {
        c, err := l.Accept()
        if err != nil {
            continue
        }
        fmt.Printf("%s\n", c.RemoteAddr())
        rpc.ServeConn(c)
    }
}
```

The `RemoteAddr()` function returns the IP address and the port number used for communicating with the RPC client. The `rpc.ServeConn()` function serves the RPC client.

Executing `RPCserver.go` and waiting for `RPCclient.go` will create the following output:

```
$ go run RPCserver.go
127.0.0.1:52289
```

Executing `RPCclient.go` will create the following output:

```
$ go run RPCclient.go localhost:1234
Reply (Multiply): -8.000000
Reply (Power): 0.250000
```

Doing low-level network programming

Although the `http.Transport` structure allows you to modify the various low-level parameters of a network connection, you can write Go code that permits you to read the raw data of network packets. There are two tricky points here:

- Network packets come in binary format, which requires you to look for specific kinds of network packets and not just any type of network packet. Put simply, when reading network packets, you should specify the protocol or protocols that you will support in your applications in advance.
- In order to send a network packet, you will have to construct it on your own.

The next utility to be shown is called `lowLevel.go`, and it will be presented in three parts. Note that `lowLevel.go` captures ICMP packets, which use the IPv4 protocol and prints their contents. Also note that working with raw network data requires root privileges for security reasons.

The first segment of `lowLevel.go` is as follows:

```
package main

import (
    "fmt"
    "net"
)
```

The second part of `lowLevel.go` contains the following Go code:

```
func main() {
    netaddr, err := net.ResolveIPAddr("ip4", "127.0.0.1")
    if err != nil {
        fmt.Println(err)
        return
    }
    conn, err := net.ListenIP("ip4:icmp", netaddr)
    if err != nil {
        fmt.Println(err)
        return
    }
```

The ICMP protocol is specified in the second part of the first parameter (`ip4:icmp`) of `net.ListenIP()`. Moreover, the `ip4` part tells the utility to capture IPv4 traffic only.

The remaining part of `lowLevel.go` contains the following Go code:

```
    buffer := make([]byte, 1024)
    n, _, err := conn.ReadFrom(buffer)
    if err != nil {
        fmt.Println(err)
        return
    }

    fmt.Printf("% X\n", buffer[0:n])
}
```

The preceding code tells `lowLevel.go` to read just a single network packet because there is no `for` loop.

Network Programming – Building Servers and Clients

The ICMP protocol is used by the `ping(1)` and `traceroute(1)` utilities, so one way to create ICMP traffic is to use one of these two tools. The network traffic will be generated using the following commands on both Unix machines while `lowLevel.go` is already running:

```
$ ping -c 5 localhost
PING localhost (127.0.0.1): 56 data bytes
64 bytes from 127.0.0.1: icmp_seq=0 ttl=64 time=0.037 ms
64 bytes from 127.0.0.1: icmp_seq=1 ttl=64 time=0.038 ms
64 bytes from 127.0.0.1: icmp_seq=2 ttl=64 time=0.117 ms
64 bytes from 127.0.0.1: icmp_seq=3 ttl=64 time=0.052 ms
64 bytes from 127.0.0.1: icmp_seq=4 ttl=64 time=0.049 ms
--- localhost ping statistics ---
5 packets transmitted, 5 packets received, 0.0% packet loss
round-trip min/avg/max/stddev = 0.037/0.059/0.117/0.030 ms
$ traceroute localhost
traceroute to localhost (127.0.0.1), 64 hops max, 52 byte packets
 1  localhost (127.0.0.1)  0.255 ms  0.048 ms  0.067 ms
```

Executing `lowLevel.go` on a macOS High Sierra machine with root privileges will produce the following output:

```
$ sudo go run lowLevel.go
03 03 CD DA 00 00 00 00 45 00 34 00 B4 0F 00 00 01 11 00 00 7F 00 00 01 7F
00 00 01 B4 0E 82 9B 00 20 00 00
$ sudo go run lowLevel.go
00 00 0B 3B 20 34 00 00 5A CB 5C 15 00 04 32 A9 08 09 0A 0B 0C 0D 0E 0F 10
11 12 13 14 15 16 17 18 19 1A 1B 1C 1D 1E 1F 20 21 22 23 24 25 26 27 28 29
2A 2B 2C 2D 2E 2F 30 31 32 33 34 35 36 37
```

The first output example is generated by the `ping(1)` command, whereas the second output example is generated by the `traceroute(1)` command.

Running `lowLevel.go` on a Debian Linux machine will generate the following output:

```
$ uname -a
Linux mail 4.14.12-x86_64-linode92 #1 SMP Fri Jan 5 15:34:44 UTC 2018
x86_64 GNU/Linux
# go run lowLevel.go
08 00 61 DD 3F BA 00 01 9A 5D CB 5A 00 00 00 00 26 DC 0B 00 00 00 00 00 10
11 12 13 14 15 16 17 18 19 1A 1B 1C 1D 1E 1F 20 21 22 23 24 25 26 27 28 29
2A 2B 2C 2D 2E 2F 30 31 32 33 34 35 36 37
# go run lowLevel.go
03 03 BB B8 00 00 00 00 45 00 00 3C CD 8D 00 00 01 11 EE 21 7F 00 00 01 7F
00 00 01 CB 40 82 9A 00 28 FE 3B 40 41 42 43 44 45 46 47 48 49 4A 4B 4C 4D
4E 4F 50 51 52 53 54 55 56 57 58 59 5A 5B 5C 5D 5E 5F
```

The output of the `uname(1)` command prints useful information about the Linux system. Note that on modern Linux machines you should execute `ping(1)` with the `-4` flag in order to tell it to use the IPv4 protocol.

Grabbing raw ICMP network data

In this subsection, you will learn how to use the `syscall` package to capture raw ICMP network data and `syscall.SetsockoptInt()` in order to set the options of a socket. Keep in mind that sending raw ICMP data is much more difficult, as you will have to construct the raw network packets on your own. The name of the utility is `syscallNet.go`, and it will be shown in four parts.

The first part of `syscallNet.go` is as follows:

```
package main

import (
    "fmt"
    "os"
    "syscall"
)
```

The second segment of `syscallNet.go` contains the following Go code:

```
func main() {
    fd, err := syscall.Socket(syscall.AF_INET, syscall.SOCK_RAW, syscall.IPPROTO_ICMP)
    if err != nil {
        fmt.Println("Error in syscall.Socket:", err)
        return
    }

    f := os.NewFile(uintptr(fd), "captureICMP")
    if f == nil {
        fmt.Println("Error in os.NewFile:", err)
        return
    }
```

The `syscall.AF_INET` parameter tells `syscall.Socket()` that you want to work with IPv4. The `syscall.SOCK_RAW` parameter is what makes the generated socket a raw socket. The last parameter, which is `syscall.IPPROTO_ICMP`, tells `syscall.Socket()` that you are interested in ICMP traffic only.

The third part of `syscallNet.go` is as follows:

```
    err = syscall.SetsockoptInt(fd, syscall.SOL_SOCKET, syscall.SO_RCVBUF,
256)
    if err != nil {
        fmt.Println("Error in syscall.Socket:", err)
        return
    }
```

The call to `syscall.SetsockoptInt()` sets the size of the receive buffer of the socket to 256. The `syscall.SOL_SOCKET` parameter is for stating that you want to work on the socket layer level.

The remaining Go code of `syscallNet.go` is as follows:

```
    for {
        buf := make([]byte, 1024)
        numRead, err := f.Read(buf)
        if err != nil {
            fmt.Println(err)
        }
        fmt.Printf("% X\n", buf[:numRead])
    }
}
```

Due to the `for` loop, `syscallNet.go` will keep capturing ICMP network packets until you terminate it manually.

Executing `syscallNet.go` on a macOS High Sierra machine will produce the following type of output:

```
$ sudo go run syscallNet.go
45 00 40 00 BC B6 00 00 40 01 00 00 7F 00 00 01 7F 00 00 01 00 00 3F 36 71
45 00 00 5A CB 6A 90 00 0B 9F 1A 08 09 0A 0B 0C 0D 0E 0F 10 11 12 13 14 15
16 17 18 19 1A 1B 1C 1D 1E 1F 20 21 22 23 24 25 26 27 28 29 2A 2B 2C 2D 2E
2F 30 31 32 33 34 35 36 37
45 00 40 00 62 FB 00 00 40 01 00 00 7F 00 00 01 7F 00 00 01 00 00 31 EF 71
45 00 01 5A CB 6A 91 00 0B AC 5F 08 09 0A 0B 0C 0D 0E 0F 10 11 12 13 14 15
16 17 18 19 1A 1B 1C 1D 1E 1F 20 21 22 23 24 25 26 27 28 29 2A 2B 2C 2D 2E
2F 30 31 32 33 34 35 36 37
45 00 40 00 9A 5F 00 00 40 01 00 00 7F 00 00 01 7F 00 00 01 00 00 1D D6 71
45 00 02 5A CB 6A 92 00 0B C0 76 08 09 0A 0B 0C 0D 0E 0F 10 11 12 13 14 15
16 17 18 19 1A 1B 1C 1D 1E 1F 20 21 22 23 24 25 26 27 28 29 2A 2B 2C 2D 2E
2F 30 31 32 33 34 35 36 37
45 00 40 00 6E 0D 00 00 40 01 00 00 7F 00 00 01 7F 00 00 01 00 00 09 CF 71
45 00 03 5A CB 6A 93 00 0B D4 7B 08 09 0A 0B 0C 0D 0E 0F 10 11 12 13 14 15
16 17 18 19 1A 1B 1C 1D 1E 1F 20 21 22 23 24 25 26 27 28 29 2A 2B 2C 2D 2E
```

```
2F 30 31 32 33 34 35 36 37
45 00 40 00 3A 07 00 00 40 01 00 00 7F 00 00 01 7F 00 00 01 00 00 FE 9C 71
45 00 04 5A CB 6A 94 00 0B DF AB 08 09 0A 0B 0C 0D 0E 0F 10 11 12 13 14 15
16 17 18 19 1A 1B 1C 1D 1E 1F 20 21 22 23 24 25 26 27 28 29 2A 2B 2C 2D 2E
2F 30 31 32 33 34 35 36 37
45 00 24 00 45 55 00 00 40 01 00 00 7F 00 00 01 7F 00 00 01 03 03 AB 12 00
00 00 00 45 00 34 00 C5 73 00 00 01 11 00 00 7F 00 00 01 7F 00 00 01 C5 72
82 9B 00 20 00 00
45 00 24 00 E8 1E 00 00 40 01 00 00 7F 00 00 01 7F 00 00 01 03 03 AB 10 00
00 00 00 45 00 34 00 C5 74 00 00 01 11 00 00 7F 00 00 01 7F 00 00 01 C5 72
82 9C 00 20 00 00
45 00 24 00 2A 4B 00 00 40 01 00 00 7F 00 00 01 7F 00 00 01 03 03 AB 0E 00
00 00 00 45 00 34 00 C5 75 00 00 01 11 00 00 7F 00 00 01 7F 00 00 01 C5 72
82 9D 00 20 00 00
```

Running `syscallNet.go` on a Debian Linux machine will generate the following output:

```
# go run syscallNet.go
45 00 00 54 7F E9 40 00 40 01 BC BD 7F 00 00 01 7F 00 00 01 08 00 6F 07 53
E3 00 01 FA 6A CB 5A 00 00 00 00 AA 7B 06 00 00 00 00 00 10 11 12 13 14 15
16 17 18 19 1A 1B 1C 1D 1E 1F 20 21 22 23 24 25 26 27 28 29 2A 2B 2C 2D 2E
2F 30 31 32 33 34 35 36 37
45 00 00 54 7F EA 00 00 40 01 FC BC 7F 00 00 01 7F 00 00 01 00 00 77 07 53
E3 00 01 FA 6A CB 5A 00 00 00 00 AA 7B 06 00 00 00 00 00 10 11 12 13 14 15
16 17 18 19 1A 1B 1C 1D 1E 1F 20 21 22 23 24 25 26 27 28 29 2A 2B 2C 2D 2E
2F 30 31 32 33 34 35 36 37
45 C0 00 44 68 54 00 00 34 01 8B 8E 86 77 DC 57 6D 4A C1 FD 03 0A 8F 27 00
00 00 00 45 00 00 28 40 4F 40 00 34 06 74 6A 6D 4A C1 FD 86 77 DC 57 B0 B8
DD 96 00 00 00 00 52 F1 AB DA 50 14 00 00 90 9E 00 00
45 00 00 54 80 4E 40 00 40 01 BC 58 7F 00 00 01 7F 00 00 01 08 00 7E 01 53
FB 00 02 FB 6A CB 5A 00 00 00 00 9A 80 06 00 00 00 00 00 10 11 12 13 14 15
16 17 18 19 1A 1B 1C 1D 1E 1F 20 21 22 23 24 25 26 27 28 29 2A 2B 2C 2D 2E
2F 30 31 32 33 34 35 36 37
45 00 00 54 80 4F 00 00 40 01 FC 57 7F 00 00 01 7F 00 00 01 00 00 86 01 53
E3 00 02 FB 6A CB 5A 00 00 00 00 9A 80 06 00 00 00 00 00 10 11 12 13 14 15
16 17 18 19 1A 1B 1C 1D 1E 1F 20 21 22 23 24 25 26 27 28 29 2A 2B 2C 2D 2E
2F 30 31 32 33 34 35 36 37
45 00 00 54 80 9B 40 00 40 01 BC 0B 7F 00 00 01 7F 00 00 01 08 00 93 EC 53
E3 00 03 FC 6A CB 5A 00 00 00 00 83 94 06 00 00 00 00 00 10 11 12 13 14 15
16 17 18 19 1A 1B 1C 1D 1E 1F 20 21 22 23 24 25 26 27 28 29 2A 2B 2C 2D 2E
2F 30 31 32 33 34 35 36 37
45 00 00 54 80 9C 00 00 40 01 FC 0A 7F 00 00 01 7F 00 00 01 00 00 9B EC 53
E3 00 03 FC 6A CB 5A 00 00 00 00 83 94 06 00 00 00 00 00 10 11 12 13 14 15
16 17 18 19 1A 1B 1C 1D 1E 1F 20 21 22 23 24 25 26 27 28 29 2A 2B 2C 2D 2E
2F 30 31 32 33 34 35 36 37
45 C0 00 44 68 55 00 00 34 01 8B 8D 86 77 DC 57 6D 4A C1 FD 03 0A 8F 27 00
00 00 00 45 00 00 28 40 D1 40 00 34 06 73 E8 6D 4A C1 FD 86 77 DC 57 8E 8E
DD 96 00 00 00 00 6C 6E D3 36 50 14 00 00 71 EF 00 00
```

```
45 00 00 54 80 F8 40 00 40 01 BB AE 7F 00 00 01 7F 00 00 01 08 00 F2 E7 53
E3 00 04 FD 6A CB 5A 00 00 00 00 23 98 06 00 00 00 00 00 10 11 12 13 14 15
16 17 18 19 1A 1B 1C 1D 1E 1F 20 21 22 23 24 25 26 27 28 29 2A 2B 2C 2D 2E
2F 30 31 32 33 34 35 36 37
45 00 00 54 80 F9 00 00 40 01 FB AD 7F 00 00 01 7F 00 00 01 00 00 FA E7 53
E3 00 04 FD 6A CB 5A 00 00 00 00 23 98 06 00 00 00 00 00 10 11 12 13 14 15
16 17 18 19 1A 1B 1C 1D 1E 1F 20 21 22 23 24 25 26 27 28 29 2A 2B 2C 2D 2E
2F 30 31 32 33 34 35 36 37
45 00 00 54 82 0D 40 00 40 01 BA 99 7F 00 00 01 7F 00 00 01 08 00 4A 82 53
E3 00 05 FE 6A CB 5A 00 00 00 00 CA FC 06 00 00 00 00 00 10 11 12 13 14 15
16 17 18 19 1A 1B 1C 1D 1E 1F 20 21 22 23 24 25 26 27 28 29 2A 2B 2C 2D 2E
2F 30 31 32 33 34 35 36 37
45 00 00 54 82 0E 00 00 40 01 FA 98 7F 00 00 01 7F 00 00 01 00 00 52 82 53
E3 00 05 FE 6A CB 5A 00 00 00 00 CA FC 06 00 00 00 00 00 10 11 12 13 14 15
16 17 18 19 1A 1B 1C 1D 1E 1F 20 21 22 23 24 25 26 27 28 29 2A 2B 2C 2D 2E
2F 30 31 32 33 34 35 36 37
45 C0 00 44 68 56 00 00 34 01 8B 8C 86 77 DC 57 6D 4A C1 FD 03 0A 8F 27 00
00 00 00 45 00 00 28 41 74 40 00 34 06 73 45 6D 4A C1 FD 86 77 DC 57 2E 9B
DD 96 00 00 00 00 C3 D6 44 57 50 14 00 00 09 5A 00 00
45 C0 00 44 68 57 00 00 34 01 8B 8B 86 77 DC 57 6D 4A C1 FD 03 0A 8F 27 00
00 00 00 45 00 00 28 44 27 40 00 33 06 71 92 6D 4A C1 FD 86 77 DC 57 C5 C2
DD 96 00 00 00 00 CF DD DB BE 50 14 00 00 CE C3 00 00
45 C0 00 58 94 B4 00 00 40 01 E7 2E 7F 00 00 01 7F 00 00 01 03 03 F1 DA 00
00 00 00 45 00 00 3C 85 E1 00 00 01 11 35 CE 7F 00 00 01 7F 00 00 01 95 1E
82 9A 00 28 FE 3B 40 41 42 43 44 45 46 47 48 49 4A 4B 4C 4D 4E 4F 50 51 52
53 54 55 56 57 58 59 5A 5B 5C 5D 5E 5F
45 C0 00 58 94 B5 00 00 40 01 E7 2D 7F 00 00 01 7F 00 00 01 03 03 F9 EA 00
00 00 00 45 00 00 3C 85 E2 00 00 01 11 35 CD 7F 00 00 01 7F 00 00 01 8D 0D
82 9B 00 28 FE 3B 40 41 42 43 44 45 46 47 48 49 4A 4B 4C 4D 4E 4F 50 51 52
53 54 55 56 57 58 59 5A 5B 5C 5D 5E 5F
45 C0 00 58 94 B6 00 00 40 01 E7 2C 7F 00 00 01 7F 00 00 01 03 03 D2 EB 00
00 00 00 45 00 00 3C 85 E3 00 00 01 11 35 CC 7F 00 00 01 7F 00 00 01 B4 0B
82 9C 00 28 FE 3B 40 41 42 43 44 45 46 47 48 49 4A 4B 4C 4D 4E 4F 50 51 52
53 54 55 56 57 58 59 5A 5B 5C 5D 5E 5F
45 C0 00 58 94 B7 00 00 40 01 E7 2B 7F 00 00 01 7F 00 00 01 03 03 D6 AC 00
00 00 00 45 00 00 3C 85 E4 00 00 02 11 34 CB 7F 00 00 01 7F 00 00 01 B0 49
82 9D 00 28 FE 3B 40 41 42 43 44 45 46 47 48 49 4A 4B 4C 4D 4E 4F 50 51 52
53 54 55 56 57 58 59 5A 5B 5C 5D 5E 5F
45 C0 00 58 94 B8 00 00 40 01 E7 2A 7F 00 00 01 7F 00 00 01 03 03 F1 B4 00
00 00 00 45 00 00 3C 85 E5 00 00 02 11 34 CA 7F 00 00 01 7F 00 00 01 95 40
82 9E 00 28 FE 3B 40 41 42 43 44 45 46 47 48 49 4A 4B 4C 4D 4E 4F 50 51 52
53 54 55 56 57 58 59 5A 5B 5C 5D 5E 5F
45 C0 00 58 94 B9 00 00 40 01 E7 29 7F 00 00 01 7F 00 00 01 03 03 CD 43 00
00 00 00 45 00 00 3C 85 E6 00 00 02 11 34 C9 7F 00 00 01 7F 00 00 01 B9 B0
82 9F 00 28 FE 3B 40 41 42 43 44 45 46 47 48 49 4A 4B 4C 4D 4E 4F 50 51 52
53 54 55 56 57 58 59 5A 5B 5C 5D 5E 5F
45 C0 00 58 94 BA 00 00 40 01 E7 28 7F 00 00 01 7F 00 00 01 03 03 9D 8F 00
00 00 00 45 00 00 3C 85 E7 00 00 03 11 33 C8 7F 00 00 01 7F 00 00 01 E9 63
```

```
82 A0 00 28 FE 3B 40 41 42 43 44 45 46 47 48 49 4A 4B 4C 4D 4E 4F 50 51 52
53 54 55 56 57 58 59 5A 5B 5C 5D 5E 5F
45 C0 00 58 94 BB 00 00 40 01 E7 27 7F 00 00 01 7F 00 00 01 03 03 A3 13 00
00 00 00 45 00 00 3C 85 E8 00 00 03 11 33 C7 7F 00 00 01 7F 00 00 01 E3 DE
82 A1 00 28 FE 3B 40 41 42 43 44 45 46 47 48 49 4A 4B 4C 4D 4E 4F 50 51 52
53 54 55 56 57 58 59 5A 5B 5C 5D 5E 5F
45 C0 00 58 94 BC 00 00 40 01 E7 26 7F 00 00 01 7F 00 00 01 03 03 D4 66 00
00 00 00 45 00 00 3C 85 E9 00 00 03 11 33 C6 7F 00 00 01 7F 00 00 01 B2 8A
82 A2 00 28 FE 3B 40 41 42 43 44 45 46 47 48 49 4A 4B 4C 4D 4E 4F 50 51 52
53 54 55 56 57 58 59 5A 5B 5C 5D 5E 5F
45 C0 00 58 94 BD 00 00 40 01 E7 25 7F 00 00 01 7F 00 00 01 03 03 A6 8D 00
00 00 00 45 00 00 3C 85 EA 00 00 04 11 32 C5 7F 00 00 01 7F 00 00 01 E0 62
82 A3 00 28 FE 3B 40 41 42 43 44 45 46 47 48 49 4A 4B 4C 4D 4E 4F 50 51 52
53 54 55 56 57 58 59 5A 5B 5C 5D 5E 5F
45 C0 00 58 94 BE 00 00 40 01 E7 24 7F 00 00 01 7F 00 00 01 03 03 F1 C6 00
00 00 00 45 00 00 3C 85 EB 00 00 04 11 32 C4 7F 00 00 01 7F 00 00 01 95 28
82 A4 00 28 FE 3B 40 41 42 43 44 45 46 47 48 49 4A 4B 4C 4D 4E 4F 50 51 52
53 54 55 56 57 58 59 5A 5B 5C 5D 5E 5F
45 C0 00 58 94 BF 00 00 40 01 E7 23 7F 00 00 01 7F 00 00 01 03 03 A3 FE 00
00 00 00 45 00 00 3C 85 EC 00 00 04 11 32 C3 7F 00 00 01 7F 00 00 01 E2 EF
82 A5 00 28 FE 3B 40 41 42 43 44 45 46 47 48 49 4A 4B 4C 4D 4E 4F 50 51 52
53 54 55 56 57 58 59 5A 5B 5C 5D 5E 5F
45 C0 00 58 94 C0 00 00 40 01 E7 22 7F 00 00 01 7F 00 00 01 03 03 B9 AA 00
00 00 00 45 00 00 3C 85 ED 00 00 05 11 31 C2 7F 00 00 01 7F 00 00 01 CD 42
82 A6 00 28 FE 3B 40 41 42 43 44 45 46 47 48 49 4A 4B 4C 4D 4E 4F 50 51 52
53 54 55 56 57 58 59 5A 5B 5C 5D 5E 5F
45 C0 00 58 94 C1 00 00 40 01 E7 21 7F 00 00 01 7F 00 00 01 03 03 B3 B7 00
00 00 00 45 00 00 3C 85 EE 00 00 05 11 31 C1 7F 00 00 01 7F 00 00 01 D3 34
82 A7 00 28 FE 3B 40 41 42 43 44 45 46 47 48 49 4A 4B 4C 4D 4E 4F 50 51 52
53 54 55 56 57 58 59 5A 5B 5C 5D 5E 5F
45 C0 00 58 94 C2 00 00 40 01 E7 20 7F 00 00 01 7F 00 00 01 03 03 F2 62 00
00 00 00 45 00 00 3C 85 EF 00 00 05 11 31 C0 7F 00 00 01 7F 00 00 01 94 88
82 A8 00 28 FE 3B 40 41 42 43 44 45 46 47 48 49 4A 4B 4C 4D 4E 4F 50 51 52
53 54 55 56 57 58 59 5A 5B 5C 5D 5E 5F
45 C0 00 58 94 C3 00 00 40 01 E7 1F 7F 00 00 01 7F 00 00 01 03 03 DD BE 00
00 00 00 45 00 00 3C 85 F0 00 00 06 11 30 BF 7F 00 00 01 7F 00 00 01 A9 2B
82 A9 00 28 FE 3B 40 41 42 43 44 45 46 47 48 49 4A 4B 4C 4D 4E 4F 50 51 52
53 54 55 56 57 58 59 5A 5B 5C 5D 5E 5F
```

Where to go next?

Philosophically speaking, no programming book can ever be perfect and neither is this one! Did I leave some Go topics out? Absolutely yes! Why? Because there are always more topics to cover in a book, so if I tried to cover everything, the book would never be ready for publication! This situation is somehow analogous to the specifications of a program—you can always add new and exciting features, but if you do not freeze its specifications, the program will always be in development and it will never be ready for your intended audience! Some of the Go topics I missed might even appear in the second edition of this book!

The good thing is that after reading this book, you will be ready to learn on your own, which is the biggest benefit that you can get from any good computer programming book. The main purpose of this book is to help you learn how to program in Go and gain some experience in doing so. However, there is no substitute for trying things out on your own, and failing often becomes the only way to learn a programming language, so you must keep developing nontrivial things.

So, this is the end of yet another Go book for me, but the beginning of a journey for you! Thanks for buying this book. You are now ready to start writing your own software in Go and learning new things!

Additional resources

Take a look at the following resources:

- Visit the documentation of the `net` standard Go package, which can be found at `https://golang.org/pkg/net/`. This is one of the biggest documentation found in the Go documentation.
- Although this book talked about **RPC**, it did not talk about **gRPC**, which is an open source, high-performance RPC framework. A package with the Go language implementation of gRPC can be found at `https://github.com/grpc/grpc-go`.
- The ICMP protocol for IPv4 is defined in RFC 792. It can be found in many places, including `https://tools.ietf.org/html/rfc792`.
- **WebSocket** is a protocol for two-way communication between a client and a remote host. There is a WebSocket implementation for Go at `https://github.com/gorilla/websocket`. You can learn more about WebSocket at `http://www.rfc-editor.org/rfc/rfc6455.txt`.

- If you are really into network programming and you want to be able to work with RAW TCP packets, you mind find interesting and helpful information and tools in the gopacket library, which can be found at `https://github.com/google/gopacket`.
- The `raw` package, which is located at `https://github.com/mdlayher/raw` allows you to read and write data at the device driver level for a network device.

Exercises

- Develop a **FTP client** in Go.
- Next try to develop a FTP server in Go. Is it more difficult to implement the FTP client or the FTP server? Why?
- Try to implement a Go version of the `nc(1)` utility. The secret when programming such fairly complex utilities is to start with a version that implements the basic functionality of the original utility before trying to support every possible option.
- Modify `TCPserver.go` so that it returns the date in one network packet and the time in another.
- Modify `TCPserver.go` so that it can serve multiple clients in a sequential way. Note that this is not the same as being able to serve multiple requests concurrently. Put simply, use a `for` loop so that the `Accept()` call can be executed multiple times.
- TCP servers, such as `fiboTCP.go`, tend to terminate when they receive a given signal, so add signal handling code to `fiboTCP.go`, as you learned in *Chapter 8, Telling a Unix System What to Do*.
- Modify `kvTCP.go` so that the `save()` function is protected using `sync.Mutex`. Is this required?
- Develop your own small web server in Go using a plain TCP instead of using the `http.ListenAndServe()` function.

Summary

This chapter addressed many interesting things including developing UDP and TCP clients and servers, which are applications that work over TCP/IP computer networks.

As this is the last chapter of this book, I would like to congratulate and thank you for choosing this book. I hope that you found it useful and that you will continue using it as a reference throughout your Go journey.

Soli Deo gloria

Other Books You May Enjoy

If you enjoyed this book, you may be interested in these other books by Packt:

Go Systems Programming
Mihalis Tsoukalos

ISBN: 978-1-78712-564-3

- Explore the Go language from the standpoint of a developer conversant with Unix, Linux, and so on
- Understand Goroutines, the lightweight threads used for systems and concurrent applications
- Learn how to translate Unix and Linux systems code in C to Golang code
- How to write fast and lightweight server code
- Dive into concurrency with Go
- Write low-level networking code

Other Books You May Enjoy

Distributed Computing with Go
V.N. Nikhil Anurag

ISBN: 978-1-78712-538-4

- Gain proficiency with concurrency and parallelism in Go
- Learn how to test your application using Go's standard library
- Learn industry best practices with technologies such as REST, OpenAPI, Docker, and so on
- Design and build a distributed search engine
- Learn strategies on how to design a system for web scale

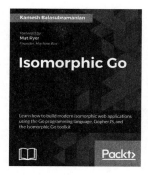

Isomorphic Go
Kamesh Balasubramanian

ISBN: 978-1-78839-418-5

- Create Go programs inside the web browser using GopherJS
- Render isomorphic templates on both the client side and the server side
- Perform end-to-end application routing for greater search engine discoverability and an enhanced user experience
- Implement isomorphic handoff to seamlessly transition state between the web server and the web browser
- Build real-time web application functionality with websockets to promote user collaboration
- Create reusable components (cogs) that are rendered using the virtual DOM
- Deploy an Isomorphic Go application for production use

Other Books You May Enjoy

Security with Go
John Daniel Leon

ISBN: 978-1-78862-791-7

- Learn the basic concepts and principles of secure programming
- Write secure Golang programs and applications
- Understand classic patterns of attack
- Write Golang scripts to defend against network-level attacks
- Learn how to use Golang security packages
- Apply and explore cryptographic methods and packages
- Learn the art of defending against brute force attacks
- Secure web and cloud applications

Leave a review - let other readers know what you think

Please share your thoughts on this book with others by leaving a review on the site that you bought it from. If you purchased the book from Amazon, please leave us an honest review on this book's Amazon page. This is vital so that other potential readers can see and use your unbiased opinion to make purchasing decisions, we can understand what our customers think about our products, and our authors can see your feedback on the title that they have worked with Packt to create. It will only take a few minutes of your time, but is valuable to other potential customers, our authors, and Packt. Thank you!

Index

*

* character 227

=

= operator 23

A

abstract classes 279
abstract types 262
acyclic graphs 176
algorithm complexity 176
anonymous functions 221
Apache 127
Apache HTTP server benchmarking tool 503

B

basic data types, Go
  about 89
  arrays 93
  constants 113
  loops 90
  maps 110
  pointers 118
  slices 97
BenchmarkFibo() function 458, 459
benchmarking
  about 452
  example 452
beta version, Go
  installing 423, 424
big endian byte order 298
Big O notation 176
binary format
  using 301
binary tree
  about 177
  advantages 180, 181
  balanced tree 177
  depth of a node 177
  depth of a tree 177
  implementing, in Go 178, 179, 180
  leaf 177
  root of a tree 177
  unbalanced tree 177
black set 50
blank identifier 27
buffered channels 376, 377, 378, 386
buffered file input and output 290
buffered writing
  benchmarking 459, 461, 462, 463
bufio package 290
build command 85
byte slice 102
bytes package
  using 314, 315

C

C code
  about 66
  calling from Go 59
  calling from Go, same file used 59, 60
  calling from Go, separate files used 60
  example 61
  mixing, with Go code 63, 64
C++ 440
cat(1) utility
  implementing 323, 324, 325
channel of channels
  about 380
  using 381, 382, 383
channel

about 53, 341, 352, 375
  as function parameters 356
  reading from 354
  receive-only channel 357
  send-only channel 356
  unidirectional channel 356
  writing to 352, 353, 355
circular list 203
closures 221
code optimization 421, 425, 426
code profiling
  about 421, 426
  convenient external package 435
  net/http/pprof standard Go package 427
  simple example 427, 428, 429, 433, 434
code testing 421
composite types 132
concurrency 343
concurrent TCP server
  developing 551, 552, 553, 554
  handy concurrent TCP server 556, 558, 559, 561
constant generator iota 115, 116, 117
container package
  about 203
  container/heap, using 204, 205, 206
  container/list, using 207, 208, 209
  container/ring, using 209, 210
context package
  about 403
  advanced example 407, 409, 411, 412
  using 403, 404, 405, 406, 407
  Worker pool 412
continuation 364
continuation stealing 364
copy() function 102
CPU profiling 427
cross-compilation 465, 467
CSV files
  reading 302, 303, 305
custom Go package
  compiling 233
  developing 231, 232
  init() function 233, 234, 235
  private functions 233
  private variables 233
custom interfaces
  developing 265
cyclic graphs 176

D

daemon processes 284
data race condition 397, 400
data
  loading, on disk 308, 309, 310, 311
  saving, on disk 308, 309, 310, 311
dates
  dealing with 121, 122
  format, changing 127, 128
  parsing 125
  working with 125
deadlocks 367
Debian Linux 8
debug parse tree 83
defer keyword 67, 68, 69
deferred functions 67
deserialization 309
Directed Acyclic Graph 176
Directed Graph 176
directives 11
directory trees
  traversing 325, 326, 327
DNS (Domain Name System) 490
DNS lookups
  MX records of domain, obtaining 494, 495
  NS records of domain, obtaining 492, 493, 494
  performing 490, 492
DNS.go utility 490
documentation
  generating 469, 470, 472, 473, 474, 475
DOT 440
doubly linked list
  about 191, 192
  advantages 196
  implementing, in Go 193, 194, 195
dtrace tool 74, 75
Dynamic Programming technique 553

E

eBPF (enhanced Berkeley Packet Filter)
  about 328
  reference links 329
  using 329
Echo service 544
encapsulation 233
Erlang programming language 343
error handling
  about 38, 40, 41, 42
  error data type 38, 40
error output 28
example functions
  creating 467, 469

F

fair scheduling strategy 364
Fibonacci sequence 452
file permissions 316, 317
file
  writing to 305, 306, 307, 308
finite automaton 140
flag package
  about 284
  using 285, 286, 287, 288, 289
fmt.Println() function
  about 244
  implementing 244, 245, 246
  URL 244
for loop 90
forgetMutex.go
  using 389
fork-join concurrency model 364
format specifier 20
functions
  about 220
  anonymous functions 221
  multiple values, returning 221, 222, 223
  other functions as parameters, accepting 229, 230
  other functions, returning 228, 229
  pointers, returning 226, 227
  return values, naming 223, 224, 225
  with pointer parameters 225, 226

G

garbage collection safe-point 53
Garbage Collection
  about 10, 47, 49
  exploring 53, 54, 55
  tricolor algorithm 50, 51, 52, 53
  unsafe code 55, 56
  unsafe package 57
Go arrays
  about 93, 94
  disadvantages 97
  multi-dimensional arrays 94, 95, 96
Go assembler 78, 79
Go channels
  about 375, 376
  buffered channels 376
  channel of channels 380
  nil channels 379
  order of execution, specifying for goroutines 383, 384, 385, 386
  signal channel 376
Go code, Go package
  exploring, of net/url package 236, 237, 238
  reading 236
  viewing, of log/syslog package 238, 239
Go code
  benchmarking 452
  compiling 13
  enhancing, with data structures 175
  example 62
  executing 14
  practical advices 86
  simple benchmarking example 452, 453, 455, 457
  testing 447
  tests, writing for 448, 450, 451
  unreachable Go code, finding 464, 465
Go compiler 46, 47
Go constants
  about 113, 114
  constant generator iota 115, 116, 117
Go documentation server
  URL 467
Go environment 76, 77, 78

[583]

Go execution tracer 442
Go functions
  calling, from C code 64
Go garbage collector 442
Go interfaces
  about 262, 263
  switch statement, using 268, 269, 270
  using 266, 267, 268
Go loops
  about 90
  examples, for loops 91, 93
  for loop 90
  range keyword 91
  while loop 91
Go maps
  about 110, 111, 112
  nil map, storing to 112, 113
Go node 79
Go package
  about 64, 65, 220
  custom Go package, developing 231, 232
  downloading 17, 18, 19
  Go code, reading 236
  resources 257
  rules, for creating 239, 241
Go playground
  URL 467
Go pointers
  about 118, 120
  dereferencing 118
Go profiler
  web interface 437
Go race detector 398
Go rules
  about 14
  curly braces, formatting 16, 17
  Go package, using 15, 16
Go scheduler 343, 364, 366
Go slices
  about 97
  byte slices 102
  capacity 100
  copy() function 102, 104
  example 105, 106, 107
  expanding automatically 100
  length 100
  multi-dimensional slices 105
  operations, performing on 98, 99, 100
  sorting, with sort.slice() 108
go tool trace utility
  about 441
  using 442, 443, 444, 445, 446
Go version 1.10
  about 422
  changes 423
  versus, Go version 1.9 422
Go
  about 8
  advantages 9, 10
  basic data types 89
  beta version, installing 423
  binary trees 177
  composite types 132
  concepts 8
  disadvantages 11
  doubly linked list 191
  error handling 38
  features 9
  functions 220
  hash tables 181
  history 8, 9
  key value store, developing 168
  linked list 186
  object-oriented programming 277, 278
  queues 196
  RC version, installing 423
  stacks 200
  URL 9
  web client, creating 520
  web server, creating 496
gob format 309
godoc utility 12
GOMAXPROCS environment variable 365, 366, 367
goroutine
  about 341, 342, 344
  Add() and Done() calls, not agreeing 350
  creating 344, 345
  multiple goroutines, creating 346, 347
  timing out 370, 371, 372, 373, 374, 375

waiting to finish 348, 349
grammar 140
graphs
  about 176
  acyclic graphs 176
  cyclic graphs 176
Graphviz 440
gray set 50
group ID
  finding 337
gRPC 574

H

handler function 496
hash table
  about 110, 181
  advantages 186
  implementing, in Go 182, 183, 184
  lookup functionality, implementing 185
HTML output
  constructing 249, 250, 251, 252, 253, 254, 255, 256
HTML template 246
HTTP connections
  timeout period on server side, setting 529, 531
  timeout, from client side 531
  timing out 526
HTTP handlers
  testing 517, 518, 519, 520
HTTP tracing 514, 515, 517
http.Request type 481
http.Response structure 480
http.RoundTripper 480
http.Transport type 482
Hugo utility 533

I

inheritance 279
init() function 233, 234, 235
io.Reader interface 290
io.Writer interface 290
IP (Internet Protocol) 483
IPv4 484
IPv6 484

J

Java 55
JSON format 309

K

kernel processes 284
key value store
  developing 168, 169, 171, 172

L

Last In First Out (LIFO) 67
latency 54
linked list
  about 186
  advantages 191
  implementing, in Go 187, 188, 190
  node, removing from 187
little endian byte order 298
log files, writing to
  about 30
  information sending 32, 33, 34
  log servers 31
  logging levels 31
log levels, writing to
  logging facilities 31
log.Fatal() 35, 36
log.Panic() 36, 37
log/syslog package
  code, viewing 238, 239
logging facility 31
logging level 31
logical processors 365
low-level network programming
  performing 566, 567, 568
  raw ICMP network data, grabbing 569, 570

M

macOS High Sierra 8
map key 110
mark-and-sweep algorithm 54
memory profiling 427
method overriding 280
modulo operator 183

[585]

monitor goroutine 395
multi-dimensional arrays 94
multi-dimensional slices 105
mutator 51
mutual exclusion variable 386
MX records of domain 494

N

name servers, domain 492
nc(1) command-line utility 484
net package 480
net standard Go package 538
net/http package 480
net/url package
 code, exploring 236, 237, 238
netCapabilities.go utility 488
netConfig.go utility 485
network interfaces
 configuration, reading 485, 486, 487, 488, 490
new keyword
 using 137
Nginx 127
nil channels
 about 379
 using 379, 380
node trees 79, 80, 83, 85
nodes 176

O

object file 46
object-oriented programming (OOP) 11, 259
online references 43
output
 printing 19, 20, 21

P

panic() function
 about 69, 70, 71
 using 71, 72
parallelism 343
pattern matching
 about 139, 140
 advanced example 143, 144, 145
 example 143
 simple example 140, 141
Perl 440
Pi
 calculating, with accuracy 165, 166, 167
pipeline 357, 358
pragmas 11
preprocessor 10, 11
private functions 233
private variables 233
process 284, 342
process ID 284, 330
program 284
programming languages
 Alef 8
 C 8
 Oberon 8
 Pascal 8
ptraceRegs.go utility 329
Python 440

Q

queues
 about 196, 197
 implementing, in Go 197, 199, 200

R

race conditions
 catching 397, 398, 400, 401, 402
random numbers
 generating 211, 212, 213, 214
 random strings, generating 214, 215, 216
range keyword 91
re-slicing 99
read-only channel 357
receive-only channel 357
recognizer 140
recover() function 69, 70, 71
reflection
 about 270, 271
 advanced example 273, 274, 275, 276
 advantages 276
 disadvantages 277
 object-oriented programming 279
 simple reflection example 271, 273

regular expressions (regexp)
  about 139, 140
  advanced example 143, 144, 145
  IPv4 addresses, matching 146, 147, 148, 149, 150
  simple example 140, 141, 143
Release Candidate (RC) version, Go
  installing 423, 424
Remote Procedure Call (RPC) 562
RFC850 format 128
RPC client 563, 564
RPC server 564, 566
Ruby 440

S

scheduler 364
select keyword
  about 367
  using 368, 369, 370
semaphores 376
send-only channel 356
serialization 309
SetDeadline() function 528
shared memory
  about 386
  using goroutines 395, 396, 397
shared variables 386
short assignment statement 23
SIGINT signal 317
signal channel
  about 376
  using 383
slice literals 98
sort.Slice() function
  using 108
SQLite3
  commands 257
stacks
  about 200
  implementing, in Go 200, 201, 203
stalling join 364
standard output
  about 21
  using 21, 23
static linking 10

stop-the-world garbage collector 54
strace tool 73, 74
stream encoding 309
strings package
  using 312, 313, 314
strings
  about 151, 152, 153
  rune 154
  strings standard Go package 157, 159, 160
  Unicode package 156
structure literal 132
structures
  about 132, 133, 134
  new keyword, using 137
  pointers 135, 136
switch statement 161, 163, 164
sync.Mutex type 387
sync.RWMutex type
  about 391
  using 392, 393
syscall package
  about 241, 242, 243, 244
  fmt.Println() function 244, 245, 246
  using 331
syscall.PtraceRegs 329, 331
system calls
  about 242
  tracing 331, 332, 333, 335
systems programming 283

T

task 364
TCP (Transmission Control Protocol) 483
TCP client
  about 538, 540
  destination port 538
  different version 540, 542
  source port 538
TCP packets 483
TCP server
  about 542, 543
  different version 544, 545, 546
TCP/IP 483
templates 246
text files

exact amount of data, reading 299, 301
reading 291
reading, character by character 295
reading, from /dev/random 297, 298
reading, line by line 291, 293
reading, word by word 293, 295
text output
    generating 247, 248, 249
text template 246
thread 342
times
    dealing with 121, 122
    format, changing 127, 128
    parsing 123, 125
    working with 123
traceSyscall.go utility
    using 332
tricolor algorithm 50
tricolor mark-and-sweep algorithm 50
tshark 533
tuples 137, 138, 139
type assertion 263, 265
type methods 259, 260, 261

U

UDP (User Datagram Protocol) 483
UDP client
    developing 547, 549
UDP server
    developing 549, 551
unbuffered file input and output 290
Unicode 10
unidirectional channels 356
Unix epoch time 121
Unix operating system 8
Unix pipes
    programming 323
Unix processes 284
Unix signals
    all signals, handling 320, 321, 322
    handling 317, 318
    two signals, handling 318, 319, 320
Unix standard error 28
Unix utilities

about 72, 73
    dtrace tool 74, 75
    strace tool 73, 74
Unix
    stderr 19
    stdin 19
    stdout 19
unsafe code 55, 56
unsafe package
    about 57
    example 57, 58, 59
user ID
    finding 336
user input
    acquiring 23
    command line arguments, working with 26, 27
    standard input, reading from 24, 25
user processes 284

W

web client, in Go
    creating 520, 521
    making advanced 522, 523, 524, 525
web interface, Go profiler
    about 437
    profiling example 437, 438, 439
web server
    creating 496, 497, 498, 499
    HTTP server, profiling 499, 500, 502, 503
website, Go
    creating 504, 507, 508, 510, 512, 513, 514
WebSocket 574
while loop 91
white set 50
Wireshark 533
work stealing strategy 364
Worker pools
    about 412
    implementing 413, 414, 416, 417
write barrier 50
write-only channel 357

X

XML format 309